Liz Baird

The emergence of print in late f
Italy gave a crucial new importance to the ....
of texts, who could strongly influence the interpretation and status of literary works by determining the form and context in which they would be read. Brian Richardson examines the Renaissance production, circulation and reception of texts by writers including Dante, Petrarch, Boccaccio and Ariosto, as well as popular works of entertainment. In so doing he sheds light on the impact of the new printing and editing methods on Renaissance culture.

Bought: BL, 2.6.08.

PRINT CULTURE IN
RENAISSANCE ITALY

CAMBRIDGE STUDIES IN PUBLISHING AND PRINTING HISTORY

GENERAL EDITORS

Terry Belanger and David McKitterick

TITLES PUBLISHED

The Provincial Book Trade in Eighteenth-Century England
by John Feather
Lewis Carroll and the House of Macmillan
edited by Morton N. Cohen & Anita Gandolfo
The Correspondence of Robert Dodsley 1733–1764
edited by James E. Tierney
Book Production and Publication in Britain 1375–1475
edited by Jeremy Griffiths and Derek Pearsall
Before Copyright: the French Book-Privilege System 1486–1526
by Elizabeth Armstrong
The Making of Johnson's Dictionary, 1746–1773
by Allen Reddick
Cheap Bibles: nineteenth-century publishing and the British and Foreign Bible Society
by Leslie Howsam
Print Culture in Renaissance Italy: the editor and the vernacular text, 1470–1600
by Brian Richardson

PRINT CULTURE IN RENAISSANCE ITALY

THE EDITOR AND
THE VERNACULAR TEXT, 1470–1600

BRIAN RICHARDSON
University of Leeds

PUBLISHED BY THE PRESS SYNDICATE OF THE UNIVERSITY OF CAMBRIDGE
The Pitt Building, Trumpington Street, Cambridge, United Kingdom

CAMBRIDGE UNIVERSITY PRESS
The Edinburgh Building, Cambridge CB2 2RU, UK
40 West 20th Street, New York NY 10011–4211, USA
477 Williamstown Road, Port Melbourne, VIC 3207, Australia
Ruiz de Alarcón 13, 28014 Madrid, Spain
Dock House, The Waterfront, Cape Town 8001, South Africa

http://www.cambridge.org

© Cambridge University Press 1994

This book is in copyright. Subject to statutory exception
and to the provisions of relevant collective licensing agreements,
no reproduction of any part may take place without
the written permission of Cambridge University Press.

First published 1994
Reprinted 1996
First paperback edition 2004

A catalogue record for this book is available from the British Library

Library of Congress cataloguing in publication data

Richardson, Brian.
Print culture in Renaissance Italy: the editor
and the vernacular text, 1470–1600 / Brian Richardson.
p. cm. – (Cambridge studies in publishing and printing history)
Includes bibliographical references
ISBN 0 521 42032 6 (hardback)
1. Printing – Italy – History – 15th century.
2. Printing – Italy – History – 16th century.
3. Bibliography – Italy – Venice – Early printed books.
4. Bibliography – Italy – Florence – Early printed books.
5. Editing – History – 15th century.
6. Editing – History – 16th century. 7. Transmission of texts.
I. Title. II. Series.
Z115.R53   1994
686.2′0945′09024 – DC20   93-30907 CIP

ISBN 0 521 42032 6 hardback
ISBN 0 521 89302 X paperback

Transferred to digital printing 2004

CONTENTS

Preface		*page* ix
Acknowledgements		xiv
Abbreviations		xv
Note on transcription		xvi
1	Printers, authors and the rise of the editor	1
2	Editors and their methods	19
3	Humanists, friars and others: editing in Venice and Florence, 1470–1500	28
4	Bembo and his influence, 1501–1530	48
5	Venetian editors and 'the grammatical norm', 1501–1530	64
6	Standardization and scholarship: editing in Florence, 1501–1530	79
7	Towards a wider readership: editing in Venice, 1531–1545	90
8	The editor triumphant: editing in Venice, 1546–1560	109
9	In search of a cultural identity: editing in Florence, 1531–1560	127
10	Piety and elegance: editing in Venice, 1561–1600	140
11	A 'true and living image': editing in Florence, 1561–1600	155
Conclusion		182
Notes		189
Select bibliography		235
Index of Italian editions 1470–1600		242
Index of manuscripts and annotated copies		251
General index		252

PREFACE

The edition of Dante's *Commedia* printed in Venice in 1477 has on its last page a sonnet which identifies in turn the four people responsible for the volume. First of all there is the author, 'Dante Alleghieri Fiorentin poeta'. Then comes the commentator, named as Benvenuto da Imola. These two are followed by a certain Cristoforo Berardi of Pesaro, described as the 'unworthy corrector' of the work 'in so far as he understood its subject' ('indegno correctore | per quanto intese di quella i subietti'). Last of all comes the German printer, Windelin of Speyer.

Of the four roles represented in this epilogue, those of author, commentator and printer need no explanation. Less familiar, though, is the third figure, the 'correctore'. What were his probable functions in the relationship between Dante's poem and the assumed fifth character, the reader of Windelin's volume? Berardi's position in the list suggests that his tasks were carried out before those of the printer. He would, first of all, have prepared for Windelin copy-texts of the *Commedia* and of its commentary using one or more sources, most probably introducing his own punctuation and correcting aspects of the language as he saw fit. However, Berardi's duties extended beyond that of the textual critic, for he had overall responsibility for the contents of the volume. He included summaries of each canto and a life of the author: items intended, like the commentary, to help readers to find their way around the poem and to understand it. He no doubt also composed the colophon-sonnet, since only he could reasonably have described himself as 'unworthy'.

There was probably more than a grain of truth in this self-assessment, for the sonnet is a clumsy one. Yet a 'correctore' or editor could justifiably present himself as having played a key role in the production of a book such as this Dante because of his supervision of the text and because of his provision or creation of supplementary material to assist the reader. These were also among the main features which would affect the reception of the book and thus determine its commercial fortunes. Whether, therefore, one is considering vernacular print culture in the Italian Renaissance from the perspective of the printer or from that of the consumer, the contribution of the editor deserves close examination. He is the central figure of the present book, and his aims and methods form its principal subject.

The study of editors also touches on broader cultural issues on which print had a significant influence, even if not an exclusive one. One of the themes running through this book is that of the contribution of editors to the evolution of vernacular scholarship (that is, the establishment of reliable literary texts and their understanding) as a worthy partner to classical scholarship. The two aspects of the editorial function which we have already identified – the correction of the text and the provision of help for the reader – had their roots in long-established practices in the editing of classical texts. These date back at least to the third and second centuries BC, when the scholars of the Museum in Alexandria produced a standardized text of Homer, established a more helpful system of spelling, punctuation and accentuation, and wrote works to explain and comment on literary texts. From their example sprang a tradition of emendation and exegesis which passed to ancient Rome and thence to the medieval world. It was still flourishing in the fourteenth century in very different contexts. In northern Greece, for example, the Byzantine scholar Triclinius was emending classical Greek poetic texts and compiling new sets of interpretative notes or scholia. In papal Avignon, Petrarch was piecing together, correcting and annotating the most complete and accurate edition then known of the Roman historian Livy. And these skills were already being applied outside classical literature. To this same tradition belonged the task undertaken by Giovanni Boccaccio in Florence later in this century: that of editing all the vernacular poetry of his compatriot Dante and commenting on the early canti of the *Commedia*. After Boccaccio's time and for much of the fifteenth century, more attention was paid by humanists to classical texts, and vernacular literature was not treated with the same seriousness. In the last third of the fifteenth century, however, the gap between the two disciplines began to narrow: the status of the vernacular rose once more, interest in its history grew, and its older texts began to be treated with the same care that might be applied to a classical text. One of the outstanding instances, though it had nothing to do with print, was the compilation, probably in 1476 and by the foremost classical scholar of his age, Angelo Poliziano, of the collection of medieval Tuscan verse called the 'Raccolta aragonese'. Poliziano is not known to have edited any texts, classical or vernacular, for printing, but his influence can be detected in the best vernacular editions of the late Quattrocento and early Cinquecento, for instance those of Pietro Bembo. Later in the Cinquecento, the influence of the most rigorous traditions of classical editing is evident in the work of the Florentine Vincenzio Borghini, a friend of the great classical scholar Piero Vettori. It is no coincidence that the most scholarly vernacular editions were printed by two presses which specialized in classical Greek and Latin works: that of Aldo Manuzio in Venice and that of the Giunta family in Florence. We shall, though, also see a striking continuity between vernacular editing and the less reputable methods of those editing classical texts for the printing press, illustrated for instance by E. J. Kenney in *The Classical Text* or in Martin

Lowry's studies of Nicholas Jenson and Aldo Manuzio. Among the practices which aroused scorn even then were the failure to treat authoritative sources with the respect they deserved, the overuse of conjecture as a means of emendation, unscrupulous rewriting of passages which seemed obscure, vagueness about sources, even dishonest claims about access to non-existent manuscripts, and (though this was normally the fault of printers) an imprudently hasty rate of work.

There was, however, one major respect in which the tasks of vernacular and classical editors differed. When printing began, the vernacular (one cannot yet call it Italian) still had to go through a process of standardization. Italy was politically disunited, so there was no capital city to impose a linguistic norm. The Tuscan of the three great fourteenth-century authors, Dante, Petrarch and Boccaccio, provided a natural model for those writers who wished to rise above the confines of their regional tongue, but up to the end of the fifteenth century nobody thought of imitating it more than partially. And the rise of humanism in the fifteenth century meant that Latin became another powerful influence, particularly on the spelling, syntax and vocabulary of the vernacular. Thus in the second half of the fifteenth century most writers were using a language which involved some sort of compromise between their own spoken usage and Trecento Tuscan, while also containing many Latinizing elements. This was the case even with Tuscans themselves, since their everyday language had changed in several respects since the Trecento. But then, in the early Cinquecento, the view began to gain acceptance that vernacular literature could not become a worthy alternative and successor to humanistic Latin literature unless writers strove to learn and re-create the language of what was acknowledged to be the highest point of its tradition.

The move towards the selection and adoption of a Trecento Tuscan standard was naturally reinforced by the new pressures and opportunities of print culture. This is by its nature a meeting point of economic and cultural forces, and the work of editors was shaped not only by the world of letters to which they belonged by background and inclination but also by the commercial motives of printers. Among the features which distinguished print culture from manuscript culture were, firstly, a large initial investment in labour, in materials (the press itself, metal type, and especially paper), and in transport, with a consequent need to recoup this capital outlay as quickly as possible; secondly, the production of hundreds or thousands of theoretically identical copies simultaneously, aimed at a relatively wide audience rather than at one which would inevitably be restricted; and, thirdly, a cheaper unit cost, which made it easier for those who were relatively poor (and therefore probably less well educated) to afford books. If a printer was to be more successful than his competitors, then, careful thought had to be given to the needs and expectations of a varied and widespread public. One of the most important ingredients in the success of an edition would be its language. A linguistic colouring from outside Tuscany, or more precisely from outside Trecento Tuscany, might be acceptable to readers if a work were circulating in its region of origin, but it was very likely to restrict the market for a work elsewhere. Thus editors, with their pivotal role in this two-way relationship between the printing industry and the reading public, increasingly assumed the power to adapt the original text so that it was better suited to success in the new medium, to the mutual benefit of printers and of the purchasers of their products. Editors also had to think of the needs of inexperienced readers, those who needed more help in understanding texts and in learning to imitate those Tuscan works which were seen as models of correct usage. The annotations and glossaries which they compiled for such readers played their part in the spread of Tuscan among those who might not have used it before.

A second theme underlying this book is therefore that of the contribution of editors to the process of the establishment and spread of a Tuscan-based vernacular. It must be said that their efforts were, here too, often erratic: editors came from a wide cross-section of literary society and, like Berardi, did not always have a great deal of expertise in Tuscan usage. When grammars of the new language began to be printed, editors did not apply their precepts immediately or meticulously. But even their improvisation and inconsistency have an importance for us, since they allow us to measure the rate at which everyday linguistic practice was changing and the extent to which it deviated from the lofty precepts of grammarians such as Fortunio and Bembo.

Another of the broader features of Renaissance Italy which the study of editors can help to illustrate is the cultural rivalry, the divergences and mutual influences, between the various cities of the peninsula. This book focuses on two cities whose distinctive approaches to the written word, to fine art and to politics gave rise to much discussion in the Cinquecento: Florence and Venice. Such a narrowing of scope has meant the virtual exclusion of centres such as Rome, Milan, Bologna and Naples, which were nevertheless important and deserve further study, but which were not so influential in setting editorial standards. It is true that Florentines tended towards stubborn cultural isolation and that their printed books were often of interest only to a local readership. But Florence's proud independence, based on her long-established leadership in vernacular culture and her prominence in humanistic studies, helped her to maintain a tradition of vernacular scholarship which no other city could rival, and she did prove able in the course of the Cinquecento to learn from the innovations of Venetian culture. Venice's importance as a centre of editing also stemmed partly from the quality of her scholarly tradition. The Venetian state was at the vanguard of the new literature, based on the study and imitation of the Tuscan Trecento, which came to dominate Italy in the first decades of the sixteenth century, and one of the products of this study of earlier literature was a new, more painstaking attitude towards editing it. Allied to this, though, was a printing industry which was far larger and more prolific than that of Florence or of any other Italian city and which provided a medium through which the linguistic and editorial principles evolved in Venice could spread to other parts of Italy. The opportunities provided by the city's presses attracted men of letters from all over the peninsula, and this reinforced the tendency for Venetian editing to avoid the narrow patriotism which could restrict the success of Florentine books. Yet the emphasis on producing books in quantity and for a wide readership meant that Venetian editing could sometimes imitate the initiatives of other centres rather passively and even risked sacrificing the high editorial standards of which she showed herself capable early in the sixteenth century. A further story which this book sets out to tell, then, is how editors in Florence and Venice influenced or emulated each other at certain points, yet still retained an approach to their task which was characteristic of the cultural centre in which they worked.

ACKNOWLEDGEMENTS

My thanks must go first of all to Conor Fahy for his generous advice from the earliest stages of my research and throughout the writing of this book. Richard Andrews was also a constant source of help. David McKitterick and Kevin Taylor offered useful comments as the book was drafted. Giulio Lepschy gave valuable support and advice. I have a long-standing debt of gratitude to Carlo Dionisotti for his teaching and for pointing me to some overlooked sources. Among the others who have provided guidance and information are Robert Black, Mark Davie, Bernard O'Donoghue, Dennis Rhodes, Christina Roaf and Diego Zancani.

Paolo Trovato has shown many kindnesses to me during the years in which we have been working in parallel on the subject of editors, and I thank him too most warmly for his generous spirit of collaboration. His recent book, *Con ogni diligenza corretto: la stampa e le revisioni editoriali dei testi letterari italiani (1470–1570)*, is everything one would expect of him: incisive, perceptive, meticulous, constantly interesting and informative. My book inevitably overlaps with his to some extent, but I hope that it will prove complementary in certain areas, such as the work of editors in Florence and the efforts of editors to provide additional material which would help readers to use and understand key texts.

Generous encouragement and tolerance has come from my family: from my father, who sadly did not live to see the completion of a project which he followed with close interest, and my mother; and from Catherine, Sophie, Alice and Laura.

I am very grateful to the Leverhulme Trust for financial support which allowed me to take study leave in order to complete this book, and to the University of Leeds for a contribution towards publishing costs.

ABBREVIATIONS

AIS	Karl Jaberg and Jacob Jud, *Sprach- und Sachatlas Italiens und der Südschweiz*, 8 vols. (Zofingen, Ringier, 1928–40)
ASI	*Archivio storico italiano*
BLF	Biblioteca Laurenziana, Florence
BMC	*Catalogue of Books Printed in the Fifteenth Century Now in the British Museum*, 9 vols. (London, 1909–49)
BMF	Biblioteca Marucelliana, Florence
BMV	Biblioteca Nazionale Marciana, Venice
BNF	Biblioteca Nazionale Centrale, Florence
BRF	Biblioteca Riccardiana, Florence
DBI	*Dizionario biografico degli Italiani* (Rome, Istituto della Enciclopedia Italiana, 1960–)
ED	*Enciclopedia dantesca*, 5 vols. (Rome, Istituto della Enciclopedia Italiana, 1970–6)
GSLI	*Giornale storico della letteratura italiana*
GW	*Gesamtkatalog der Wiegendrucke* (Leipzig, Hiersemann, 1925–)
IGI	*Indice generale degli incunaboli*, 6 vols. (Rome, Libreria dello Stato, 1943–81)
IMU	*Italia medioevale e umanistica*
IS	*Italian Studies*
LB	*La Bibliofilia*
SB	*Studi sul Boccaccio*
SFI	*Studi di filologia italiana*

NOTE ON TRANSCRIPTION

I have edited quotations from medieval and Renaissance Italian sources in a fairly conservative manner, while attempting to make them easily comprehensible. Abbreviations have been expanded and the consonant *v* has been distinguished from the vowel and semivowel *u*. Words have normally been separated and accentuation, capitalization and punctuation have been modified, except in a very few cases where it seemed preferable to leave the text in its original state in order to show exactly what changes an editor made or to illustrate an unusual system of accentuation.

1 · PRINTERS, AUTHORS AND THE RISE OF THE EDITOR

THE FIRST CENTURY OF PRINTING IN ITALY saw a steady increase in the importance of the editor of vernacular texts. For the three decades after the appearance of the first dated vernacular book in 1470, the fact that an editor had been at work was not always considered worth mentioning, nor was particular attention drawn to it. Often editors were not identified; and even when they were, their names might be hidden away in a dedicatory letter or at the end of a book, in some concluding verses or in a colophon. However, around the start of the sixteenth century, this situation began to change. First of all, title pages of vernacular books announced the fact that they had been revised or that they contained additional material. The first page of the Venetian Petrarch of 1500, for example, said that the poems had been 'newly corrected' ('correcti novamente'). As competition between Venetian editions of Petrarch grew more intense, the next step was to use title pages to identify editors by name. One would only have learned the identity of the editor of the 1500 Petrarch, a certain Nicolò Peranzone, from reading the preface or the colophon. In 1508, though, the edition printed by Bartolomeo de Zanni included Peranzone's name on the title page, pointing out that he had corrected the poems 'with many shrewd and excellent additions' ('con molte acute et excellente additione').[1] This kind of prominent identification of the editor, alongside the author and the printer, gradually became normal practice. By the end of the sixteenth century, books could even occasionally contain portraits of editors.[2]

But editors, unlike compositors or pressmen, were not indispensable to printers. Their employment was a potential source of delay and expense which was no doubt avoided whenever a printer felt that it was sufficient merely to reproduce a manuscript or an earlier printed edition without alteration. What, then, were the benefits which outweighed these disadvantages and which, as the printing industry grew, meant that the role of the editor was drawn increasingly to the public's attention?

To a certain extent, the demand for editors came from living authors. Some of these, of course, took an active part in correcting their own texts both before and during printing.[3] But, in a climate of growing linguistic orthodoxy, with the Tuscan of the great writers of the Trecento becoming generally accepted as a literary norm, and with more attention being paid to punctuation and

1

spelling, those who were uncertain of their usage might ask someone else to check their work before printing. And the degree of correctness of an edition was not just a matter of personal satisfaction for authors: it could also have a bearing on any income they hoped to earn from the printing of their works. An early example of a writer having a work scrutinized specifically for printing is provided by Nicolò Malermi's translation of the *Legenda aurea* by Iacopo da Varazze, which was printed in Venice in 1475 only after the Venetian Malermi had had his language revised by a Florentine.[4] When Giorgio Interiano of Genoa sent his *Vita et sito de Zychi* to Aldo Manuzio to be printed in 1502, he gave it, said Aldo, on condition that it should be emended wherever necessary, though in the event Aldo corrected only the spelling, leaving everything else as it stood for the sake of authenticity.[5] By the mid sixteenth century, an editor could act on behalf of an author to see that a work was printed correctly. Pietro Aretino was delighted to accept Lodovico Dolce's offer to 'rewrite' ('riscrivere') the second book of his *Lettere* in 1541: it was important, he wrote to Dolce on 1 September, that a work should be well written and well punctuated, and he gave Dolce permission to add to or delete from the book as he thought fit. Bernardo Tasso found it natural to use Dolce and Girolamo Ruscelli as his agents when he wished to have some of his *Rime* printed in Venice.[6] Ruscelli was also one of those approached by a doctor, Giuseppe Pallavicino, who wanted to have a collection of his own letters corrected for printing. When Ruscelli proved unable to undertake the task, Pallavicino was grateful to be able to entrust the letters to another leading editor, Francesco Sansovino, in 1565 in the hope that Sansovino would be able to 'enrich' them ('con isperanza ch'ella ... le possa arricchire').[7] Bernardo Cappello concluded his collection of *Rime* (1560) with a sonnet claiming that he would hardly dare publish his work without the contribution of his editor, Dionigi Atanagi, calling him a new Aristarchus who gave polish and adornment to the wisest writers of the age ('O novello Aristarco, e 'n questa etate | et lima e fregio a' più saggi scrittori').

But the persons who most frequently employed editors were printers and publishers. There were various reasons for which they might decide to have a text edited. In the first place, they might want to improve its readability. If its language was considered unacceptable to the readers at which the printed edition was aimed, it was very likely that an editor would be asked to adapt it in order to meet his audience's demands. He could also be expected to improve punctuation and accentuation. Secondly, the printer or publisher would want the edition to be as complete as possible, especially if there were rival editions competing for the market. He would want to be sure that there were no gaps in the text. If the work had been printed before, he might want to attract purchasers by having some sort of supplement added. This might be additional text or some ancillary material which would make the text easier to understand and easier to consult: a table of contents, for example, or an index of proper names and of subjects, or an index of the first lines of a collection of poems. If the text was important enough, an editor might be asked to add one or more commentaries, or perhaps even to write a new one, or to add diagrams and illustrations which would act as a visual commentary on the text. He might include (or again, he might himself compile) material concerning the author's life and writings or exegetic material such as a glossary or linguistic notes.

Publishers of literary works would have had a third important consideration in mind. Such works were to be learned from or imitated as well as enjoyed. Their readers therefore needed to be reassured not only that the edition which they were buying was useful to them because of the range of its contents but also that its text was 'corretto': 'correct', that is, and usually also 'corrected' by an editor. Those producing the book therefore had to project an image of themselves as caring about accuracy, and the best way of doing this was to claim in the preliminary or concluding matter of a book that careful revision had brought it back to its pristine form.

In practice, such revisions tended to achieve not authenticity but conformity with contemporary standards. Nevertheless, printers and editors did not tire of using letters to the readers or dedications in order to compose, with differing degrees of honesty, variations on the same commonplace: that the carelessness, ignorance or greed of those responsible for the text up to now (whether scribes or printers) had been responsible for the degradation of the text, but that, thanks to the generosity and care of the present printer, the work was now in its original state. Among examples of the contrast drawn by editors between earlier printers and their current employer's striving for perfection, one can cite Francesco Tanzio's letter to the reader in his Milanese Petrarch of 1494. One could, he wrote, easily remedy the corruption to which printed books were liable, 'if printers had their books very diligently revised and corrected beforehand by men who were learned and expert in the subject-matter with which those books deal which they wish to print' ('se gl'impressori faccessono prima diligentissimamente rivedere e corregere da huomini docti et experti in quella facultate qual tractano essi libri che vogliono imprimere'). Unfortunately, he went on, few printers were willing to employ editors, because of 'their insatiable avarice' ('la insatiabile loro avaritia'), but his own printer, Ulrich Scinzenzeler, was an exception, putting honour before gain.[8] Similarly, Pre' Marsilio of Fossombrone, in the dedication of a Petrarch printed in Venice in 1513, applauded the invention of printing but lamented the lack of care, due to 'base gain' ('vil guadagno'), which had led to the corruption of both Latin and vernacular texts; his printer, though, Bernardino Stagnino, had a different motivation, being 'most desirous of every perfection in such a work' ('de ogni in tale opera perfectione desidera<n>tissimo'). In the middle of the sixteenth century, Lodovico Dolce and Lodovico Domenichi praised the care which Gabriele Giolito took over

3

the texts which he printed (Dolce in his Ariosto, Domenichi in his Petrarch), and Girolamo Ruscelli wrote of Giovanni Griffio's 'affection for scholars' ('affectione alli studiosi') and 'diligence in printing' ('diligenza nell'impressione') in one of his Petrarchs of 1554. Ruscelli also praised Melchior Sessa the younger effusively in 1558 because of the printer's generosity in bringing classical and the best vernacular works to the greatest degree of perfection achievable in printing and 'in providing them with annotations, explanations and other such things for the benefit and convenience of those who study them' ('con aiutarli d'annotationi, di dichiarationi, e d'altre cose sì fatte per utile e comodo de gli studiosi').[9]

A comparable strategy was to include a letter or poem written in the name of the long-dead author, but probably composed by an editor, extolling the printer's concern for the correctness of the text. This material would no doubt be intended to attract purchasers and could be reproduced unscrupulously by different printers with minimal changes. In a Latin and vernacular edition of the *Aesopus moralisatus* printed in Verona in 1479, a sonnet attributed to Aesop credited an anonymous 'correttore' for removing the errors (blamed on scribes) which previously afflicted both versions of the text, and pointed out how vulnerable to change and needful of correction were works destined mainly for children and the uneducated. In 1487 the Dalmatian Bonino Bonini reprinted the work in Brescia, merely inserting his own name in the sonnet so as to get credit for the revision.[10] An example from Florence is the *Decameron* of 1516, which was preceded by a letter to the reader purporting to be from Boccaccio. The writer says that the printer Filippo di Giunta, having collected several texts copied from the original, has newly printed the work, using the judgement of several learned Florentines, so that he has brought it back to the state in which it left the author's hands.

Finally, the employment of an editor could help the printer or publisher to stress that a book was up to date.[11] Newness was an important selling point in this period. Title pages often pointed out that a book had been 'nuovamente stampato': the Florentine *Decameron* just mentioned is one of many examples. If this information was considered important, it would have been even more useful to be able to claim that an edition came fresh from the hands of an editor. Thus, for instance, the title pages of Ariosto's *Orlando furioso* in the 1530s and 1540s often had phrases such as 'very recently printed and corrected' ('novissamente stampato e corretto') or 'very recently restored to its integrity' ('novissamente alla sua integrità ridotto').

Printers as different as Bonini and Filippo di Giunta, then, wanted to be seen to be satisfying their readers' demand for texts edited well and recently. But editing was not something that the great majority of printers could do for themselves. In general, printers were businessmen and craftsmen, only rarely men of letters. In any case, the running of the operations and finances of a printing house created a multitude of other demands on their time. It is true that in some exceptional cases a printer might edit his own texts. For instance, the printer (and priest) Boneto Locatelli of Bergamo corrected the *Rationale divinorum officiorum* of Guillaume Durand in 1491. We shall see in chapter 4 that Aldo Manuzio edited Petrarch in 1514. Some editors nominally made the transition to printing in their own right, though the extent of their involvement on the technical side must remain open to doubt.[12] But the fact that a printer claimed responsibility and credit for a text did not necessarily mean that no separate editor was involved. We know, for instance, that Vincenzio Borghini wrote at least two prefaces signed by members of the Giunta family.[13] From the beginning of printing in Italy and throughout the Cinquecento, printers who wished to attain eminence in their craft needed a network of editorial contacts among local intellectuals.

This was especially necessary for the academic and erudite market, at which many Italian presses were aiming. The strong humanist tradition in Italy meant that the reading public was justified in looking for respectable standards of scholarship in the printing of the Latin and Greek classics or in legal, philosophical, theological and medical works. Printers of such texts would therefore draw attention to their collaboration with men of letters. An advertisement for the company formed by the French printer Jenson in 1473 or soon after claimed that 'the excellent Nicholas Jenson has readers skilled in both tongues [i.e. Latin and Greek] and seeks out the best-known experts, not merely one of them but several ... so that in his texts there remains nothing to be supplied or deleted'.[14] And scholars were well aware of the way printers depended on them: a year later, dedicating his edition of Valla's Latin translation of Herodotus printed by Jacques Le Rouge in Venice, Benedetto Brugnoli was confident enough to assert that printers could not produce books properly on their own but needed others, not least those who would revise and correct copy-text for them ('qui exemplaria recognoscant atque emendent').[15] The Venetian state was particularly conscious of the relation between its prestige and a high standard of printing of classical texts. It could exert a degree of compulsion on printers to produce correct Latin texts: two book-privileges, which gave the exclusive right to publish an edition within its territory for a certain number of years, were granted in 1504 and 1514 respectively on condition that the legal texts concerned were diligently corrected ('castigati').[16] On 30 January 1516 the Consiglio dei Dieci took the remarkable decision to charge the scholar Andrea Navagero to check all classical works ('de recognoscer tute le opere de humanità') which were to be printed in Venice.[17]

Even when producing texts in the vernacular, still considered by many to be inferior to Latin, printers often needed to employ men of letters as editors. This was not just because many early printers were foreigners, but also because the editorial problems here could be just as great as with a Latin text.

First of all, there was a growing demand for printed editions of the Trecento classics: but it was difficult for printers to produce accurate ones, since the texts had become corrupt over the years. Furthermore, it was important to be able to satisfy the demand from many readers for commentaries and other aids to understanding these texts. As for works written outside Tuscany or since the Trecento, it might be felt desirable to adapt these at least to some extent in order to bring them more closely in line with Trecento Tuscan, whose importance as a model, even if only a partial model, was becoming accepted by an increasing number of users of the vernacular. But that was no easy matter. Tuscan was more or less distant from the native dialects of the great majority of those who worked for presses outside Florence, and even a knowledge of contemporary Tuscan did not make one an expert in the Tuscan of the Trecento. Latin grammars were readily available, but no grammar of Tuscan was printed until 1516. All these tasks, then, called for skills of a sort which most printers did not possess.

Since printers were often accused, as we have seen, of sacrificing quality in order to save money, they did not want to seem to be employing editors simply in order to increase their profits. They preferred to claim that their concern for accuracy arose out of an altruistic desire to benefit their public, with no expense being spared in this honourable cause. Thus the Venetian printer Nicolò Zoppino said, in a letter to the reader in an edition of 1529, that he would endeavour to make his future works welcome and praiseworthy to every reader both for their spelling and for their simple Tuscan language, regardless of any necessary expense and effort incurred ('sforzaromi in ciò che per l'avenire l'opre nostre et per la nuova ortographia et per la semplice lengua toscana grate et lodevoli appresso di ciascun lettore si restino, non mi curando di spesa né di fatica alcuna che gire vi potesse').[18] But no press could survive on idealism alone. A hard-headed businessman such as Zoppino would only have been prepared to spend money on the correction and usefulness of texts as long as he believed that this outlay was also going to promote sales. Even if printers did not like to say this, editors were naturally ready to remind them that a well-edited text could help a book to be a success rather than a failure on the bookstalls. In 1470 an editor of Latin texts, Giovanni Andrea Bussi, claimed that his prefaces allowed printers to put a higher price on their products.[19] Presenting Pulci's *Morgante* in the Venetian edition of 1502, Niccolò Massetti of Modena said that the work had become incomprehensible; he had corrected many verses 'so that the printer can sell it' ('perché 'l stampator puossa essa vendere').[20] Twenty years later, Cassiodoro Ticinese wrote of how Zoppino had asked him to correct and revise ('corrigendo ... rivedere') the works of the Florentine poet Girolamo Benivieni, pointing out that much profit would accrue to the printer as well as no little praise to the editor ('affermando che molto allui di emolumento et a me non pocha lode de ciò risulterebbe').[21]

During the first century of printing, therefore, an investment in editing came to be seen as one of the keys to success both by ambitious printers and publishers and by authors. A reputable printing house would want to win readers with the superior correctness and usefulness of its books, as well as with their typographical elegance. Authors would want to ensure that their works were reproduced accurately and would be received well, often because they hoped for direct or indirect financial benefit. But to what sort of persons did printers and authors turn when they wanted a text edited, and what relationships existed in practice between editors and their employers?

Printers in Venice and Florence came from a wide variety of social, cultural and geographical backgrounds. Nicholas Jenson, for example, was a Frenchman, probably a metalworker by profession; Aldo Manuzio, born near Rome, began his career as a teacher; Filippo di Giunta came of a Florentine family of weavers and was a stationer before turning his hand to printing. Most printers were laymen, but the clergy too had close connections with printing in Italy from the very beginning. Some presses were run by members of the secular clergy, others by friars: the Franciscans had a press at the Frari in Venice, a Dominican ran the press of Sant'Iacopo di Ripoli in Florence, where nuns helped to compose the type. Women could also be involved to some extent in the financial side of printing after the death of their husbands, as one sees from examples such as those of the widows of John of Speyer, Giorgio Rusconi and Melchior Sessa the younger.

Those who edited vernacular texts were just as diverse in their origins, save that they were almost all Italian and apparently always male.[22] At one end of the social scale were three of the most learned vernacular philologists of the Cinquecento, Pietro Bembo, Vincenzio Borghini and Lionardo Salviati, all members of distinguished Venetian or Florentine families. Nicolò Delfino, editor of the *Decameron* in 1516, was another Venetian patrician and served his state in various offices. Alessandro Vellutello was a member of a patrician family of Lucca. Others could claim respectability on account of their education or career. Girolamo Centone, Castorio Laurario, Lucio Paolo Rosello, Lodovico Domenichi and Francesco Sansovino all studied law in Padua (and, in Sansovino's case, also in Bologna) before working as editors for Venetian printers. Tizzone Gaetano may have given diplomatic and military service to the brothers Federigo and Pirro Gonzaga before becoming an editor in Venice in the 1520s.[23] Marco Guazzo was a soldier as well as a writer and editor. Editors might also come from the ranks of the regular or secular clergy: Franciscans were actively connected with the Venetian printing industry in its early years, and editors in Venice and Florence in the second half of the Cinquecento included members of the Dominican, Carmelite, Benedictine and Camaldolese orders.

Editors generally had some source of income other than the printing press.

Some were teachers, like Colombino Veronese, who helped two Germans print Dante's *Commedia* in Mantua in 1472 and was one of the first vernacular editors to be identified by name. Gerolamo Squarzafico and Antonio Brucioli also worked part-time for printers and part-time as teachers in Venice.[24] Members of the secular clergy might also come into this category. Pre' Marsilio, for instance, said in the Venetian Petrarch of 1513 that he had been fitting his editorial work into the gaps left by his priestly and teaching duties ('le mie sacerdotali e scolastice occupationi').[25] Church and school took up most of the time of Antonio Craverio, who worked for Giovanni Giolito in Turin in the 1530s.[26] Another cleric who was teaching Latin in 1587 and worked as a corrector for the Venetian press was Ascanio Guidotti from Naples.[27] In the fifteenth century, at least, some scribes helped printers by editing texts for them.[28] But the growth in the importance of editors in Venice from the 1530s onwards meant that a handful of secular men of letters were for long periods able to make a living there mainly or even exclusively from activities connected with the press, combining editing with work as translators, anthologists and sometimes authors. Even then it was exceptional for editors to work regularly for just one printer, as did Lucio Paolo Rosello for Gregorio de Gregori in 1522-3 or as did Dolce for Gabriele Giolito after 1542. Most were freelances, working for different employers as the opportunity arose.

In Florence, editors were probably mostly of local origin. But those working in Venice, by far the largest centre of printing in Italy, came from all over Italy, with many employees being drawn eastwards from Padua and beyond, and others northwards from the Marche or further south. To give a few examples, Rosello was from Padua, Domenichi from Piacenza, Squarzafico from Piedmont, Peranzone, Pre' Marsilio and Atanagi from the Marche, Sebastiano Manilio from Rome, Tizzone Gaetano from Pofi, near Frosinone; Ruscelli was born in Viterbo and had travelled as far afield as Milan, Rome and Naples. Tuscans were well represented, helped no doubt by their linguistic advantage. Vellutello, as we have seen, was from Lucca; Brucioli and Anton Francesco Doni were Florentines; Francesco Sansovino lived mainly outside Tuscany but was brought up by the Florentine architect Iacopo Sansovino; Francesco Baldelli came from Cortona and Tomaso Porcacchi was from Castiglione Aretino. Only Bembo and Dolce, of the best-known editors, were from Venice itself. The picture in the Venetian printing industry, at least, is one of an open society where differences of regional and social origin had little or no importance.

What was the position of an editor in relation to those with whom he was collaborating? In order to gather sufficient evidence for an answer to this question, we need to broaden our horizons to include Latin works and cities other than Venice and Florence; it seems fair to assume that, in the present context, the experience of editors dealing with different languages and in different centres were similar.

It needs to be said first that an editor could occasionally be an employer rather than an employee. He might be acting as publisher, instigating and financially promoting the printing of a book. In Venice, editors as well as printers, publishers and authors requested book-privileges from the state.[29] An editor might have a work printed wholly at his own expense, or might bear a proportion of the costs. Piero Donato Guadagnoni entered into a partnership with a printer and a notary in Umbria in 1470, according to which he was to provide a corrected legal text and capital of 150 ducats, and was to receive a quarter of the profits.[30] In 1472, Antonio Moretto and Squarzafico appear to have financed a Venetian edition of the letters of Leonardo Bruni, and Francesco Caimi financed his edition of Petrarch's *De vita solitaria* in Milan in 1498.[31] In Padua the French printer Pierre Maufer borrowed money, in order to help him finance editions, from people who were going to be remunerated for their services to him: 25 ducats in 1476 from Giacomo Bordegazzi, working probably as proof-corrector, and 100 ducats in 1477 from the lecturer Giovanni Pietro Carari and a colleague, Paolo Varisco, who were working for him as editors.[32] Sebastiano Manilio was granted a book-privilege for the 1494 edition of his translation of Seneca's letters and shared the costs (as the colophon says) with the printers Stefano and Bernardino di Nalli.[33] The colophons of some incunabula printed in Venice by the company of Ratdolt, Maler and Löslein say that the first two printed the books together with Löslein as corrector and partner ('socius'), though one cannot be sure that this referred to a financial partnership. Innocente Ziletti, teacher and later bookseller, is described as 'helper and partner' ('adiutor sociusque') in the 1476 edition of the Italian version of Petrarch's *De viris illustribus*.[34] Carlo Bembo obtained book-privileges for the Petrarch and Dante of 1501–2 edited by his brother Pietro and printed by Aldo Manuzio, and Pietro may well have been in financial partnership with Aldo.[35] Alessandro Vellutello financed his edition of Virgil (1534), his Dante (1544), and (jointly with Giovanni Giolito) the fourth edition of his Petrarch (1538).[36] A medical lecturer from the Bolognese Studio, acting as partner ('consocius'), editor and accountant to a printer in Modena in 1475, lived in the printer's house and received money for clothing expenses.[37] This situation was reversed in 1553–5 when Ruscelli had the printer Plinio Pietrasanta living in his house in Venice and paid him a salary.[38]

Most often, though, an editor would work at the request of a printer, a publisher, or an author. His function was that of an intermediary between the author's study and the printing house, but he remained to some extent aloof from the latter and normally prepared his copy-text away from the press. Tomaso Porcacchi wrote in his dedication of Bembo's *Asolani* in 1571 that he had been spending the previous summer in the country villa of a marquis near

Verona, writing in the cooler hours of the day and at other times discussing virtue, when letters arrived from the printer Giolito in Venice asking him to contribute to some planned new editions of Bembo's works. This self-portrait of an editor working in idyllic, scholarly seclusion, part of a courtly world into which printers entered cap in hand, was doubtless much idealized. But we do have some evidence of a clear distinction between the editor or editors who established the text, working outside the printing house, and the more humble copy-editor who added the final touches and saw the book through the press. Two Latin canon law texts printed by Nicholas Jenson in Venice in 1474 and 1475 were edited for him in Padua by Alessandro Nievo (da Nevo) of Vicenza, who delegated the routine work to a pupil of his, Pietro Albignani da Trezzo of Brescia. Letters by Albignani at the end of these volumes show that someone else, a certain Francesco Colucia from Verzino in Calabria, supervised the correct printing of the volumes in Venice on Jenson's behalf. Colucia, who had already edited Rutilius Taurus for the French printer in 1472, was asked by Albignani on both occasions to ensure that the pressmen did not stray from the path he had set out and that due care was taken over spelling and punctuation.[39] In 1479 Albignani edited these and other texts for the printer John of Cologne. Albignani was still working in Padua, still had (for at least one edition) a contact in Venice, a priest called Bartolomeo Pozzi (Puteus), and was still uncertain enough of the compositors to ask Pozzi to ensure that they followed his text closely.[40]

This separation of spheres of influence, combined with the rapidity with which composition of type and imposition were followed by printing, could well mean that the editor was not involved in proof-correction. This was evidently the case, for instance, when Gerolamo Squarzafico complained in 1484 that his Venetian edition of Petrarch was not printed or corrected as he would have wished; or when Francesco Tanzio claimed in his Milanese Petrarch of 1494 that any mistakes would be those which compositors almost inevitably made (he did not expect to see proofs, therefore, nor did he think anyone else would correct them effectively); or when Griffio wrote in the supplement to the *Decameron* which he printed for Valgrisi in Venice, 1552, that the house of the editor, Girolamo Ruscelli, was so far away from the press that it was not possible to consult him whenever the printers were unable to decipher his original.[41] Later, Tomaso Garzoni distinguished 'il correttore' from the proof-reader ('lo scontratore') in his list of those involved in book-production.[42] However, Pre' Marsilio described in 1513 how he acceded to Bernardino Stagnino's request that he should correct proofs on the day of printing ('son condesceso a la requisitione et instantia de meser Bernardino Stagnino ... a dover correggere ... e libri ne la sua officina a la giornata da stamparsi') as well as preparing Petrarch's *Trionfi* for the press. Filippo Beroaldo the younger assured readers of his edition of Tacitus's *Annales* (Rome, 1515) that he had proof-read the printed sheets together with the original manuscript and noted any misprints in his errata (f. 74v). Atanagi corrected first proofs at home in Venice in 1562, then visited the printer to check a second set of proofs, as well as sending a set to the absent author for his revisions.[43] He made it clear that his visits to the printing house were very inconvenient, as if they called for some extra reward for his services. But wherever proof-reading took place, it is natural to suppose that editors often had the dual role of preparation and correction. They could also have an interest in the layout of the text and might give the printer instructions on this.[44]

The appearance of Alessandro Nievo's name in the two legal texts just mentioned, when their detailed editing was apparently carried out by a pupil, suggests that printers might occasionally use the name of a distinguished local figure in order to lend prestige to an edition, even though the person concerned had made little or no contribution to the preparation of the text. One such figure, at the other extreme from the sort of editor (such as Pre' Marsilio) who entered the hurly-burly of the printing house or who undertook hack work in order to make ends meet, may have been Tommaso Sclarici dal Gambero, a retired teacher of law from the University of Bologna with some reputation as a poet.[45] An edition of Petrarch printed in this city by Francesco Griffo in 1516 includes a Latin letter from Sclarici which describes how the printer had come to him in a state of anxiety over the validity of the text which he was printing, and how Sclarici duly gave him reassurance. By printing this letter, Griffo could show potential readers that his Petrarch, approved by a local academic who was also a poet, was preferable to rival editions from other cities.

But the relations between editors and pressmen were not always smooth. Fifteenth-century humanists had often viewed printers with distaste, and even those who worked as editors did not hesitate to bite the hands that fed them. This was partly for reasons to do with scholarship, because printers were often unlettered and allegedly debased learning. But in some cases it must also have been out of social snobbery: printing was described by a character in a mid-sixteenth-century dialogue of Doni as a 'mechanical and sordid' or 'plebeian' business in which many were only motivated by a desire for gain.[46] Moreover, some earlier printers had a reputation for rowdiness.[47] We have already seen some instances of editors criticizing earlier printers, and we shall see in the next chapter some criticisms voiced by editors about the way in which printers did not always follow the text which they were given. There was a tendency for editors to brand the whole race of printers as ignorant and mean, sometimes linking their ignorance with that of previous editors. Francesco Tanzio lamented, in his Petrarch of 1494, the way in which printed texts degenerated (like manuscript ones) from edition to edition. He blamed this on the fact that printers were all unlettered ('illitterati') as well as on their stinginess in not employing editors. Nicolò Peranzone complained in his 1500 edition of the *Trionfi* that Petrarch's works had become incorrect through the ignorance of both 'correctori' and printers. In 1516 Castorio Laurario, writing in a Venetian edition of Boccaccio's *Corbaccio*, said that the avarice of printers, always wanting to save metal, made it impossible for editors to avoid mistakes and hence criticism.[48] The anonymous editor of the Florentine *Decameron* of the same year condemned the greed and ignorance of earlier printers of the text, as well as the audacity of earlier (Venetian) editors in believing that they knew Florentine better than the Florentines. Lodovico Dolce followed a similar line in his editions of the *Decameron* (in the dedication to Bembo in 1541 and in the letter to the readers in 1552). The avarice of printers, he wrote, and the ignorance of those responsible, both pressmen and 'correttori', meant that there were few authors in any language who had not needed, or who did not still need, long and very diligent emendation ('castigatione'). Most people now thought only of gain, wrote Ruscelli in his edition of Pandolfo Collenuccio's *Compendio dell'historie del regno di Napoli* (1552), and bookmen had a proverb that 'good and bad books alike, they all sell, and so the less one spends on them, the more property remains in the house of those who have them made' ('i libri così tristi come buoni, tutti si vendono, e però quanto meno spesa si fa in essi, tanta più robba resta in casa di chi gli fa fare'). He thought it was time for states to take as much care over regulating printing as they did over regulating cloth-dyeing and other trades.

In view of such criticisms, it is not surprising that printers did not always feel warmly towards editors. For obvious reasons, it was not easy for them to express their views, but there are exceptions, rare cases of printers taking their revenge on an editor. Griffio's note prefacing the errata in Ruscelli's *Decameron* of 1552, mentioned above, complained that Ruscelli's original was so difficult to read that even the editor himself could hardly understand it in places. Ruscelli could be understanding about the problems which his corrections caused for printers: in his heavily-revised edition of Collenuccio's *Compendio*, he introduced a note on the printing errors with the comment that it was impossible to avoid them 'in such a confusion of crossings out and replacings' ('in tanta confusione di cassature e rimettiture'). But Ruscelli evidently gained a reputation for intolerance. Bitter criticism of him is found in the comments which accompany the long list of errors in the *Del modo di comporre in versi* printed by Giovanni Battista and Melchior Sessa in 1559. The two brothers said that they were giving the list, even though most of the errors were self-evident, so as not to give Ruscelli reason to make one of his usual attacks at the end of books against 'us poor printers'; he wanted them to lend him their hands in order to attack them by making them print words against themselves ('per non dar causa al S. Ruscelli di far contra noi poveri stampatori qualche invettiva, come suol far bene spesso in fine de' libri, e vuole che noi stessi gli prestiamo le mani per darci de' pugni, facendoci a noi medesmi stampar parole contra di noi').[49]

What of the ways in which editors were seen by those contemporary men of letters who did not work for printers? Some of their reactions were very critical. In the very first years of printing in Italy, Niccolò Perotti lamented the corruption which the new art had brought to classical texts. He blamed 'not so much the ignorance of printers as the negligence, to use no more harsh a term, of those who set themselves up as correctors and masters of the books of the ancients' ('huius autem rei causa est non tam inscitia eorum qui imprimunt quam negligentia, ne quid gravius dicam, quorumdam qui se correctores ac magistros veterum librorum constituunt'). These men, he said, thought that what they did not understand sufficiently was wrong and consequently corrected that which needed no correction.[50] Over half a century later, Nicolò Liburnio wrote that Italy had been flooded with printers to such an extent that only a few presses, such as Aldo's, produced Latin and Tuscan texts which were well edited ('debitamente castigati'). His solution to the problem of correctness was for rulers to order books to be sold at higher cost; in this way (presumably because better editors could be afforded) books would not have to pass through such rough hands on their way into print.[51] But writers might also accord to editors a rather grudging acceptance, seeing them as a necessary evil. After Paolo Giovio had read the manuscript of Vasari's biographies of artists early in 1548, he wrote to warn the author that, if he wanted the printing to be free of faults, he would have 'to drink the chalice of paying an assiduous and diligent editor' ('bere il calice di pagare un correttore assiduo et diligente').[52]

A similar contrast between attitudes towards editors is shown by two of the best-known authors of the later Cinquecento, Torquato Tasso and his father Bernardo. Torquato's attitude is a negative one, but for reasons different from Perotti's: he believed that writers were most worthily employed as secretaries of princes and should avoid menial association with the printing industry. In one of his dialogues, *Il Minturno*, he uses as interlocutors the poet Antonio Minturno and the editor Girolamo Ruscelli, who had edited the works of Minturno in 1559 and who had helped Bernardo and himself in pleading their case to the Spanish king, Philip II.[53] Minturno remarks on Ruscelli's prudence in separating himself from the common people and schoolteaching ('le scuole de' fanciulli') and in raising himself instead to working with statesmen; he is at a loss to understand why Ruscelli has not been imitated by other men of letters working for printers in Venice. Likewise, Ruscelli portrays the period of drudgery which he spent in the service of Venetian printers as an unwelcome parenthesis in his life, forced on him for want of worthier employment. In Rome, he says, his talents had received nothing more tangible than praise; as for Naples, society was too bound up in fashionable trivia. For a short space, therefore, Ruscelli earned his living by attending to the correction of books in Venice and by ensuring that his books were the most beautiful and the best understood. But, fortunately, says Tasso, he was able to give up this hack work and return to the Kingdom of Naples to concern himself principally with matters of state.

It is true that the life of a courtier was seen as an attractive, if elusive, alternative to work as an editor.[54] Nevertheless, Tasso's portrayal of Ruscelli's editorial career as brief and unwelcome is a considerable deformation of the truth. For one thing, the short period which Ruscelli is said to have spent in Venice lasted for over a decade and a half, from about 1549 or 1550 until his death in 1566. And in the dedication of his *Del modo di comporre* Ruscelli describes his debt to Venice, and his affection and admiration for the city, in terms which appear to be genuine enough.

Tasso is also ignoring the importance which editors had by then acquired both for printers and for authors. For an example to complement the opinion of Giovio quoted above, one need only look at some of the letters from Torquato's father Bernardo to Ruscelli and Dolce. In April 1554 Bernardo wrote to Dolce that he particularly wanted the first four books of his *Rime* to be printed by Gabriele Giolito, rather than by any other printer, so that Dolce's diligence might make up for any faults in them. He sent the manuscript to Dolce in October, giving him authority to alter not just matters of form ('la scrittura') but also sentences and words. The edition was duly printed by Giolito in 1555. But Bernardo was to regret the trust which he had placed in Dolce. In 1556 he wrote to another printer working in Venice, Giacomo Giglio, to complain about Giolito's edition. He had asked Dolce, he says, to check that 'la scrittura' was well punctuated and well corrected; however, Dolce allowed the work to appear full of confusion and errors which were not in the copy-text. Bernardo therefore turned to Giglio to enquire about the possibility of a further edition, relying this time on Dolce's rival, and Giglio's partner, Ruscelli to act as his editor. A year later Giglio had died, but Bernardo still pleaded with Ruscelli to help him have his works reprinted. Dolce, he said, had (among other things) removed the titles which he had given in order to elucidate the destination and meaning of his poems, and had removed some poems, apparently for political reasons. He could not afford to have the poems reprinted at his own expense; all he asked for was twenty free presentation copies. Later in 1557 he wrote again to Ruscelli to inquire about the cost of printing 2,000 copies of his epic *Amadigi*, so that he could find the necessary sum of money.[55] Because of one editor's ill-considered interventions, then, Bernardo requested the mediation of another editor in order that his work might be printed correctly. In the event, he was reconciled with Dolce, who provided a letter in defence of the *Amadigi* when this was printed in 1560.[56] It is clear that editors were now acting as agents who could arrange for a work to be printed and (unless the author was able and willing to supervise the printing process himself) had also acquired considerable control over the author's text, making suggestions for improvements (as Bernardo asked Ruscelli to do with the *Amadigi*), preparing the manuscript for publication, and presenting the text to the public in a favourable light. Torquato Tasso's Minturno might despise the profession of editor, associated as it was with commerce and the schoolroom; but Bernardo Tasso knew that in practice, and perilous though the experience could be, authors might look to editors to ensure the success of their works.

Editors, like printers, sometimes asserted that their motives were philanthropic. Giovanni Andrea Bussi, editor of classical texts for the first printers in Italy, Sweynheim and Pannartz, wrote in his edition of Aulus Gellius that he wanted to make all men of the Latin world and himself more learned, and said that he was not paid for his work.[57] A claim on the part of editors that they wanted 'to be of benefit' became a commonplace. Thus Squarzafico wrote in 1477 that he had edited Asconius's commentaries on Cicero's speeches 'for the public benefit' ('pro utilitate publica'). In a book-privilege request of 1512, fra Giocondo claimed to have edited Latin texts to the common benefit of students and to have laboured and purchased old copies with a desire to help everyone ('zovare ad ognuno'). And Ruscelli wrote at the end of his life that readers could recognize in his writings an overwhelming desire to help each one ('giovar a ciascuno') as far as he could and to win their goodwill, and that this was dearer to him than 'all worldly and transitory wealth' ('ogni mondana e transitoria ricchezza').[58] Pre' Marsilio claimed in his Petrarch of 1513 that he was constrained to work for Stagnino not so much by the need to earn a living ('dal cottidiano vivere') as by the natural instinct of being born into the world not for himself alone ('dal naturale instinto de non esser a me solo nel mondo nasciuto').

Altruism apart, editors also worked, of course, for prestige and financial gain. We saw earlier that some relatively wealthy editors invested in their own publications and would have received a certain number of copies or a share in profits. Those working for others would also have expected some tangible regard. Bussi's claim about working to bring enlightenment is belied by his gratitude to Sixtus IV for appointments which, he hoped, would enable him to give up the chores of editing.[59] Printers certainly anticipated that the debit column of their balance sheets would often have to contain entries relating to an outlay on editorial work. An agreement drawn up in 1507 for collaboration between four Venetian printing houses anticipated that the company would have to meet the costs of buying copy-texts, correcting them, making indexes, and 'all other expenses pertaining to the usefulness of the works' ('tute altre spexe pertinente a la utilitade de le opere').[60]

A variety of terms were used in employing editors. Documents from the early days of printing in Padua show that one variable was the means of payment, with some university teachers being paid or asking to be paid for their work solely in cash and other editors being paid in kind, and that another variable was the length of contract, since payment could be arranged for a specific edition or for a certain length of time. Maufer contracted in 1476 to give books to the value of 30 ducats a year to a doctor of arts, Giacomo Bordegazzi, in payment for his correction of any books produced by Maufer's four or five presses.[61] Giovanni Pietro Carari agreed to provide Maufer with a corrected manuscript of a commentary on Avicenna, giving him each day the number of pages necessary to keep four presses working. Printing took place in 1477 and lasted for six months. In exchange, Carari was to have sixty-two copies of the finished work, forty as financial backer and twenty-two as editor.[62] With the market price at 4½ ducats, he could have earned 99 ducats as editor alone. This was a more than respectable sum, worth more than the recompense of between 30 and 48 ducats plus board and lodging which a compositor might hope to receive in a year, and comparable with the 50 to 100 ducats which was the annual pay of a schoolmaster or university teacher of average reputation (though outstanding teachers could earn much more).[63] Bernardo Machiavelli (father of Niccolò) undertook to compile an index of geographical names in Livy's history for a Florentine printer in 1475, and in return could keep the printed copy from which he worked.[64] Books could also be exchanged for other goods rather than for cash: Squarzafico, who had a reputation as a tippler, tried to sell the histories of Livy to a nobleman of Treviso in 1471 for three *carri* of wine.[65] And editors doubtless presented some copies as gifts to potential benefactors.

Not all arrangements for payment were generous, and some haggling might be involved. A mere two copies of the *Speculum iudiciale* by Guillaume Durand were to be given in 1478–9 by the German printer Johannes Herbort to Francesco Brevio, who had provided the corrected text; but Brevio did not get even these, since the printer alleged that the text was incorrect. Some other pieces of evidence from the same period relate to *correctores*: maybe editors, or proof-correctors, or men acting in both capacities. In 1478, Matteo Albricci claimed from a printer called Giacomo (perhaps Jacques Le Rouge) two copies, worth 6 ducats each, of three works which he had helped to correct, first with one and then with two collaborators. He said in evidence that it was customary to give to correctors at least three copies of the books which they had revised. Another dispute concerned the fixing of a monthly salary. In 1479 a law scholar who had been working as a 'corrector' in Venice, Giovanni Nicolò Polletti, asked 6 ducats per month for his correction of a text by Avicenna. The printer offered him 4½ ducats (a wage comparable with that of a teacher), but they failed to reach an agreement. After a period of reflection, Polletti agreed to the offer but found to his anger that someone else had been employed in his place.[66]

On other occasions a combination of remuneration in cash and kind was used. In Padua, the printer Bono Gallo agreed to pay 10 ducats to the lecturer Nicoletto Vernia in 1474 for the correction of a work on Aristotle by Gaetano di Thiene, and was also to give Vernia twelve copies of the book.[67] There are two similar cases from Bologna. The two 'correctors' of a legal text were to receive 120 ducats each, a copy of the book in question and a copy of another legal text. This seems a large sum when compared with the printer's salary of 3 ducats a month, later raised to 5; but the book consisted of three large folio volumes, and over two years elapsed between the signing of the contract in 1473 and the completion of the book in 1475. In the following year the editor of a medical text, Matteo Moretti, a lecturer at the Bolognese Studio, was offered twenty copies of the book of which he was to provide a corrected exemplar for printing or, alternatively, a cash equivalent according to the price fixed for the sale of the book in Bologna. But he also had the option (which he took up) of making a financial contribution to the capital necessary to print the volume and of receiving, from the profits, a share corresponding to the amount he had invested. However, any recompense was to be forfeited if his exemplar was found to be incorrect.[68] A Milanese agreement of 1474 stipulated that the Pisan humanist Bonaccorso was to correct copies of three Latin historical works for the printer Filippo da Lavagna; in return, Bonaccorso was to receive 60 ducats, at the rate of 4 *lire imperiali* per ducat, and two copies of each of the books. A new agreement was drawn up between the two parties in 1475, valid this time for a whole year and renewable. Bonaccorso was to correct all works of rhetoric and poetry to be printed by Filippo with his three presses in return for an annual payment of 240 *lire imperiali*, to be paid in three-monthly instalments, and two copies of each book corrected by him. This was substantially more than the 120 or 144 *lire* paid to apprentice printers and compositors in 1500; but the agreement was evidently of benefit to the printer as well, since it lasted until 1477.[69]

The evidence from the sixteenth century shows similar combinations of methods of payment. Fra Giocondo wrote to Aldo in 1514 to suggest that he should be given one or two copies of the volume containing Perotti's *Cornucopiae* and Latin grammarians in recognition of his work on it. He also reminded Aldo of their agreement about his editorial work, pointing out that for his preparation of Columella he was to receive 10 ducats and ten copies.[70] An arrangement of the same sort was used by Bernardo di Giunta in Florence in 1518. Instead of regarding the unit of work as one book, however, he made a contract to give the editor Mariano Tucci an annual payment of 20 florins and books to the value of 8 florins (the florin was similar in value to the ducat).[71] Towards the end of the century Bernardo's sons Filippo and Iacopo offered Lionardo Salviati the choice between three methods of payment for his edition of the *Decameron*, printed in 1582: 200 *scudi* in cash (100 immediately and the rest after six months), the *scudo* being worth rather less than the florin which it had replaced earlier in the Cinquecento; the forming of a joint company with profits to be shared equally; or the sum of two *carlini* (about one fifth of a *scudo*) for each copy sold.[72] But printers did not only provide tangible benefits of this sort: a potential indirect method of payment was the editor's right to dedicate the work to a patron or pupil or to some other influential figure from whom he hoped to gain money or other favours. Angelo Riccio, for instance, accepted the task of 'revising the sheets' ('rivedere i fogli') for the fourth edition of the *Rime* of Giuliano Goselini (1581) on condition that he could dedicate it to whoever he chose; and Francesco Melchiori of Oderzo was given the same right when he was chosen by Goselini to oversee the fifth edition (which came out in 1588, the year after the poet's death).[73]

Dionigi Atanagi's letters from Venice in the early 1560s tend to stress the risks involved in editing without a clear contract. In 1562 he was working on credit for the patrician Mocenigo family. In March he was expecting at least some wine and flour and hoping for some cash. He referred pointedly to the way in which he had been let down by Iacopo Zane, whose poems he edited in that year: Atanagi must have been expecting to receive a proportion of the books, for he complained that Zane had sold them all himself. It was the last time, he said, that he would serve a gentleman without a pledge in his hand ('senza il pegno in mano'). Yet this failed to arouse any pity in the Mocenigo household, and by May he had still received nothing. Atanagi also complained of lack of luck with his dedications. He wrote in 1561 that the last three had been in vain and that, if the fourth (for the collection of verse on the death of Irene di Spilimbergo) were unsuccessful, he would give up 'dedicating to masters' ('il dedicare a signori').[74]

The financial and other rewards for editing, then, were far from certain but could be quite considerable. The size of the payment offered in some of the contracts mentioned confirms what was said earlier about the value to printers of an investment in the correctness of their texts.[75] Nevertheless, the occupation was an irregular and precarious one, and it had a relatively low standing in the eyes of other men of letters. In most cases, editing will have been just a way of earning extra money; only a few can have taken it on for motives of pure scholarship. This is something which we must bear in mind as we go on to look more closely at editors' working methods.

2 · EDITORS AND THEIR METHODS

WE HAVE BEEN CONSIDERING SO FAR THE RELATIONSHIPS between editors and those for and with whom they worked. We now need to look at the factors which shaped their methods and results. Firstly, we can examine some contemporary attitudes towards the transmission of written texts by considering the standards for the copying of vernacular texts which had been established in the manuscript age and by assessing the nature of the guidance which classical textual criticism offered to vernacular editors when printing began. Secondly, we can look at practical constraints arising specifically from the aims and methods of the printing industry.

From the earliest days of the circulation of vernacular texts in manuscript, scribes would often deliberately adapt the language or even the contents of their exemplars according to personal tastes. It is worth analysing briefly the ways in which they treated their sources, since their methods provided the natural model for editors when texts began to circulate in printed as well as in handwritten form.

One influence for change was that of the place of copying of a text, especially when this differed from its place of composition. A well-known example is that of the Tuscanization of the poets of the thirteenth-century Sicilian school which is found in the three earliest surviving Tuscan *canzonieri*, and which was very probably already present in their archetype.[1] Tuscan authors of the same century were themselves subject to linguistic adaptation. This could occur even within Tuscany, as with the Pisan-Lucchese colouring given to the letters of Guittone d'Arezzo in the Laurenziano Rediano 9. It also occurred when Tuscan poetry or prose was 'exported' in the Trecento and brought closer to the usage of other regions.[2] A similar process could affect the circulation of texts within the north or the south of Italy: one Bolognese text underwent change in the Veneto, for instance, and a Sicilian one took on Calabrese or Neapolitan features.[3] In the late Quattrocento, the *Arcadia* of the Neapolitan author Iacopo Sannazaro was Tuscanized in a manuscript written by a scribe who was probably Tuscan. But there could be a deepening of regional colour, as one finds in a Neapolitan manuscript of the same work.[4]

Adaptations could also result from differences in linguistic fashion between periods of time or between cultural milieux. The mid-Quattrocento manuscript which is the oldest surviving copy of Dino Compagni's *Cronica*, written between about 1310 and 1312, contains linguistic features which originated only in the Quattrocento.[5] Certain manuscripts of Dante's *Commedia*, the 'gruppo del Cento' written in Florence between about 1340 and 1360, present characteristics due both to changes in usage since Dante's time and to the interest shown in the work by the Florentine merchant class.[6] Then, particularly in the Quattrocento, scribes tended increasingly to Tuscanize texts from outside Tuscany. Behind this trend lay the combined prestige of Dante, Petrarch and Boccaccio, helped by the nature of the Tuscan dialect itself, which shared characteristics of both northern and southern dialects and was relatively close to Latin, the major language of learning. (The Tuscan of the Quattrocento, on the other hand, had no authors of anything like comparable prestige and was in any case harder to imitate because it had acquired a greater range of alternative forms since the Trecento.) An example of the spread of Tuscan is provided by a *canzone* written in the Trecento by Antonio da Ferrara. In its original form, the language is a hybrid combination of Tuscan and northern Italian with additional Latin influences; but in manuscripts of the fifteenth and sixteenth centuries the poem is Tuscanized.[7] For the moment, though, influences other than Tuscan could still be felt. A fifteenth-century manuscript of the Trecento Sicilian *Istoria di Eneas*, probably copied on its island of origin, has Tuscan as well as Sicilian features but also contains forms imported from Aragonese Naples.[8]

Another characteristic of the manuscript transmission of vernacular texts was one shared with all types of copying: the tendency to try to correct anything thought, rightly or wrongly, to be a material error. In some cases this was done by drawing on other manuscript sources. Such a process of contamination affected Dante's *Commedia* from a very early stage. The pressure of demand on copyists, who were not always equal to the task of reproducing a long and complex work, led to errors which in turn gave rise to the need to seek corrections from independent manuscripts.[9] When Boccaccio compiled his collections of Dante's poetry, he attempted to eliminate errors through this eclectic form of correction. But he also made use of what was then, and was to remain for many centuries, an even more common source of editorial 'improvements': arbitrary conjectures based on his own taste.[10] Of course, since linguistic tastes change, this procedure was particularly risky. For example, over a century after the death of Cavalcanti, someone in the humanist Florence of the early Quattrocento saw that a line from one of his sonnets, 'di farne mercé allei non tardo', was one syllable short of the required eleven, and tried to remedy this by changing it to 'di lei gratificar giamai non tardo'.[11] This emendation now seems clumsy because a Latinism such as *gratificare* is incongruous in Duecento lyric poetry. But such sensitivity to the independence of different phases in the evolution of the vernacular was acquired only gradually in the Cinquecento. At the time when printing began, the literary languages of the past and present were still seen as part of the same continuum, so that what was currently fashionable could be superimposed on the usage of an earlier period.

The further a work was, in its readers' perception, from the literary canon, the more subject it was to reworking. The *Decameron* was not generally ranked as highly as Boccaccio's other prose works until the Cinquecento, and its manuscript tradition is characterized, as Branca has shown, by the creative copying of amateur scribes from the mercantile milieu who felt free to add episodes from other sources or to emphasize or suppress Boccaccio's comments on the characteristics of the inhabitants of certain cities. Similarly, friars would alter the text of the translation of Aesop's fables to suit their own prejudices: for example, the swallow, coloured black and white like a Dominican, was accompanied by complimentary or contemptuous adjectives according to whether the copyist came from this order or was a Franciscan.[12]

Latin and Greek texts were naturally treated with more respect. Nevertheless, classical textual criticism was still in its infancy when printing began; even for the most serious of vernacular editors, it provided no ready-made method to borrow. A major practical obstacle to the development of any form of textual criticism, classical and vernacular alike, was the difficulty of tracing and getting access to manuscripts or even to early printed editions in order to use them as sources. It would have been hard to know what was available in one's own city, let alone in other cities and in other states; and, even if one knew where books were, they might be jealously guarded by their owners.[13] This situation helps to explain why source material was often poor. Yet, as a few examples will show, even classical editors used their sources in ways which today can seem haphazard.

The editor or editors of the Aldine printing of Aristotle's *Metaphysics* (1497) used their Quattrocento manuscript 'in a very free and easy fashion', as Martin Lowry puts it, and the exemplar of the *Nicomachean Ethics* (1498) was followed equally erratically.[14] For the letters of the younger Pliny (1508), Aldo had used his contacts in order to obtain what was probably a manuscript of the fifth or sixth century, so ancient in appearance that he presumed that it dated from Pliny's time; but he found it hard to read and used it carelessly and inconsistently.[15] If editors were using more than one source, they did so eclectically and arbitrarily rather than according to any systematic method. Examples can be found among Aldo's Greek editions, such as the Sophocles of 1502, the Euripides of 1503, and Plutarch's *Moralia* of 1509 edited by Demetrius Ducas. For the last of these a thirteenth-century manuscript was used, but apparently only after part of the work had already been printed; the inference is, in Lowry's words, 'that the process of criticism and emendation did not precede that of printing but advanced jerkily alongside it, step by alternating step', and that there was no coherent recension.[16]

Such casual use of sources was often compounded by excessive recourse to conjecture in an effort to improve on these sources. Bussi's edition of Pliny's *Natural History* (Rome, 1470), for instance, contains alterations to its manuscript source which were at best unnecessary, at worst violent.[17] The unidentified scholar who edited Caesar's *De bello gallico* for Jenson's Venetian printing of 1471 was faced by a passage in which an infinitive had been misplaced; he tried to make sense of it by adding a phrase of his own but inevitably only made the text more corrupt.[18] At the other extreme from such interventions, though, an editor might print an evidently confused text without attempting to resolve the problem.[19]

Contemporary works in the classical languages were less likely to need emendation but suffered, if anything, from even more cavalier treatment by editors. Aldo Manuzio's edition of the Greek grammar of Constantine Lascaris merged two sources, representing slightly different drafts of the work, into a single hybrid version.[20] Giovanni Tortelli's *De orthographia* had not fared any better in Rome in 1471. As Luisa Capoduro has shown, the editor, almost certainly an Augustinian friar from Genoa called Adamo da Montaldo, completed quotations left unfinished by Tortelli and added new ones; he inserted his own observations, some inspired by local Genoese knowledge; and he added ideas on spelling which could contradict those of Tortelli, as when the author suggested the spelling *Creta* but the editor then proposed *Crheta*.[21]

It must also be said, though, that here and there one can detect elements of an approach to classical textual criticism which offered hope of better things to come. One of the last Greek works printed by Aldo, the lexicon of Hesychius (1514), has been praised for the skill, linguistic knowledge and shrewd emendations of its Cretan editor, Marco Musuro.[22] There were, too, some sound statements of editorial principle, even if they were not necessarily translated into practice. We have already seen, in chapter 1 (p. 13), how Perotti warned in the early 1470s about the abuse of arbitrary correction by editors of Latin texts. He went on to argue that it would be prudent to charge someone by papal authority with the duty of regulating printing and of employing someone moderately skilled to examine and emend all the formes ('tabellas') of texts when they were about to be printed and to see that the original editors ('correctores') did not dare to conjecture anything rashly. He urged these editors, when correction was necessary, to keep to a limit which he had defined earlier in the letter: they should add nothing of their own but emend either from other copies, or from the same author's meaning expressed more clearly elsewhere, or from the authority of the writer who was being used as a source, or from the clear imitation of somebody, or from the evidence of Greek. Giorgio Merula voiced similar feelings in 1472, criticizing editors for following their own instincts rather than relying on the evidence of the available sources (though ironically he made this attack in the preface of an edition of the *Scriptores rei rusticae* which contained many blemishes).[23] Around 1490 the two leading scholars of Florence and Venice, Angelo Poliziano and

Ermolao Barbaro, showed their contemporaries the importance of respect for the readings of old manuscripts.[24] Poliziano pointed out in his first *Miscellanea* (chapter 57) that old codices keep vestiges of the true reading, which are subsequently obliterated by scribes. His connoisseurship of such things was seen by most people as doctrinaire and illogical: even Matteo Bosso, who had lent Poliziano an old copy of Ausonius, wrote around the turn of the century that 'Poliziano is wont to judge codices like wines, by age rather than by reason' ('solet enim Policianus codices, quasi vina, magis vetustate quam ratione probare').[25] Poliziano was also unusual for his age in his understanding of the need to see the genealogical interrelations of groups of manuscripts, the care with which he identified the manuscripts which he cited, and his awareness of the history of linguistic usage. In the second *Miscellanea* (chapter 5), he set out a hierarchy of evidence to be used in emendation, similar to Perotti's list: first the authority of an old source, then the evidence of a suitable writer, and lastly the sense, as long as this was not established arbitrarily ('veteris auctoritas codicis', 'scriptoris idonei testimonium', 'sensus' which was not 'undecunque decerptus').[26]

If care in using one's sources and wariness about using conjecture were exceptional among editors of Latin and Greek texts, one must not expect to find different standards applied to vernacular texts, naturally less prestigious objects of study. In later chapters we shall see many examples of the use and abuse of conjecture. Editors of printed vernacular texts could also be extremely vague about their sources, using perhaps a phrase such as 'un libro antico' without telling us approximately how old it was or even whether the book was manuscript or printed. One sometimes has the impression that, if they name the owner of a 'libro' containing a particular reading, this is done to show the range of their social contacts and to flatter the person concerned. And, in the transfer to print, the vernacular text was even more malleable than the classical text, more liable to be adapted to suit the tastes and interests of the moment. As in manuscript copying, regional colouring could still be altered; old-fashioned usage could still be replaced by something more up to date; references that were unwelcome, for instance for political reasons, could still be removed.

Lack of respect for earlier states of a text, whether classical or vernacular, can also be seen in the treatment of manuscripts used as printers' copy. Their normal fate, wrote Aldo Manuzio, was to be torn apart and die in order that the new-born text might emerge.[27] The risk of damage was recognized in the contract concerning Aldo's edition of the *Epistole* of St Catherine of Siena: he had to promise to return undamaged the manuscript and printed texts which he had received for use as copy-text, or else pay their value in cash.[28] The few exemplars that have survived show that editors and printers had no compunction about using early manuscripts as their copy-text, even though this meant inserting, in ink, corrections or instructions for divisions into chapters, pages, and so on. This method seems to have been preferred unless the original was difficult to follow or in need of completion or correction from other sources, or unless the editor showed exceptional reverence for it.[29] Manuscripts were not generally seen as museum pieces any more than the language of the texts they contained was seen as immutable. The decision to print direct from an old manuscript, rather than to have a new one copied out, could, of course, be intended to achieve accuracy, not just to save time and money, but the fact remained that the source might be damaged or even sacrificed in the interests of the printed text.

While printing brought obvious benefits as regards the wider diffusion of texts, it also brought circumstances which could make the establishment of reliable texts even more difficult than it had been in the manuscript age. Printers were working under pressure of time in order to recoup as quickly as possible their financial outlay in paper, type and labour. They had limited supplies of type, and thus it might not be possible to leave formes set up long enough for careful proof-reading. Printing often began before editing had been completed, and it then proceded at a rapid pace. One must remember that printers could be producing more than one work concurrently on their presses, so that there might be pauses in the production of a book; but even so, editors often lamented the speed at which they had to work. 'Revised with disorderly haste', in what spare time his work as a lawyer allowed, with the printers clamouring for their text, was Girolamo Bologni's description of Caesar's *Commentarii* in his edition of 1480, printed in Treviso.[30] Gerolamo Squarzafico had to race to keep ahead of the Venetian pressmen when he was completing a commentary on Petrarch in 1484. In Florence, an anonymous editor apologized to a publisher in 1492 that he had corrected Landino's *Formulario di lettere e di orationi volgari* only 'as far as the short time granted by the swift printer, greedy for gain, could allow' ('quanto el breve tempo dal veloce impressore, cupido del guadagno, concesso poteva patire'). Pre' Marsilio complained, in the Venetian Petrarch of 1513 already mentioned, that one of his problems in establishing a correct text was that he was allowed very little time by the printers who were breathing down his neck in their haste for copy. On top of his duties as a priest and teacher, he was obliged to have a corrected gathering of his copy-text ready every two days. Since each of these gatherings probably consisted of sixteen pages of closely-printed text (sixty-five lines per page), the priest would indeed have had to work rapidly, and evidently had no chance to establish any consistent method before printing began.[31] But things could have been worse. For a Petrarch printed two years later, Ottavio degli Stefani was asked by Alessandro Paganini to revise the text only after it had been printed; all he could do was to compile, over five days, a list of corrections which was then printed in a supplementary gathering.[32] Such lists of errors and corrections are commonly found at the end of books and were evidently used even in less desperate circumstances. Other signs of haste and inconsistency are the use of stop-press corrections even in the most carefully edited of texts, and the way in which dedications and other important material could be added after some sheets had been printed.[33] Moreover, editors were not encouraged to build up expertise in particular authors or periods: they had to be prepared to turn their hands to different genres and might also have to move between editing texts in Latin (both classical and neo-Latin) and the vernacular.[34] If they were working anonymously, there was an even greater chance that their task would be badly done. Another cause of inconsistency could be the involvement of more than one editor in a large work.[35]

These factors encouraged, but were probably not the only cause of, a feeling evident from time to time among both printers and editors that a work in print was not something definitive but an open text which readers might continue to edit for themselves. Lists of errors and colophons often contain invitations from the printer to the reader to correct any mistakes that had not been noticed, as if it were taken for granted that such perfection was unattainable. Even an editor might give readers the freedom to change his version as they saw fit. A Milanese printing of the *Astronomicon* of Manilius (1489) says that Stefano Dolcino has emended the work, in so far as he was able, in three hundred places, but he left untouched what seemed uncertain: he gives carte blanche to teachers to add the rest in order to perfect the work.[36] Such a comment shows that print culture did not introduce suddenly the 'sense of closure' or finality which, as Walter Ong has suggested, it encouraged in the long term.[37] A similar invitation to correct difficult passages was issued by Filippo Beroaldo when editing Tacitus in 1515. And so with vernacular texts: Francesco Asolano, for instance, listed some variants in an edition of Boccaccio's *Decameron* (Venice, 1522) in order that readers could use them according to their judgement; the printer Gabriele Giolito wrote that variants had been put in the margins of his *Decameron* of 1546 so that each reader could choose what seemed to him most appropriate and elegant ('quello ... che più proprio e più ornato li pare'); and Tizzone Gaetano said in his dedication of the same author's *Fiammetta* (Venice, 1524) that he had left space in the margins so that expert lovers of the language could (without animosity on their part or resentment on his) bring the work back to its pristine state. Gaetano can hardly be accused of saying this just out of laziness or false modesty, given the extent of his interventions and the attention which he draws to them: rather, he saw his printed text as one possible state to which further changes could be made by pen. A particularly pessimistic note accompanied the list of errors in the Petrarch of 1532 edited by Sebastiano Fausto: the editor or printer said that it was impossible to spot all misprints and therefore provided a long list of the usual categories of error (such as confusion of similar letters, double for single consonants or vice versa, wrong punctuation or word division, missing or superfluous letters), so that readers could correct instances of these errors for themselves. One could, however, try to make a virtue out of variety in spelling. At the end of an edition of Benedetto Dell'Uva's poem *Le vergini prudenti* (Florence, Sermartelli, 1582), a note said that the presence of prepositions both separated from articles (as in *de le*) and joined with them (as in *nelle*) had been caused at first 'by the inconsistency of the correttori' ('dalla varietà de' correttori'), but had thereafter been left so as to satisfy both Tuscans, who linked the two words in both prose and verse, and outsiders ('forestieri'), who linked them in prose but not in verse.

After haste, a second factor which influenced the achievements of editors was a lack of control over the final printed text. They sometimes complained that those working in the printing house did not always follow the text given to them. We saw in chapter 1 (p. 10) the concern of Pietro Albignani lest the printers (which could mean both compositors and proof-readers) should stray from the path of his correction. Another editor wrote in the 1480s of his reluctance to have anything to do with printers because they were wont to alter and upset ('mutare ac invertere') the corrected text.[38] The priest Giorgio de Spadari added a disclaimer at the end of the Roman breviary which he edited for Jenson in 1478: any additions, changes or repetitions in the Psalms were to be ascribed not to his ignorance but to the printers, who tried to please everyone as far as they could.[39] Punctuation added by editors was particularly liable to be ignored, according to Marco Astemio, an editor working in Venice in the mid 1520s.[40] Accentuation posed similar problems: the editor of the Aldine Petrarch of 1533 said that any deviations from his new system were to be ascribed to the printers ('impressori'), who could never be rebuked sufficiently; when they corrected an error, they often added two ('à quai non se puote mai tanto sgridare, che baste; in tanto che spesso, mentre uno errore emendano, ne incorreno in due'). Compositors might misunderstand the editor's marks, or fail to incorporate some emendations through oversight.[41] If they were working through a text out of sequence, for instance when composing by forme, they might abbreviate or pad out a text in order to make it fit into the limits of a page.[42] There was also the risk of interference from the dialect of compositors during the process of reading a phrase in the copy-text, memorizing it, and setting it in type.[43] Compositors might carry on the editor's work where he had not been consistent in indicating some point of presentation, such as the use of upper-case letters or the way in which words were to be separated.[44] On the whole, it seems that compositors and proof-readers had some freedom, especially when printing began, but that their deliberate changes were relatively minor or at any rate occasional.[45] One must still be wary, of course, both of supposing that any text in which changes were introduced had been prepared by an editor, and, where an editor is named, of attributing every feature of an edition to him.

Practical considerations could also affect the quality of the copy-text. If this had to be a manuscript, there were strong reasons for choosing a recent one, even though *recentior* might mean *deterior*. The printer would save money, because a recent manuscript would be cheaper, and because it was easier for compositors and proof-correctors to read, with the result that he would have lower labour costs. A contemporary manuscript would also probably be closer in format and script to the model which the printed book was trying to resemble. However, a printed copy-text will have been preferred where possible. For compositors, this was the easiest source to follow, especially if it was in the same format as the new edition and could thus be followed page by page.[46] But the practice of copying from edition to edition was also the result of the attitudes of scholars. An illogical conservatism, as Timpanaro calls it, which persisted well into the nineteenth century, led Biblical, classical and vernacular scholars alike to see the use of manuscript sources as a departure from the authentic tradition rather than a return to it.[47] This meant that an editor was using a source which was at least one stage removed from the original and therefore had a higher probability of containing false readings. When Francesco Tanzio perceptively commented on this in his Milanese Petrarch of 1494, he used the image of water losing more of its purity the further it goes from its source ('così come l'aqua, la quale, quanto più si lontana da la sua origine, tanto più perde de la sua perfectione').

Finally, just as early printed books mimicked manuscripts in their physical appearance, so it was natural that there should be a continuity of practice from manuscript to print in the way in which the text was transmitted. Indeed, the need to sell multiple copies of a text positively encouraged the different kinds of linguistic adaptation that we saw being used when manuscripts were copied. We shall see that at first, in the Quattrocento, printed texts were adapted to make them palatable and accessible to a local readership. But, as the Venetian press and then others outgrew their regional market and began to expand, it became desirable to impose a supraregional standardization, based naturally on the Trecento Tuscan which had already provided just such a model in the age of manuscripts, and which writers in the last third of the Quattrocento were already studying with increasing attention. With the transition from manuscript to print, it thus became even more probable that what an author had written would not be seen as set in tables of stone and that an editor would at least iron out what he saw as any linguistic imperfections and perhaps even engage in some creative rewriting.

3 · HUMANISTS, FRIARS AND OTHERS: EDITING IN VENICE AND FLORENCE, 1470–1500

WE SAW IN CHAPTER 1 THAT, IF PRINTERS were to specialize in Latin texts of the sort which demanded scholarly expertise, they needed to be able to draw support from men of letters in their community. The output of printers in Venice shows that they were, from the start, notably successful in using such resources. Scholderer's estimates for the period from 1469 to 1480 give a total of 596 Venetian editions, of which 206 (35 per cent) can be categorized as classical literature, 121 (20 per cent) as theology, 100 (17 per cent) as law, and 71 (12 per cent) as science.[1]

One of the groups whose collaboration ensured the academic strength of Venetian publishing was that of humanists working in Venice and the Veneto. Benedetto Brugnoli, Giorgio Merula and Marcantonio Sabellico, for instance, all of whom edited classical texts, were teachers in the Scuola di San Marco, a position which was especially welcome from the printers' point of view since the pupils of the school included patrician men of letters whose favour would have been a valuable asset. Other notable figures involved in editing were Ognibene da Lonigo, Giovanni Calfurnio (who, like Brugnoli, was a pupil of Ognibene), the neo-Latin poet Raffaele Zovenzoni, and Lodovico Carbone, a humanist of the Ferrarese court.[2]

Another major source of editorial assistance was the neighbouring university of Padua, some of whose teachers and students were quick to see the financial and academic advantages of cooperation with Venetian printers. We have already seen in chapter 1 how a teacher of canon law, Alessandro Nievo, and his pupil Albignani edited legal texts for Jenson in the 1470s, and how Carari and Vernia edited philosophical texts.

As for works of philosophy and theology, Venetian printers could count on help from a third source, Franciscan and Dominican friars, some of whom were lecturers in Padua.[3] Their contribution was not restricted to texts for lecturers and students but included works in the vernacular. The colophon of the *Specchio della fede* by the Franciscan friar Roberto Caracciolo (printed by Giovanni Rossi of Vercelli, 1495) shows that the work was 'examined and corrected' ('vista e corretta') by Stefano da Capua, a member of the same order.[4] The Dominicans too helped with the production of vernacular texts. When Malermi's translation of the Bible was printed by Antonio Miscomini in 1477, it was provided with chapter headings ('rubricata') by a lecturer in

Biblical studies, fra Marino of Venice.⁵ Works of St Catherine of Siena, herself a member of a Dominican order, were edited, as one would expect, by fellow Dominicans. Her *Dialogo della divina provvidenza* (Matteo Capcasa for Lucantonio Giunta, 1494) was edited by an anonymous friar ('frate N.'), and the book-privilege printed at the beginning of her *Epistole devotissime* (Aldo Manuzio, 1500) says that the letters had been collected over a period of twenty years by fra Bartolomeo da Alzano of Bergamo.⁶

But not all the texts edited in Venice by friars were devotional in nature, at least as far as the Franciscan order was concerned. We shall see later in the chapter that a Franciscan was responsible for the editions of Dante's *Commedia* dated March and November 1491 and that Petrarch's *Trionfi* were edited by a friar of the same order, Gabriele Bruno, in May of the same year. Bruno was an important figure in the Franciscan community, and it is worth pausing to look at the development of his links with printing.⁷ At some time between April 1476 and mid 1480 he became guardian of a convent in his native Venice, perhaps the Frari. He is described as minister of the Province of the Holy Land in 1491, and by 1508 he had become minister of the Province of Romania (an area corresponding with modern Greece).⁸ His first known involvement in editing came in 1480, when he read through, on behalf of the Inquisition, some Latin works by Giovanni di Dio to be printed by Jenson. In 1485 he edited for Erhard Ratdolt another spiritual work in Latin, the *Rationale divinorum officiorum* by Guillaume Durand. Dedicating it to no less a person than the General of the Franciscan Order, fra Francesco Senni, Bruno showed himself well versed in all the clichés which were editors' stock in trade. He referred to previous editions of the *Rationale* which were incorrect through the fault either of copyists or of printers, and on the other hand he mentioned his own desire to help others and his diligence. He paid a compliment to Ratdolt by making it clear that he had undertaken the work at the printer's request, and said that he had worked hard to emend the text, not only by his own efforts but helped both by the learning of other scholars and by old copies of the work ('omni studio quo potui non solum mea cura verum etiam aliorum doctorum doctrina ex veteribus etiam exemplaribus emendavi'). A few years later Bruno compiled the first-ever subject index for the Latin Bible, and this appeared in 1492 together with a Bible edited by a fellow Franciscan, Pietro Angelo di Montolmo.⁹

At the end of 1492 or early in 1493, Piero Quarengi printed Bruno's translation of a thirteenth-century Latin work by Giordano Ruffo on horses and their care. At the end of the translation appears a dedication, dated 16 December 1492, which shows the extent of Bruno's links with lay society. It is addressed by him to a member of a famous family of condottieri in the service of Venice, Gianconte Brandolini; it mentions Gianconte's brother Ettore as Bruno's special patron ('singular patrone'); and it says that the translation was undertaken at the request of Lazaro Mazzarello of Modena.¹⁰ The language of the volume shows that Bruno was quite content with a koinè influenced both by Latin and by northern dialects. The sonnet following the dedication shows Bruno's standards of language and metre to be far from Petrarchan, in spite of his experience in editing the *Trionfi*:

> Chi vol haver del cavallo bon governo,
> cognoscere sua natura e soi defecti,
> qual sian cativi e quali siano electi,
> senza fastidio leza questo quinterno.
>   Questa doctrina el studio de Salerno,
> sanare cavalli zoti e farli drecti
> e fuora de infirmtà farli perfecti,
> non sepe mai insegnar, come discerno;
>   ma ben misier Jordano cavaliere
> in questo libro ha dimonstrato in parte
> e quanto se intende in questo mestiere.
>   Et sancto Alo lassò questo su le carte
> perché l'ebe da Dio questo pensiere
> contra el diavolo nemico de quest'arte.[11]

Bruno's varied work for printers must have occupied a considerable proportion of his time in the period from 1485 to 1492. But it seems to have ended after this translation, a break which was probably symptomatic of a more general tendency for the mendicant orders to reduce their contributions to literature and to the teaching of philosophy in the first half of the Cinquecento.[12]

Printers in Venice had, then, created a solid infrastructure of editorial support and experience in both academic and religious circles. This network was built up initially to assist with the production of Latin texts; but, as it was natural for them to do, printers turned to the same sources of editorial help – minor humanists, law graduates, friars – for vernacular texts as well as for classical, legal, philosophical and theological ones. As the case of Bruno shows, however, those who had the training to scrutinize a Latin text were not necessarily well prepared to do the same with an author such as Petrarch, because the Tuscan of the great Trecento authors was not yet taught, and could not be studied, in the same way as Latin. For the same reason, it was not easy for editors to bring the language of non-Tuscan authors towards a Trecento Tuscan standard. The problem of fixing a standard was exacerbated by the interregional character of Venetian printing, with editors coming not just from the local community but from as far afield as Lombardy, Piedmont, Tuscany, the Marche and Campania. Such a Babel-like mixture may have made it easier for editors to accept the principle of a neutral linguistic norm based on a usage found only in books, but it also meant that they approached their work from very different linguistic backgrounds. Let us look first at how these men coped with the demands of the three great Trecento authors, Dante, Petrarch and Boccaccio, before going on to consider their editing of non-Tuscan texts.

Of Boccaccio's works, only those written before the *Decameron* were treated with some seriousness by intellectuals. One thus finds minor humanists named in some printed editions, though they were perhaps not involved in the editing of the texts. Five Latin couplets by Bonino Mombrizio appear at the end of the Milanese *Filocolo* of 1476, and Girolamo Bologni contributed a sonnet to the *Ameto* printed in Treviso in 1479.[13] In Venice, the only named editor of these earlier works was Gerolamo Squarzafico of Alessandria. He had been in the service of Venetian printers as early as 1471, when he worked for Windelin of Speyer, and he produced about twenty editions of works in Latin and the vernacular over a period of thirty years.[14] His dedicatory letters to two of his editions of the *Filocolo* (Milan 1478 and Venice 1503) and to the *Fiammetta* (1481) say nothing of the state of the texts or of his sources.[15] In one of the Latin works which he edited, a translation of Plutarch's *Lives* (1502), he laments the ignorance of some editors and criticizes obliquely the Roman edition of 1470, but without specifying how he has improved the text.[16] It is hard to imagine him taking anything more than a superficial interest in the text of Boccaccio, especially since his Italian is uncertain at times and (as one would expect in this period) strongly influenced in spelling and phonology by both northern Italian and Latin, although many of his verb forms conform with Trecento models. Some linguistic changes (as well as misprints) are introduced in the *Fiammetta*, tending to make the language less Tuscan and more Latinizing, but they are not systematic.[17]

This should not lead one to overlook the importance of Squarzafico's role as editor in presenting to the reader the figure of Boccaccio as a man and as a writer. In the first place, Squarzafico compiled a new life of Boccaccio which appeared with the *Filocolo* in 1472 and was reprinted several times. Secondly, his dedicatory letters encouraged readers to take an interest in Boccaccio's style. The dedication to the *Fiammetta* praised Boccaccio as the greatest vernacular prose writer. Over twenty years later, however, in the *Filocolo* of 1503, his approach was rather more critical, and a eulogy of Boccaccio's art was followed by the comment that the Tuscan's style was elegant but rather lax ('fluxo') and in need of restraint.

In spite of the interest of humanists, though, Venetian editions of Boccaccio's early works tended simply to copy a previous one from the same city. Thus the *Fiammetta* of 1491 copied Squarzafico's edition of ten years earlier; and successive editions of the *Filocolo* form a series (1481, 1488, 1497) which is derived from the Venetian edition of 1472 and in which each link depends on its immediate chronological predecessor.[18] The case of the *Filocolo* epitomizes a gradual deterioration of standards among Venetian printers as one moves into the 1490s: a deterioration which jointly affected the quality of printing, cramped and inelegant, and the philological quality of the works printed.

As for the *Decameron*, regarded by subsequent generations as Boccaccio's masterpiece, it had no named editor. The work was first printed in Venice in

1471 by Cristoforo Valdarfer, but it was the Mantuan edition of 1472 which was the ancestor of subsequent incunabula printed in the Veneto. Each of these bases its text on its immediate predecessor; only one, that of Antonio da Strada (1481), does not follow blindly all the errors of its principal source, the Vicenza edition of 1478. However, when it comes to unfamiliar words, it introduces some errors of its own, such as the *lectio facilior* 'habitatori' for 'habituri' (1 intr. 48).

From the second edition onwards, the *Decameron* contained an index with references to folio numbers. In the Vicenza 1478 edition, each story has a short title (such as 'Novella de ser Ciappelletto'), and subsequent Venetian editions follow this pattern. Those of 1492 and 1498 number the first ten stories; both use woodcuts to illustrate the stories and contain Squarzafico's life of the author, thus adding a rudimentary visual commentary and setting the work in a historical context. However, nobody yet provided an index to help readers use the work for reference.

Petrarch's vernacular poetry, the *Canzoniere* and the *Trionfi*, fared considerably better than Boccaccio's works as regards the amount of attention dedicated both to the text and to its presentation and interpretation. Wilkins has divided the incunabula into four families on the basis of the arrangement of the texts of the *Canzoniere* and *Trionfi* and of the supplementary material which the editions provide; and, for the most part, the same genealogy can be applied to their texts.[19] With two exceptions, Venetian editions belong to the third and fourth of these groups, C and D. The C family derives from the Paduan edition of 1472 and contains three further editions, those of Vicenza 1474, Venice 1477 and Venice 1482. The colophon of the 1472 edition mentioned its derivation from Petrarch's own manuscript, partly autograph, now Vatican lat. 3195, and at that time owned by a Paduan family. This is the only case in which a fifteenth-century edition of Petrarch specifies its source. Apart from some printing errors and some influence of dialect habits, the edition faithfully reproduces Petrarch's spelling and even his punctuation. Yet each of its three descendants tries to improve upon the 1472 text. It is possible that its claim to authenticity was simply thought to be false, but the lack of respect for it also suggests that Renaissance editors did not feel it necessary to be absolutely faithful even to an autograph text.[20]

The remaining eleven Venetian editions belong to Wilkins's D family, which descends from the Bolognese edition of 1475-6. This contained the text of only the first 136 poems of the *Canzoniere* (those on which Filelfo had written a commentary), together with the *Trionfi*. Some of these Venetian texts show signs of editorial revisions.[21] It is probable that the editor of the 1478 edition used another source in order to improve his text.[22] In 1484 the *Canzoniere* was restored to its entirety, using a text of the C family for poems 137-366. The 1486 edition is an important one since, as Paolo Trovato has pointed out, it is the first to contain revisions which bring it closer to Trecento Tuscan: for instance, pretonic *i* for *e* in '*me* legaro' > 'mi', 'secur' > 'sicur', and the removal of Latinizing consonantal groups (as in 'sospecto' > 'sospetto'). The next edition, of 1488, increases the number of errors and northern forms but, on the other hand, adds sonnets 350 and 351 from another source and tries to correct some of the errors of its ancestors. The editor also keeps an eye on metrical regularity, and some emendations remove an extra syllable from hypermetric lines.

No editors declare their identity fully until the early 1490s, when two names appear in the colophons of editions printed in Venice by Piero de Piasi. In 1490 the sonnets were revised ('coreti et castigati') by Girolamo Centone, a Paduan lawyer. However, Centone's text follows closely that of the previous Venetian edition, that of 1488.[23]

The revision of the *Trionfi* in the 1491–2 edition is said to have been carried out in May 1491 by fra Gabriele Bruno, whose links with printing in Venice were mentioned earlier in this chapter. His text is close to that of 1490, but at least two readings go back to earlier editions, including the Paduan edition of 1472.[24] The *Canzoniere* of 1492 included even more errors and northern forms. This edition was the source of the next two, of 1493 and 1494 respectively. The second of these was used in turn for the edition printed by Bartolomeo de Zanni in 1497 which, significantly, tried to make its language more Tuscan, though it also brought in some errors and Latinizing forms.[25]

The text of Petrarch had, then, become increasingly corrupt during the 1490s, but the Zanni edition of 1497 had at least shown some renewed attempts to bring Petrarch's language closer to its Tuscan roots. In Zanni's printing of 1500 such attempts were more pronounced. The editor was Nicolò Peranzone of Montecassiano, a teacher who later returned to his native Marche.[26] His text was derived from the edition of 1497. Many errors remained, but he attempted to correct some errors and introduced some Tuscanized forms.[27] Moreover, Peranzone was the first editor to provide any discussion of his methods, even if this was based on commonplaces. He complained in a letter that Petrarch's works had become corrupted and disordered by the ignorance of both editors ('correctori') and printers. This had led commentators astray; and so, he said, he had unselfishly undertaken the task of bringing the verses back to their 'first style' ('nel suo primo stile') and remedying corruptions in the text. He also claimed to have made some additions to the commentary. But he mentioned no manuscript or printed sources: all his corrections seem conjectural.

Venice still had some ground to make up on the standard set by one of the two Milanese editions of 1494, that printed in March by Ulrich Scinzenzeler and edited by the priest Francesco Tanzio.[28] Petrarch, Tanzio wrote in a letter to the reader, had been cured of 'old wounds' ('antiche piaghe') and thus almost 'restored to pristine health' ('restituto alla pristina sanitate'), partly through his own diligence but thanks especially to a loan of a Petrarch which the Milanese poet Gasparo Visconti had diligently corrected 'with many exemplars' ('con molti exemplari'). It is important to note, though, that his faith in Visconti's copy (it is not clear whether this was a manuscript) derives not from its intrinsic authority but only from what he considers to be an intuitive stylistic similarity between Visconti and Petrarch. Tanzi's text may have been based on a recent Venetian edition, but he departs from the text of the D family in several cases.[29]

One of the editorial problems at which Peranzone had hinted was that of the division and ordering of Petrarch's poetry. The Paduan edition of 1472 resolved some uncertainties about the order of the *Canzoniere* but had less influence in establishing the division of the collection into two parts at poem 264 ('I' vo pensando'). Here it used the phrases 'Finit vita amoris' and 'Incipit de morte amoris' and left space for the initial letter of poem 264 to be illuminated. Its three descendants did likewise. Some other incunabula marked the division only by the space for the initial, but this practice ended in any case in 1490 with Centone's text.

There were also several difficulties concerning the organization of the *Trionfi*, a complex problem which remained a thorn in the flesh of editors well into the Cinquecento. The second *capitolo* of the *Triumphus Cupidinis* in early editions was the one which began 'Era sì pieno'. The order of the third and fourth *capitoli* of this *Trionfo* varied, but 'Poscia che mia fortuna' normally preceded 'Stanco già di mirar'. 'Nel cor pien d'amarissima dolcezza', an earlier version of the first *capitolo* of the *Triumphus Fame*, was included, normally at the start of this *Trionfo* but sometimes as the final *capitolo* of the *Triumphus Mortis*. The fragment beginning 'Quanti già ne l'età matura ed acra', a discarded opening of the *Triumphus Mortis*, was at first given at the end of the *Triumphus Pudicitie*; then, in the Bolognese edition of 1475, it was grafted onto the beginning of the first *capitolo* of the *Triumphus Mortis*, which itself began 'Questa leggiadra e gloriosa donna'. The same edition included another composite *capitolo* in which the start of 'Nel cor pien' was merged with the second capitolo of the *Triumphus Mortis*, but this was not followed in later editions.

From early on, editors supplemented Petrarch's poetry with biographical items and with simple indexes. After a few years came an innovation which was to have a long-term effect on the publishing of the Trecento classics: the introduction of a commentary. This initiative was taken not in Venice but in Bologna. In 1475 Annibale Malpigli printed Petrarch's *Trionfi* together with the commentary of Bernardo Ilicino, a Sienese who had written his commentary at the court of Ferrara between about 1468 and 1470.[30] Malpigli's venture must have been successful, because in 1476 he was asked by the bookseller Sigismondo de' Libri to print a similar type of edition of the *Canzoniere*. The commentary used, and prepared for printing by Nicolò Tomasoli of Forlì, was that by Francesco Filelfo. This too had been written for a northern court, that of the Visconti in Milan, in the mid 1440s.[31] It was unfinished, breaking off at poem 136 ('Fiamma dal ciel'), but such was its importance to those responsible for the 1476 edition that the *Canzoniere* ended at the same point.

Venetian printers were characteristically quick to imitate this novelty. In 1477 there appeared commented editions of both Dante's *Commedia* and Petrarch's *Canzoniere*. The publishers of the Petrarch volume were not Venetian, however, but two Mantuans, Gaspare Siliprandi, who provided the finance, and his son Domenico. The latter had probably seen the Bolognese *Trionfi* of 1475 but not the *Canzoniere* of 1476 since, instead of Filelfo's commentary, he used one attributed to the fourteenth-century Paduan judge and metricist Antonio da Tempo and revised for the occasion: in an introductory note Domenico mentioned that it was printed with additions by a scholar who did not wish to be named, no doubt judging it better at this stage for anyone with pretensions as a humanist to remain anonymous in the context of vernacular literature.[32] This commentary was not reprinted until the sixteenth century, its place being taken by Filelfo's, incomplete though this was. The pattern for the rest of the Quattrocento was set by the Venetian printing of 1478 (whose editor, we saw above, had also taken some care over its text). It combined the texts of the *Trionfi* and the *Canzoniere* with the commentaries by Ilicino and Filelfo, and its two Flemish printers used a technique developed in the mid 1470s by Jacques Le Rouge for classical texts such as Virgil and Terence: the text was printed in the centre of the page and the commentary, set in a smaller fount, was printed beside and below the text.[33]

However, the incompleteness of the Filelfo commentary was clearly a drawback to its use. When, in 1484, Piero de Piasi had already embarked on a new edition of Petrarch, he prevailed on the ubiquitous Gerolamo Squarzafico to complete Filelfo's commentary at very short notice.[34] Even though Squarzafico borrowed material from the 'Antonio da Tempo' commentary, he finished work only three days before printing ended. Near the end he expressed his mixed feelings of pleasure and regret:

I am glad to have reached the end and to have helped the printers, by whom I was begged for it insistently. But I regret the great haste imposed on me by the importunity of the printers, who immediately printed what I wrote. And I am sorry too that it was not printed or corrected as I would have wished. But it is done now; in time, if God pleases, we will put it right.

(Mi alegro per esser giunto al fine et per havere servito gl'impressori, de quali ne sono stato molto pregato. Mi doglio possia per la grande presteza quale m'è stato necessario di usare per l'importunitade degl'impressori che di continuo come scriveva il stampavano. Et anchora mi duole per non esser impresso ni correcto come io haria vogliuto. E'gli ogimai facto; col tempo, piacendo a Dio, il riconzaremo.)

In spite of the circumstances of its completion, the Filelfo–Squarzafico commentary proved very popular: it was included in all the remaining incunabula of the *Canzoniere* and was only substituted in 1525.

The next major editorial innovation in the printing of Petrarch was the improvement of another means of assisting readers: the system of indexation. Several early editions had included an index of first lines of the *Canzoniere* in alphabetical order of the first letter of each poem and a list of the first lines of each *capitolo* of the *Trionfi*. The Paduan edition of 1472 was the first to include folio numbers in the index. There was no foliation, however, so that readers would have had to write in their own numbering, as they would have done in order to use indexes in other early printed books. Numbering of the individual poems of the *Canzoniere* was introduced in the Venetian edition of 1477 in order to provide a cross-reference to the commentary, which preceded the text, but it was kept thereafter in folio editions even when commentaries moved into the margins. As for the *Trionfi*, the Bolognese edition of 1475 had gone some way towards helping readers to find their way around the work by providing an index (in text order) to the first lines of each section of the text dealt with by the commentary. At this stage, an index of this sort was an unexpected novelty in a vernacular book, and, just in case it proved baffling, a note explained its purpose: to help the reader who wished to read in one place rather than another and to find his desired place without laborious searching. Nevertheless, this primitive system was of no great help. The first move towards a more efficient one was made in the Petrarch of 1488. This volume was innovative in various ways: each *Trionfo* was illustrated by a woodcut, foliation was introduced, and in the margins of the *Trionfi* were found, at intervals, a series of letters running through the alphabet from A to Z and then starting again.[35] The intention was evidently to provide an index referring to a particular place on a particular folio. An index was not included, presumably because it had not been completed in time, but just such an index (to names and subjects in Ilicino's commentary, rather than to the *Trionfi* themselves) was introduced in the next Venetian Petrarch, that of 1490. The *Canzoniere* had a similar but much briefer index.

No changes of presentation were made in the system of indexing in the three remaining Venetian editions of the 1490s. One has the impression in this decade of a frequent lack of initiative on the part of printers in respect of their use of editors, in contrast with the innovations of the 1470s and 1480s. A symptom of the lack of care shown by some Venetian printers in this period was the mechanical reproduction of the index to the *Trionfi* from the 1492 edition even though two pages had been transposed. In 1500, however, Nicolò Peranzone made some changes. In place of the biographical sketch at the start of Ilicino's commentary, he used the life of Petrarch attributed to Antonio da Tempo (but written in the Quattrocento) together with some other biographical material, all of which he derived from the Venetian edition of 1473.[36] He also provided new subject indexes for the commentaries of both the *Canzoniere* and the *Trionfi*, fuller and in more exact alphabetical order.

The contribution of editors is emphasized from a very early stage in Dante's

Commedia. One of the three editions of 1472, that printed in Mantua, names a local teacher, Colombino Veronese, as the person responsible for the text and the insertion of what are called the 'ordini' (possibly the title and number of each canto). The Dante printed in Venice in July 1472 was produced with less attention (no editor is named and the canti are not numbered systematically), and for the first innovations from this city one has to wait until the edition of 1477, a heavyweight volume both literally and metaphorically, edited by Cristoforo Berardi of Pesaro. The text does not follow a single previous edition, and it seems likely that Berardi's independence (as a Pesarese) from northern Italy led him to use a manuscript.[37] Each canto was followed by the relevant section of a commentary which was here attributed to Benvenuto da Imola, though it was in fact the much earlier vernacular commentary by Iacopo della Lana. Berardi provided some 'Rubriche di Dante' before each *cantica*: a list of the first three or four words of each canto, the number of the canto, and a summary of its contents (and the summary is repeated at the start of each canto). He also offered the reader supplementary historical material from nearer Dante's age than his own, including the life of Dante by Boccaccio and a *capitolo* by Iacopo di Dante.[38] Berardi's edition thus marked a new approach to the presentation in print of the major literary works in the vernacular, and it is significant that it should have appeared in Venice, the centre which in the sixteenth century did most to promote the editorial use of supplementary material.

The edition which appeared in Milan in 1478 contained the same commentary by Iacopo della Lana. The humanist Martino Paolo Nidobeato of Novara said in his dedication that he was grieved by the inaccurate state in which the text had been printed, and had therefore persuaded Guido Terzago to have it revised and the commentary added. Nidobeato altered the commentary considerably and added references to classical sources and recent historical events.[39] He gave a list of contents as in Berardi's edition of 1477 but with folio numbers for each canto. He also devised a new form of cross-reference between the text and the commentary, which here, for the first time, surrounds the text: in the margins of the text there are letters of the alphabet (from *a* to *z*, then back to *a* and so on) which correspond with letters at the start of the relevant section of the commentary. In a Latin poem which is in effect an advertisement addressed to the reader, Nidobeato praises Dante and his own edition with its reference system ('Neve sit errandum querenti singula, prorsus | signavit propriis invenienda locis'). As for the text, it is probable that the source was a manuscript rather than a single previous edition or a combination of previous editions.

Another northern edition, the Venetian *Commedia* of 1478, was edited by one 'C. Lucius Laelius'. In a concluding sonnet, he wrote of his love for Dante and said that he wanted to restore the author from neglect and to interpret him to others. As his Latinized name suggests, Laelius's approach was that of a humanist. He chose as his source Colombino's Mantuan edition of 1472, reproducing its Latin canto headings and occasionally correcting it with readings found in one or more other previous editions and perhaps with conjectures of his own. He also attempted to 'de-Tuscanize' the text according to his humanist and regional preferences. He introduced many Latinizing spellings: Dante finds himself in a 'silva obscura', a phrase which is followed by 'cum' for 'con', 'umbra' for 'ombra', 'puncto' for 'punto', and so on. Laelius waged a war against the letter z, whether after a consonant (as in 'dinanci') or between vowels (as in 'larghecia'). Northern influence is evident in phrases like 'Nel megio del camin', 'i soi fredi rivi', and in spellings like 'folgie' for 'foglie'. The *terzine* begin with an upper-case letter but are not indented, possibly so that they will resemble Latin hexameters. Fortunately, this deformed *Commedia* found no imitators.

In 1481 there appeared in Florence a new text of the *Commedia* with a commentary by Cristoforo Landino. The commentary was very influential in Venice: it immediately replaced that of Iacopo della Lana and was included in all five editions printed during the rest of the century, four in Venice and one in Brescia. These editions also use Landino's text with varying degrees of independence.[40] Two of them, both printed in 1491, contained the name of an editor, fra Pietro Mazzanti of Figline, near Florence; a Franciscan like Gabriele Bruno, who had edited Petrarch's *Trionfi* in the same city and in the same year. The colophon of the edition dated 3 March 1491 and printed by Bernardino Benagli and Matteo Capcasa claimed that the work had been 'diligently revised and emended' ('revista et emendata diligentemente') by fra Pietro, who had filled many lacunae which he had found in all previous printed editions, as one could see in these Dantes in both text and commentary ('ha posto molte cose in diversi luoghi che ha trovato mancare in tutti e Danti li quali sono stati stampadi ... como ne dicti Danthi si potrà vedere, sì in lo testo come nela iosa'). There is no index, but fra Pietro noted important references in the margins. He also made interventions in the text. Many readings characteristic of Landino's text were replaced with variants which are sometimes found in previous editions but in a few cases are new. Although the friar was Florentine, the text contains rather more northern and Latinizing forms than its main source, the Brescia edition of 1487.[41]

Another Dante attributed to fra Pietro's editorship was printed in November 1491 by Piero de Piasi. There were two important innovations. One was the use of a system of indexation identical with that of the *Trionfi* printed by de Piasi in 1490. The second, much more important, was an appendix of seventeen *canzoni* by Dante and another poem attributed to him. They were all in a very corrupt form, but this was the first time that any of Dante's lyric poems had been printed, and none were printed again for over twenty years.[42]

The two remaining Venetian incunabula of Dante (Matteo Capcasa, 1493 and Piero Quarengi, 1497) followed the March 1491 edition in listing *notabilia* in the margin; neither the index nor the appendix of the November 1491 edition were used. The 1493 text is very close to that of March 1491, while introducing a few readings from Landino's text. The 1497 text does things the other way round: it is much closer to Landino's text but borrows some readings from fra Pietro's adaptation of it.[43]

One can see, then, a clear gradation of effort in the treatment of the *Tre corone* by editors in Quattrocento Venice: little attention was given to the prose of Boccaccio, much more to the verse of Petrarch and Dante. The editions of these two poets show how important the editor's role could be already in the first three decades of vernacular printing. One can trace some close parallels between the development of the two series of editions, with the vogue for commentaries beginning in 1477 and taking new directions in 1484, the first indexes appearing in 1490 and 1491, the contributions of the Franciscan order coming in about 1491, and, in the last decade of the century, a series of attempts being made to improve the texts of both authors and of their commentators. But an important difference was that, while editors of Petrarch tended to base their improvements on a more or less approximate knowledge of literary Florentine, editors of Dante tended to draw more freely on variants from earlier editions or even from manuscripts.

Venetian editions of texts from outside Tuscany continued to have a more or less strong northern or even Venetian linguistic colouring right up to the end of the century. However, printers in Venice were soon aiming at a readership much wider than that of their own city or even of the Veneto. It has been estimated that, of the over 8,000 surviving editions of incunabula printed in Italy, more than a third, perhaps nearly one half, were printed in Venice. But less than 14 per cent of Italians lived in the Venetian dominions and only just over 1 per cent lived in the city of Venice.[44]

The search for an interregional market was no doubt a factor which encouraged editors to seek, in some texts, a blend between the influence of Tuscan and that of northern dialects. An early example is to be found in the editing of Bartolomeo Teo, from Pontecorvo in Campania. In 1478 he and the printer Filippo di Pietro published an edition of *L'Acerba*, a poem in *terza rima* written probably in the 1320s by Cecco d'Ascoli, using as their source the edition of this work which Filippo had brought out two years earlier and initially following it page for page. However, Teo now made the language rather more Tuscan in character, and there was some attempt to remedy evident corruptions. On the other hand, the revision was by no means a thorough one: many non-Florentine forms remained, and Latinizing spellings of consonantal groups were both introduced and removed.[45]

An example from the next decade is provided by the printing in Venice of a collection of fifty short stories called the *Novellino* by a southern author, Masuccio Salernitano (*c.* 1410–75). From the Neapolitan first edition of 1476, no copy of which survives, descended an edition printed in Milan in 1483 and then, independently, one printed in Venice in 1484 by Battista de Torti. The southern dialect forms of the Milanese edition show that it followed the Neapolitan edition relatively faithfully, but the Venetian version departed consistently and deliberately from its source (whose text, in the examples that follow, is assumed to be the same as that of the Milanese edition).[46] Most of the changes made in Venice were intended to make the work more comprehensible to readers who were not from southern Italy. Often Tuscan forms were introduced.[47] However, the editor was not trying to rewrite the *Novellino* as if Masuccio were Boccaccio. His alterations to southern forms were inconsistent; many Latinizing spellings remained; and he sometimes made the language northern, or even Venetian, in character.[48] Syntactic changes were rare.[49] In the area of lexis, the editor avoided a few terms which he no doubt felt to be southern regional ones, sometimes adopting a word acceptable to Tuscans, as in 'uccisaglia' > 'occision' and 'carafetta' > 'anguistara', sometimes introducing a northern term, as in 'cio' ('uncle') > 'barba', or even a local Venetian one, as in 'pasturare' ('to take to pasture') > 'pascuar'. Among other lexical changes we find '*superata* gli sarebbe' ('he would have found excessive') > 'avanzata', where the editor has eliminated a Latinism, and 'ridendo' > 'ghignando', where he has strengthened Masuccio's characterization. Another kind of strengthening, by the addition of an ennobling adjective, occurs in 'vera Croce' > 'vera e santa Croce' and 'il pirigrino ... signore' > 'lo illustre pirigrino ... signore'. But such interventions, which were to become routine in the first part of the Cinquecento, are uncommon here.

When the Venetian editor of Masuccio found a term which baffled him, his reaction was to assume that it was a mistake. He might search for a similar word regardless of sense, as in 'camorra' ('dress') > 'camera', or 'vivati' ('passengers on a ship') > 'inviati' or 'vianti'.[50] He might also substitute a word which bore no resemblance in form but which, he believed (sometimes rightly), made sense in the context. The readings of the Neapolitan edition may have been corrupt in these cases, but he made no attempt to reconstruct words which corresponded with them.[51]

In summary, then, the editor sacrificed many aspects of Masuccio's original in order to produce a text which would be easier to understand, particularly in northern Italy. Tuscanization was, as yet, only one of the devices used for this purpose, and the text had at times a distinct northern colouring. Works printed in the 1490s continued to be Tuscanized in this still (presumably deliberately) only partial way. Paolo Varisco's edition of his own translation of Guido de Cauliaco's *Chirurgia* shows only a few Tuscanizing changes between the edition of 1480 and that of 1493, even though Varisco (a Paduan practising medicine in Venice) had used a language with a strong Veneto character.[52] Of two Venetian editions of a minor work of piety, the *Monte de le orations*, one printed probably in 1493 and the other in about 1495, the later one introduces some Florentine forms along with a more orthodox spelling system. On the other hand, several non-Florentine and Latinizing forms remain, as they do in the *Hystoria del mondo falace* which, like the *Monte*, bears the name (as editor or publisher?) of a certain 'Ioannes Florentinus'.[53] Two other religious texts of a popular nature, the *Salve Regina* (probably printed about 1495-8) and the *Virtutes psalmorum* (*c*. 1495), have a characteristically Venetian colouring; there is, for instance, 'mazore', 'chi caza via da nui ogni tristeza', 'la prima mare' (= 'madre', in rhyme with 'mancare' and 'sperare'), and in the other case 'lezando' (= 'leggendo'), 'domenega', 'vertuoso psalmo sie a quella persona ch'el dixe: o per chi el vien dito de' vardarse da usar con cative persone e usar con le vertuose'.[54]

We saw earlier that the Florentine *Commedia* of 1481 provided the model for subsequent Venetian incunabula of this work, in preference to earlier Venetian editions. However, just as printers did not necessarily seek out Florentines to work for them as editors, so Florentine editions were not always chosen as models when Venetian alternatives were available. It is noteworthy that none of the six incunabula of the *Trattato* of 'Sir John Mandeville' printed after 1492 in Venice, Milan and Bologna took the Florentine edition of that year as its source. The influence of Venetian printing in the 1490s as the provider of a linguistic norm in the north of Italy for works outside the literary canon is exemplified by the fact that the two Milanese Mandevilles of 1497 were based on the Venetian edition of 1496 rather than on either of the Milanese editions of 1480 and 1496, both of which were more strongly dialectal.[55]

Looking back on the examples studied so far in this chapter, we can see that the ways in which Venetian editors dealt with the texts of Tuscan literary works had much in common with their treatment of the texts of works from elsewhere, whether of a literary or practical nature. An editor of Dante might sometimes search for alternative readings in other copies of the *Commedia*, but the predominant tendency in all texts was to use one's own judgement in order to replace readings which were obscure or archaic or not Tuscan in form. Different balances were struck between the expanding Tuscan dialect and the host dialects of the north, with Tuscan penetrating at a slower rate in works of a more popular or practical nature. These revisions did make texts available to a wider reading public, but there was evidently a price to pay as regards loss of authenticity.

Given Venice's dominance of the book market by 1500, her editors could well have continued in this vein for a long time. However, not everybody was satisfied with the failure to search for authoritative texts of works from the Tuscan literary canon, and we shall see in chapter 4 how, at the start of the new century, a reaction against current editorial practice led to a reappraisal of philological standards. We shall see too in later chapters that the way in which a standard based on Trecento Tuscan was imposed on other printed works gradually became less imprecise and half-hearted.

A comparison between the numbers of incunabula printed in Florence and those printed in Venice reveals a great disparity between the output of the two cities. If one ranks Italian printing centres in terms of the number of surviving editions which they produced, Florence reaches only fourth place: her nearly 800 incunabula put her not only far behind Venice (which printed perhaps 4,000 or more editions) but also after Rome (with perhaps 1,500 to 2,000 editions) and Milan (estimated at 1,121 editions).[56] Florentine books also tended to be less bulky than Venetian ones, and this difference in size reflects a difference in character. While many books were printed in Venice with an eye on the market of lecturers and students in the neighbouring university of Padua, Florence did not have a comparable institution within her territory: neither her own Studio nor that of Pisa (established only in 1472) was as large or as renowned. Thus there were no Florentine counterparts, for instance, of the massive legal and theological tomes produced by Jenson and others. As for vernacular texts, most of those printed in Florence were short, often works of piety, and the proportion of major literary works was relatively low: only one *Commedia*, none of Petrarch's poetry except the *Trionfi*; of Boccaccio's prose, only one *Filocolo*, one *Decameron*, no *Fiammetta*. It seems probable, too, that the proportion of books sold outside their territory of production was lower in Florence than in Venice. There is a considerable difference between the figure of something like 1.6 editions produced in the Tuscan city in 1470–1500 for every 1,000 inhabitants of its state, and the equivalent figure for Venice of about 2.9.[57] One can only draw conclusions from such statistics by making the perhaps unjustified assumptions that the presses of the two cities had similar print runs, that their books have survived in similar proportions, and that their inhabitants bought books in similar quantities. Even so, it seems fair to say that Venice expected that more of her vernacular books would be read by people from a linguistic background other than that of the city where the books were printed.

This situation helps to explain two features which distinguish vernacular editing in Florence from its Venetian equivalent well into the Cinquecento. One is that Florentine editors were more inward-looking and insular in their approach. They tended to impose the linguistic standards that they heard about them from day to day: ones that were limited, therefore, to a certain period and a certain place, not broadly based. Here the influence of the tastes of local readers was reinforced by a strongly patriotic conviction that living Florentine was in no way inferior to the Trecento variety which was being increasingly imitated in other states. A second characteristic of Florentine editing is the relative lack of prominence given to editors. Even when they were used, they very often remained anonymous. In Venice, as we saw, local humanists and friars were quite often named as editors of Latin and vernacular texts, partly because the texts themselves were important ones and partly because editors could gain some prestige, given that printing had the support of many leading statesmen, academics and clerics. In Florence, on the other hand, the works printed were often more ephemeral and printing itself did not attract the sort of patronage which helped to launch Jenson's career. Ironically, what may be the very first Florentine book, Servius's commentary on Virgil printed by Bernardo Cennini and his son Domenico in 1471-2, has a very informative colophon which tells us how another son of Bernardo, the scribe Pietro Cennini, emended the work and collated it with many very old copies, being particularly careful that nothing extraneous should be ascribed to Servius and that nothing should be excluded or lacking which ancient copies showed to belong to him ('emendavit; cum antiquissimis autem multis exemplaribus contulit, in primisque illi curae fuit ne quid alienum Servio ascriberetur, neu quid recideretur aut deesset quod Honorati esse pervetusta exemplaria demonstrarent').[58] But thereafter it is not easy to identify editors by name. A member of the clergy, a certain fra Niccolò da Tobia, was given the *Confessionale: Defecerunt* of Antonio Pierozzi (St Antoninus) 'in volgare' to be corrected for the press of the convent of Sant'Iacopo di Ripoli in 1481.[59] We know, too, that some editors were drawn from the ranks of local humanists and of scribes such as Cennini. Giorgio Antonio Vespucci, provost of the cathedral and well known both as a scribe and a teacher, is named in the colophon as the person by whose diligence the *Martyrologium* of Usuardus was 'emended and corrected' ('emendatum correctumque') in Francesco Bonaccorsi's printing of 1486.[60] Bartolomeo della Fonte (or Fonzio), another humanist and scribe, worked for the Ripoli press.[61] The most eminent of Florentine editors, though, was Cristoforo Landino, responsible for the first Florentine edition of the *Commedia*, printed in 1481 by Niccolò di Lorenzo.

With this *Commedia* (as on other occasions in the sixteenth century), Florence responded somewhat tardily but with decisive effect to the incursions into her own literary tradition made by printing centres in northern Italy. Its main novelty was the commentary specially composed by Landino, a substantial one even though it was written very rapidly in the period following the appearance in 1477-8 of the editions of Berardi and Nidobeato (in which the *Commedia* had been edited, commented on and printed by men whose geographical origins were all non-Tuscan), and against the background of the war which followed the Pazzi conspiracy of the same year and which saw Florence pitted against the might of the Papacy and the Kingdom of Naples.[62] Landino's undertaking was exceptional in several ways. He was the only major scholar to have been connected with the editing of a vernacular work in the Quattrocento, and this was the only complete commentary on either Dante or Petrarch written expressly in order to be printed (Squarzafico's extremely sketchy commentary on Petrarch was, it will be recalled, only a completion of Filelfo's). As one would expect, a fierce patriotism underlies Landino's commentary. The extensive preliminary matter includes a defence of both Dante and Florence against 'false slanderers' and a eulogy of Florentine learning, eloquence, music, painting, sculpture, jurisprudence and commerce. It was intended that this monument not just to Dante but to all Florentine achievements should be graced by an ambitious series of engravings, one at the start of each canto, based on illustrations by Botticelli; however, the series only went up to *Inferno* 19 and the number of plates printed differed from copy to copy.

Landino was also concerned to restore the language of Dante to Florence, freeing him, as he said in his dedication, from the barbarity introduced by non-Florentine commentators, so that the poet might be restored to his *patria* after a long exile. The proof of the superiority of Florentine, he says, is that no writer with intelligence or learning has tried to use anything else. As for the text of the *Commedia*, the printers appear to have been given a manuscript as their exemplar, since the edition contains a high proportion of readings not previously found in print. But Landino was clearly indifferent to textual criticism: he drew attention to some variant readings but noted these only in passing, and there are several discrepancies between the words on which Landino commented and the printed text.

Even though the *Commedia* was printed only once in fifteenth-century Florence, the 1481 printing was, as we saw earlier in this chapter, much imitated in Venice. The same cannot be said of Florentine editions of Petrarch and Boccaccio. There was no Florentine edition of the *Canzoniere*; only the *Trionfi* were printed there, on at least two occasions. The sole commentary to appear was that of Iacopo di Poggio on the *capitolo* 'Nel cor pien d'amarissima dolcezza', reprinted in 1485/6, some ten years after it had first appeared in print in Rome.[63] Any discussions on the establishment of the text took place outside the sphere of editing.[64] The one Florentine *Decameron* of the Quattrocento, which came from the Ripoli press in 1483, was based not on a manuscript but on the Milanese printing of 1476; and it replaced archaic or obscure words with more easily understandable ones in a way which was to become familiar in the next century.[65] This kind of intervention must have been common at this press. Its 1477 edition of the *Specchio di coscienza* of Antonio Pierozzi has revisions which make the language less northern and Latinizing than the Florentine edition of 1475, and the two editions of the *Fiore di virtù* printed in about 1482–3 thoroughly revised the language of their Venetian source.[66] Florentine readers were also relatively demanding about metrical regularity, to judge by the edition of Boccaccio's *Ninfale fiesolano* printed in about 1485 by Bartolomeo de' Libri. This was derived from one of the two previous editions from Venice (*GW* 4492, *c.* 1477), without the use of any new sources. As well as the usual changes intended to bring spelling and phonology into line with contemporary

Florentine taste, the editor reduced lines of twelve syllables to the correct length of eleven.[67]

There are, however, two editions of early vernacular texts in which Florentine editors showed themselves to be capable of surpassing their Venetian counterparts in their ability to obtain high-quality manuscripts and in their respect for the readings of good sources. One was Dante's *Convivio*, printed in September 1490 by Francesco Bonaccorsi. Not only was the source manuscript of very high quality, but it was reproduced conservatively even when corrupt. In the light of what we have seen of contemporary methods, this should be seen as a mark of respect rather than of indifference.[68]

The other noteworthy edition, which came from Bonaccorsi's press in the same month, contained the *Laude* of Jacopone da Todi.[69] It is to be linked, as Carlo Dionisotti has pointed out, with the movement which kept alive the tradition of the Florentine Spirituals and which must also have been behind the editions of treatises by fra Ugo Panciera printed by Miscomini and Morgiani in 1492.[70] The volume of *Laude* gives what is probably the fullest account in any fifteenth-century Italian book of the principles according to which it was edited. The 'Proemio' tells us in unusual detail about the sources used in order to ensure that the specially-copied text delivered to the printers should be as correct as possible. Many manuscripts were used, including two very old ones from the poet's own city and another written in Perugia in 1336. The editor also explains how the *laude* were arranged and how he intended to arrive at a total of 100 (though the volume, through an oversight, actually contained 102), and, exceptionally, he provides two indexes to them, one in alphabetical order of their first lines, the other in the order of printing with a summary of contents.

The 'Proemio' declares the editor's relatively conservative approach to the medieval Umbrian dialect in which the poems were written: he has kept 'the simplicity and ancient purity according to that town of Todi' ('la simplicità e purità anticha secondo quel paese di Todi'), as seemed right to several devout and spiritual persons, without changing or adding anything. (Florentines in the Cinquecento retained this liking for working as a team of editors, even if one of them had principal responsibility.) However, he wanted to help the reader to understand the language. This was, after all, a volume intended for the spiritual improvement of its readers, printed 'for the contemplation of devout persons' ('a contemplatione delle devote persone'), as the colophon tells us. In the second index of contents, therefore, he gave some explanations of linguistic features which occur often (such as *co* for *come*, *monno* for *mondo*); and he provided explanations of words which were old or from Todi, putting these in a third index which also included information on misprints and variants and thus had a threefold function as a glossary, an errata, and a kind of apparatus criticus. Here he showed an excellent command of medieval Umbrian.

The editor keeps the reader well informed about the processes of editing and of printing. The third index gives variants (sometimes two or three of them) for over fifty readings of the text, particularly for the second half of the volume, and the presence or absence of groups of lines in different manuscripts is mentioned in the text itself. In constructing his text, the editor's method was to follow one manuscript, usually the one from Perugia, as long as it provided a coherent text. But there was also a process of comparison, and the editor had recourse to other manuscripts when they appeared more correct or when they filled lacunae. It is not surprising to find that he used conjecture and that he did to some extent adapt the language of his sources in order to make it conform with the tidy linguistic and metrical patterns expected by his contemporaries. For instance, all traces of Umbrian metaphony were removed, so that forms such as *nui* and *igli* became *noi* and *egli*. He tried wherever possible to obtain a perfect rhyme at the end of each line, which meant discarding Jacopone's occasional use of assonance and running the risk of including what he admitted (in the second index) to be 'several mutilated and imperfect words' ('più parole mozze et imperfecte'), neither Tuscan nor Umbrian, such as 'quiito' (rather than 'quieto') in rhyme with 'sbandito'. Rhyme was so important that, for its sake alone, the editor could change the forms found in his source so that they went against Tuscan phonology; for example, 'questa' could become 'quista' in order that it might rhyme with 'trista'. A number of other emendations ensured that lines had a regular number of syllables, and he also preferred to avoid dialoephe (the pronunciation in two distinct syllables of a final vowel and a following initial one), hence making changes such as 'l'umilità è' > 'l'umilitate è' or 'può essere' > 'puote esser'.

Conjecture was used fairly sparingly, usually in order to make a difficult text, which was to be used in worship, as logical and as easily comprehensible as possible. If the editor decided to emend, he usually tried to keep as close as possible to the original text; so, for instance, 'et *di gran pieno* donògli conforto' became 'del gran pianto'. Correction was also based on a study of the poet's vocabulary: thus the change of 'rengratiare' to 'regratiare' introduced a *lectio difficilior* which corresponds with the author's usage elsewhere.

This edition of Jacopone's *Laude*, though clearly not impeccable in its use of sources, was remarkable in many ways for its period. The editor's concern not only to use the best available methods but to inform his readers about them is unparalleled in the Quattrocento and for many decades thereafter in both Florence and Venice. However, such an edition could not have emerged except in a climate of serious textual scholarship, vernacular as well as classical. It was a Florentine, perhaps Angelo Poliziano, the finest classical scholar of the time, who had been responsible in the mid 1470s for some pioneering editorial work when collecting Duecento Tuscan poetry for the 'Raccolta aragonese'. The editor of Jacopone was clearly aware of Poliziano's teaching in his first *Miscellanea*, which had appeared in the previous year, on the importance of the evidence of old manuscripts and the need to identify sources precisely.[71] Poliziano, in the preface (now normally attributed to him) of the 'Raccolta aragonese' and in his *Nutricia* (composed in 1486 and printed less than a year after the *Laude*), had also pointed the way towards applying techniques of classical textual scholarship to vernacular texts by presenting the first great Tuscan authors as worthy successors of the classical poets.

The 1490 edition of Jacopone's *Laude* thus shows that in Florence a fruitful dialogue between classical and vernacular scholarship had begun by the last decade of the Quattrocento. Even though the methods used in it had no immediate legacy in editions either of Jacopone's *Laude* or of other vernacular works, it gave notice of the great (though often latent) potential of Florentine vernacular editing and held out the promise of further fine achievements in the Cinquecento. More generally, the appearance of a book such as this – hardly a potential best-seller, edited anonymously and without a dedication – indicates that good editing might be inspired simply by the opportunities of the medium of print. The diffusion of a book in many copies, it must have been felt, gave more permanence and authority to writing and therefore called for a higher degree of forethought and preparation than was normally the case in the context of manuscripts.

4 · BEMBO AND HIS INFLUENCE, 1501–1530

THE EDITING OF DANTE AND PETRARCH in Venice at the end of the fifteenth century was, we saw in the previous chapter, drifting ever further from authenticity. Each successive edition was only superficially new, since each was based on a source text printed only a few years earlier. There were, though, already signs of dissatisfaction with this situation. Editors were making efforts to remove the northern patina which in the Venetian tradition had come to overlay the original Tuscan, and in 1500 Nicolò Peranzone openly voiced criticisms of the state of the text both of Petrarch and of his commentators, though his edition was in fact little better than its predecessors.

The consciousness among editors that the text of Petrarch presented a problem which urgently needed to be addressed in a fresh way was a consequence, no doubt, of a growing interest in the poet among men of letters in northern Italy. Already by the early 1490s Giovanni Aurelio Augurello was trying in the Veneto to close the gap between humanism and a serious study of the vernacular, urging other humanists to study Petrarch's Italian verse, and arguing in the early years of the Cinquecento in favour of a strict adherence to Petrarch as a poetic and linguistic model. One of those whom he had probably already influenced in this respect was the young Venetian patrician Pietro Bembo, and it was Bembo who edited the first Petrarch and the first Dante to appear in the Cinquecento, respectively in July 1501 and August 1502.[1]

These two editions appeared in a series of Latin and vernacular texts launched in April 1501 by the printer-scholar Aldo Manuzio, who had previously specialized in Greek and Latin works. The series set out to be radically and provocatively innovative. It used a completely new typeface, the first ever italic. The format was octavo, unheard of for printed texts of this kind. It accorded to Petrarch and Dante the same status as Latin classics such as Virgil and Horace, and it presented the work of all of these authors uncluttered by commentaries and other extraneous matter for the first time in some twenty to twenty-five years. This must have restricted the readership of these editions, but it allowed those who did not need help with the interpretation of the texts to approach them with a fresh mind.

Even in this context, Bembo's two contributions to the series stand out as particularly revolutionary. First of all, one must not forget the social dimension of Bembo's involvement. When one considers that the four identifiable editors of Petrarch and Dante in the previous decade were two Franciscan friars, an obscure law graduate of Padua and an immigrant from the Marche, then one can see that it was a brave step for the eldest son of a prominent Venetian family to become involved with editing, even if natural discretion led him to reveal his name only in a few special copies of the Petrarch printed on vellum and to omit it altogether from the Dante. The texts of the two authors also broke completely with recent convention. Bembo supplied the printers not with corrected printed texts but with manuscripts which he had copied out himself and which are now united in Vatican lat. 3197.

He began work on transcribing Petrarch's lyric poetry with the help of at least two manuscripts of high quality. In the context of editing in the vernacular, this collation of manuscripts was a very significant advance beyond the practice, established in Venice in recent years, of adapting a printed source. Bembo's original transcription contained some errors of substance, and some of its orthographic and phonetic usage was non-Petrarchan, even non-Tuscan.[2] But even if his text had not been taken beyond this first phase of transcription, it would have been far superior to any printed text other than the Paduan edition of 1472, which (as we have seen) had used Petrarch's own partly autograph manuscript, now Vatican lat. 3195 (V).[3] However, at a later stage, Bembo had the use of this very manuscript, then owned by the Santasofia family of Padua.[4] From a certain point, not before sonnet 337, he copied directly from V, and also collated with it what he had written previously.

A letter to readers added to some copies of the edition, whose content and style show that it was written by Bembo but which claimed to be from Aldo, says that the printed edition was taken 'letter by letter' ('a lettra per lettra') from the original. To a great extent, Bembo was indeed faithful to his source. He followed the order of V, though he ignored the rearrangement of the final poems indicated by Petrarch's renumbering in the margins, and he made a clear division of the poems into two parts, the second starting with poem 264. As a result of his collation of the text, some readings in his original transcription were emended immediately. One example from many is the adjective 'bavarico', used in 'Italia mia' (128.66) in the lament on the use of German mercenaries, instead of its banal corruption 'barbarico'. But there were definite limits to his fidelity to Petrarch's manuscript. Some readings of V were merely noted, for later consideration, in the outer part of the outer margins and were indicated with a 'P'. The inner part of the outer margins was reserved for definite corrections to the text, as opposed to variants to be pondered or merely noted. Subsequently, only a few of the variants marked with a 'P' were accepted.[5] Some other discordant readings of V were ignored altogether.[6] Bembo's final manuscript text was thus by no means as scrupulous a copy of V as Bembo's letter might suggest, at least to a modern reader.

Some of the differences appear to be errors of transcription, due to a lapse of concentration or to an echo of another poem having come into Bembo's mind; three of them even go against the rhyme-scheme. Three errors were corrected in the printed edition, either in the text or in the errata. Some further differences, though, were deliberate alterations to Petrarch's manuscript. Two involved adding a word. In 3.14 'a voi armata' became 'et a voi armata', and in 224.4 'un lungo error' became 's'un lungo error'. At 105.45 Bembo changed his original 'a passo' (= V) to 'a pasco', a conjecture similar to the 'al pasco' of the more recent incunabula. He inserted the conjunction in 'pacificato e humile' (114.13), where there is a larger than usual gap between words in Petrarch's manuscript. He kept his 'restate' rather than Petrarch's 'ristate' at 161.14; he cautiously preferred 'risentir', the reading found in most incunabula (spelled 'ris-' or 'res-'), to Petrarch's admittedly unique Gallicism 'retentir' at 219.2; and he kept 'Et Laura mia' against Petrarch's Latinizing 'Laurea mia' at 225.10. These appear to be cases where Bembo thought that he could improve on the original, perhaps because he believed there had been an error of transcription in V like the 'et' which he rightly corrected to 'è' at 278.4.[7]

Most of the many other discrepancies between V and the 1501 edition affected form rather than sense. The motivation for them was Bembo's desire to regularize Petrarch's language, a desire which led him to remove some of the variety inherent in the poet's usage and to alter the usage of his source where it disagreed with his own tastes.[8] Here his editorial procedure, so insistent on authenticity in other respects, was dominated by his concern to establish grammatical and rhetorical norms, in a period when he had already begun 'alcune notazioni della lingua', as he announced in a letter of 2 September 1500, and when he was composing his *Asolani*, the first work by a non-Tuscan which was intended to be written entirely in Tuscan and predominantly in its Trecento form.[9] Bembo's version of the *Canzoniere* was more consistent in its usage and thus offered a model which was easier for himself and others to imitate. The rejected variants marked 'P' included some Latinizing spellings which must have seemed too like those which had become fashionable in the fifteenth century, though he liked to retain initial *h* (as in 'huomo' rather than 'uomo'), the one letter that could never affect pronunciation. Bembo also preferred not to adopt some alternatives which differed from his own more modern ones phonetically, morphologically (for instance certain noun, pronoun and verb forms) or sometimes syntactically.[10]

Bembo, himself a poet, had an ear which was alert to euphony and to questions of metrical regularity. Petrarch often used synaloephe (the running together for metrical purposes of a final vowel with a following initial vowel, although both are still heard); but Bembo sometimes avoids it, eliding or truncating final vowels and thus producing a more free-flowing line. However, there are a few examples in which he restores a final vowel. In other cases he truncates a word in order to avoid a hypermetric line.[11] The elimination of the final vowel in 'impressioni' (34.11, marked 'P') gave a line with a more orthodox stress on the sixth syllable, 'di queste impressïon l'aere disgombra'.

Petrarch's *Trionfi* were presented in an even more innovative form. Bembo established an arrangement of the *capitoli* which is still followed in modern editions. In the *Triumphus Cupidinis*, he began to copy out 'Era sì pieno' after 'Nel tempo', as if he was going to follow the conventional order of the incunabula ('Nel tempo', 'Era sì pieno', 'Poscia che', 'Stanco già'). But then he changed his mind and moved 'Stanco già' to second place. The *Triumphus Pudicitie* was probably going to consist of two *capitoli*, since 'Quando ad un giogo' was at first entitled 'Capitolo i'. This title was crossed out, however, and the fragment 'Quanti già ne l'età matura ed acra' was not transcribed. 'Nel cor pien', normally the first *capitolo* of the *Triumphus Fame*, was also omitted (to the consternation of many readers), so that this *Trionfo* only had three *capitoli* ('Da poi che morte', 'Pien d'infinita', 'Io non sapea'). The text of these *capitoli* is also different from any printed previously. Bembo's readings are generally found in one or more of the earliest editions (Venice 1470 and 1473, Padua 1472), but he probably used at least one manuscript, since he included some readings not found in earlier printed editions.[12] It is possible that he may have had access to autographs of the *Trionfi* then in Padua.[13]

Bembo's Petrarch was also novel in its careful use of punctuation, capitalization and accentuation. At first, a modern reader may not notice this aspect of the edition, because Bembo's practice corresponds so closely with our usage. But contemporaries were forcefully struck by Bembo's liberal use of signs to break up and terminate sentences (not just the colon and full stop but also the comma, semicolon and question mark) as well as by his use, for the first time in the printing of the vernacular, of the apostrophe (to indicate word division where a vowel had been lost through elision or aphaeresis) and, to a much lesser extent, of the grave accent (to distinguish the verb *è* from the conjunction *et* or *e*, and rarely to indicate stress in ambiguous cases such as *empiè* and *pièta*).[14] The introduction of this new system of punctuation meant that Bembo had to commit himself continually to interpreting the text to the reader in a particular way, choosing between alternatives (such as 'l'aura' or 'Laura') where previously there was ambiguity. Assuming that his options were the correct ones, and once readers became accustomed to his system, the new punctuation must have been a considerable help towards the rapid understanding of the text. The interpretative function of punctuation was something about which Aldo too felt strongly: the errors corrected at the end of his editions often relate to punctuation alone, and the printer noted in the dedication to his Horace of 1509 that punctuation can fulfil the role of a commentary.[15]

The letter to the readers defends Bembo's edition against the criticisms which his own usage and his text had aroused. One reading cited is 'se non se' (22.2), to which many earlier editions had preferred the simplified 'se non' or 'se non che'. This, Bembo argues, is an archaic but genuine form, supported by other occurrences in Tuscan prose and by forms found in contemporary popular regional speech. Bembo adds his approval, on stylistic grounds, of the practice of scattering some old forms among the more common ones, like stars shining in the heavens. But he is not basing his case on his own tastes: if critics disapprove of such readings, he goes on, they should reprove Petrarch rather than himself, because this is what is written in the autograph manuscript. The authority of this manuscript is compared with that which an autograph of Virgil would have. Bembo does not go into details about his *Trionfi*; however, he does not doubt that right-thinking readers will not criticize him for removing parts which do not belong and for ordering them correctly, as he believes, for the first time. Finally, he announces that his Dante, no less correct, will follow shortly and will be all the more welcome because the text of the *Commedia* has become infinitely more corrupt than that of Petrarch.

The Dante duly followed, though only in August 1502. It was as innovatory as the Petrarch. It had the same external appearance: octavo format, italic type, no commentary. The title, *Le terze rime di Dante*, was original, and the text was not based principally on an earlier printed edition. Bembo copied out the whole text himself between July 1501 and July 1502 (he evidently had a longer period of time at his disposal than most editors, working to a printer's orders, would have had), using two main sources. One was a mid-Trecento manuscript with a distinguished history, now Vatican lat. 3199 (Vat). This was probably the manuscript sent by Boccaccio to Petrarch in the 1350s, and it was now owned by Bembo's father Bernardo.[16] It remained Bembo's principal source throughout, and one hardly needs to stress the importance for the history of the text of the *Commedia*, and for vernacular philology in general, of Bembo's restoration of the work to a stage so much nearer to its source than the contaminated texts of the incunabula. But he was naturally even less subservient to this manuscript than he had been to Petrarch's own version of the *Canzoniere*. He incorporated a number of readings from another volume in his father's library, the Florentine printing of 1481, a copy of which had been sent to Bernardo by Landino. Several of its readings were incorporated as Bembo transcribed his text.[17] As with his Petrarch, he regularized punctuation, spelling, phonology and morphology.[18]

Bembo also introduced readings from other sources. Some emendations were probably suggested by the consultation of other manuscripts and incunabula.[19] He probably made some sparing use of conjecture when he felt that syntax or metre required it.[20] In preferring the literary synonym 'cangiò' to the more everyday 'mutò' of Vat and other early manuscripts at *Inferno* 13.144, he may have followed his personal taste, which leant away from Dante's eclecticism and towards Petrarch's more restricted and lofty vocabulary.

As with the Petrarch section of Vatican lat. 3197, so in the first two *cantiche* of the *Commedia* Bembo noted several variants in the margin without using them. Two of these came from a source described as 'vet[us]' (long-standing), which can be identified with Landino's text: 'Acor' for 'Acam' at *Purgatorio* 20.109, and 'seniori' for 'signori' at *Purgatorio* 29.83. But the great majority of such variants came from his father's manuscript, sometimes described as the 'ant[iquus]' (ancient and, by implication, venerable, though Bembo was by no means overawed by it).[21]

In several cases from *Inferno* 3 onwards, though less often than with his Petrarch, Bembo had second thoughts about his text. Some changes were made during transcription, within a line; others were written above the line or in the margin and seem to be the result of a thorough revision intended both to eliminate errors and to reconsider his original choices (ff. 116r and 127r are marked 'lustranda', 'to be reviewed'). Several emendations take Bembo's text away from that of Vat. Sometimes he prefers a reading which is that of Landino (but may also be found in other sources).[22] On other occasions his new reading is not found in Landino's text.[23] However, there are also many changes in which he inserts the reading of Vat after having originally written an alternative, and here Landino's text is sometimes on the side of the original reading, sometimes in agreement with Vat.[24]

Bembo could, then, make the same kind of arbitrary interventions as his contemporaries. It would have been extraordinary if this had not been the case, especially as he was an editor who was also a fledgling grammarian and a writer in search of a model for his own style. But those of his emendations which are, as far as one can tell, unsupported by any authority usually affected only minor matters of linguistic regularity, rarely syntax or lexis. And the philological care which he bestowed on Petrarch and Dante was unprecedented. He was the first to bridge the gap between the best contemporary classical scholarship and the application of its methods to editions of the great Tuscan authors. Bembo's use of a printed copy of Dante and a manuscript from his father's library followed the example of the great Angelo Poliziano, who had collated his printed copy of Terence with Bernardo Bembo's ancient manuscript of this author in June 1491 alongside Pietro, then only twenty-one years old.[25] Bembo was fortunate, to say the least, to have had such connections and to have had access to certain manuscripts through Paduan contacts and his father's library. But what his good fortune gave him was (to use a concept of Machiavelli) only an opportunity. It was Bembo's prodigious philological skill and his sheer determination which enabled him to make the most of this opportunity and to set new standards in the editing of the great Tuscan authors of the Trecento.

The first editorial reaction to Bembo's Aldine editions came in the Petrarch printed in Fano in 1503 by Girolamo Soncino. The letter addressed to the readers in the printer's name criticized the Aldine Petrarch for ordering both the *Canzoniere* and the *Trionfi* wrongly, compared with 'old copies' ('antiqui exemplari'), and referred to three manuscripts owned by local people, including the humanist Lorenzo Astemio.[26] It was suggested that the source for the Aldine edition either was not autograph or else was not Petrarch's final version (and the extent of Petrarch's work of revision was stressed). There are several respects in which the editor follows texts printed before the Aldine. The order of the second and third sonnets of the *Canzoniere* is inverted with respect to Bembo's edition and there is no division before poem 264. Although the Fano text and its punctuation are generally close to that of Bembo, it has some conservative features, preferring for instance to spell some elided and apocopated forms in full, as in 'l'havesse io' for 'l'havessio' and 'io sono' for 'i sono'. It numbers sonnets and *canzoni* in separate series, whereas the poems in the Aldine *Canzoniere* were unnumbered, as in the original manuscript. Aldo's order is abandoned in the *Trionfi*. Instead, the *Triumphus Cupidinis* follows the usual arrangement of the incunabula (with 'Era sì pieno' the second *capitolo* and 'Poscia che mia fortuna' third); the first *capitolo* of the *Triumphus Mortis* is the version in which 'Quanti già' is merged with 'Questa leggiadra'; and the *capitolo* which had been omitted from the Aldine ('Nel cor pien') was included at the start of the *Triumphus Fame*. But, against these retrograde steps, there was the interesting addition of two *rime disperse*: the *canzone* 'Quel c'ha nostra natura in sé più degno', found in the 'antico libro' belonging to Astemio, and the *ballata* 'Nova bellezza in abito gentile'. This appendix was, as we shall see, to inspire much rivalry.

Soncino's letter to the readers ended with the confident statement that his edition was by far the most correct so far printed. This claim glossed over the debt to Bembo's edition of the *Canzoniere* and showed a lack of appreciation of his work on the *Trionfi*; but at least it was recognized in Fano that any serious challenge to the authority of the Aldine edition had to be based on manuscript evidence. This was not yet the case in Florence. The first three editions of the *Canzoniere* to appear in the city were printed in 1504, 1510 and 1515 by Filippo di Giunta and apparently edited by a certain Francesco Alfieri.[27] The improvements which he claimed to have made were purely linguistic ones. He said in the first of his editions, using the same medical metaphor as Tanzio in 1494, that he had healed the wounds inflicted on Petrarch outside Tuscany and restored him to health ('alla propria sanitade'). Bembo's use of accents was extended to indicate the stress on several oxytone words. In 1510 Alfieri increased his independence from the Venetian Petrarch by restoring the *capitolo* 'Nel cor pien' to the start of the *Triumphus Fame* and by including both of the *rime disperse* which had appeared in the Fano edition as well as in Gregorio de Gregori's Venetian edition of 1508. Alfieri repeated the phrase that the first of these poems had been found 'in an old book' ('in un antico libro') but failed to specify that it had not been found by him. Florentine publishing evidently did not yet have sufficient ammunition with which to respond to the Venetian challenge with any original initiatives of its own.

As if to prove that a prophet's fame is never in his own country, Bembo's edition of Petrarch at first had limited influence in Venice itself. Only a minority of Venetian editions in this period had no commentary. There appeared in 1503 a folio edition, printed by Albertino da Lissona, which followed the pattern of Peranzone's 1500 edition but incorporated, for the *Canzoniere*, the 'Antonio da Tempo' commentary (previously published in 1477) in addition to the Filelfo–Squarzafico commentary. Ilicino's commentary was printed with the *Trionfi*. This type of edition with three commentaries continued to be popular in Venice and Milan for the next two decades.[28] There were some modifications in appearance: quarto rather than folio format was introduced by Gregorio de Gregori in 1508, and italic rather than roman type was used by Stagnino in 1513. The last folio edition was printed in 1515 by Agostino de Zanni, with a text which adopted some of Bembo's corrections but left the *Canzoniere* undivided and still read, for instance, 'barbarico' for 'bavarico' in 'Italia mia'.

Even among Petrarchs without a commentary, Bembo's new text met with considerable resistance. Lazaro de' Soardi's 1511 edition set out to be different from the Aldine both in its appearance (an original semigothic typeface with each page of type set in a woodcut border) and in its text.[29] A letter to the readers claimed that the edition was more accurate than its predecessors and also fuller than some (including the Aldine) because it had an appendix ('gionta') consisting of the two poems from Soncino's edition of 1503, the fragment 'Quanti già nell'età matura ed acra' which here begins the *Triumphus Mortis* and merges into 'Questa leggiadra e gloriosa donna', and, as the first *capitolo* of the *Triumphus Fame*, 'Nel cor pien'. The 1501 text of the *Canzoniere* was followed almost exactly and the apostrophe was used. However, the work was not divided into two parts as it had been in the Aldine, and the order followed in the *Trionfi* was that of Soncino's edition. This 1511 edition can be seen, then, as an attempt to achieve a compromise between the Aldine edition and its popular rival from Fano.

Two years later came the edition of Pre' Marsilio, printed by Bernardino Stagnino. A superficial glance might lead one to conclude that the priest was following in Bembo's footsteps. The title-page announced that the *Canzoniere* and *Trionfi* had been diligently corrected and restored to their original state and that they had been printed in the 'cursive letter' used by Aldo. The *Trionfi* were prefaced by a sonnet addressed to Bembo, paying homage to his restoration of the *Canzoniere* and the *Trionfi*; furthermore, the colophon of the *Trionfi* affirmed that the author's original was used. But this was just window-dressing. The edition preferred the Fano text to the Aldine and included all three commentaries together with the 'Antonio da Tempo' life of Petrarch. Pre' Marsilio's letter to Lodovico Barbarigo said that his main efforts as editor

55

were spent on improving the text of Ilicino's commentary. Bembo's Petrarch was still too innovative for most: no other printers or editors were yet prepared to follow it all the way, especially as regards the text of the *Trionfi* and the discarding of commentaries.

In August 1514 Aldo printed another Petrarch, apparently edited by himself. He took the opportunity to make some alterations to Bembo's text: he moved the division of the two parts of the *Canzoniere* to poem 267 ('Oimè il bel viso', the first to refer explicitly to Laura's death), corrected typographical errors, altered some other details and introduced minor changes in the system of punctuation.[30] But Aldo's main innovations came in reaction both to the appendix of two 'rime disperse' in the rival italic editions and to the hostility aroused by the 1501 *Trionfi*. He compiled an appendix of his own which, in its final version, included a letter from Aldo to his readers, the controversial *capitolo* 'Nel cor pien', eight poems by Petrarch, and the three *canzoni* whose first lines appear in Petrarch's 'Lasso me, ch'i' non so in qual parte pieghi' (*Canzoniere* 70): Cavalcanti's 'Donna me prega', Dante's 'Così nel mio parlar', and Cino da Pistoia's 'La dolce vista'.[31] This appendix was an event of considerable importance in the history of the study of medieval Italian verse. It contained the first surviving edition of Cavalcanti's great *canzone*, the first respectable text of Dante's (which had been printed in 1491 by Piero de Piasi but in a very corrupt version of which Aldo's text is independent), and the first printed edition of any poem by Cino. Aldo's letter to the readers (written by himself, as its language shows) constituted the first detailed examination of the question of the composition and order of the *Triumphus Fame* and indeed the first case study of the evolution of any of Petrarch's texts. Aldo showed his solidarity with Bembo's judgement in omitting 'Nel cor pien' from the *Trionfi*. If he included it now in the appendix, he wrote, it was not because he had changed his mind but so that readers could see why it had been excluded in 1501. He argued convincingly that it would be wrong to include this *capitolo* before the others of the *Triumphus Fame*, since it duplicated material found in 'Da poi che morte' and 'Pien d'infinita'. To substitute it for 'Da poi che morte' would similarly disturb the logical organization of Petrarch's subject-matter. There was also a stylistic argument in favour of seeing 'Nel cor pien' as a discarded draft. Petrarch showed fine judgement in rejecting it as written in too dense and dry a manner and in then distributing its material into two *capitoli* written in a more ornate style. It was for these reasons that Bembo and he (the responsibility is jointly shared) had omitted the *capitolo* in 1501. Aldo also emphasized the process by which Petrarch revised his works over a long period of time and rejected some versions such as the *capitolo* in question; but this and the other rejected poems would help everyone to recognize Petrarch's own criteria of acceptability. We can see here early evidence of a trend which was to grow in strength during the course of the Cinquecento in Venice: the provision of material which would help readers to become good judges of style.

The Aldine appendix had an immediate impact. Its poems were reproduced between 1515 and 1526 in fourteen other editions of Petrarch printed in Venice, Florence and elsewhere. In Florence, Aldo's edition was significantly given a less hostile reception than Bembo's of 1501. When Francesco Alfieri produced the third of his editions for Filippo di Giunta in April 1515, there were still signs of rivalry with Venice and Fano, and he stubbornly left 'Nel cor pien' at the start of the *Triumphus Fame*. But he included all the other material in the Aldine appendix, adding to it 'Nova bellezza' from Soncino's appendix, and his preface paid tribute to Aldo, who had died in February, though he mentioned only Aldo's services to Latin and Greek letters, not those to the vernacular.[32]

Some discordant notes came from other sources, however. In Venice, it was Alfieri's latest edition rather than Aldo's which was imitated by Alessandro Paganini in the same month in which the Giuntina was published, April 1515.[33] Paganini's Petrarch was intended to inaugurate a new series in twenty-fours format which came out between about April 1515 and June 1516. In his dedication of Sannazaro's *Arcadia*, one of the first of the series, he set out to establish the credentials of his press as a rival to the Aldine by stressing the novelty of his format, the new italic type design, and the attention paid to textual accuracy; and he made a point of mentioning the editor of the *Arcadia*, Ottavio degli Stefani.[34]

In a Petrarch printed in Bologna in September 1516, the influence of the 1514 Aldine was combined with that of the 1515 Giuntina. The limited use of the Aldine could be blamed to some extent on personal animosity: the printer was Francesco Griffo, who had been Aldo's type-cutter but had quarrelled with him and gone to Fano, where he cut a new italic fount for Soncino's Petrarch. Griffo had now moved back to his native Bologna and had set up as a printer in his own right, with plans to issue his own series of editions of Tuscan and Latin authors in twenty-fours, using Paganino's latest initiative as a model.[35] This showed a lack of originality, but Griffo's intention to emulate and rival Aldo's combination of craftsmanship and scholarship was evident both from the ambitious nature of this scheme and from his enumeration of the technical and editorial tasks which he said he had to perform as a vigilant printer, straightening lines badly laid out by some incompetent compositor and correcting the innumerable errors which, through the ignorance of the times and of scribes, had arisen in learned poems and historians.[36] As we saw earlier (p. 11), Griffo included a letter of recommendation from Tommaso Sclarici dal Gambero which sought to justify Griffo's edition in relation to the Aldine edition of 1514. Sclarici wrote that Aldo had removed many errors from the Latin works which he printed, but that he could not approve Aldo's 'new emendation' of Petrarch because Aldo was alien from the idiom of Tuscan poetry. He went on to say that the recent appearance of certain maxims ('dictamenta') had led Griffo to come to him in a state of anxiety, asking for Sclarici's opinion on whether his efforts as printer had been wasted; and Paolo Trovato has convincingly identified the work in question with the new grammar of Gian Francesco Fortunio, printed earlier in September.[37] Sclarici concluded that he could recommend what Griffo had produced, for two reasons. Firstly, the text had been harmed as little as possible 'by unusual accents' ('ab insuetis accentibus'): in other words, it had none of the Aldine apostrophes and accents. Secondly, it had been produced 'by collating the books of those who think rightly' ('collatis recte intelligentium libris'), namely the editions of Fano and Florence. Griffo's text was in fact derived principally from the Aldine text, but the order of the second and third *capitoli* of the *Triumphus Cupidinis* was reversed and 'Nel cor pien' was included, though at the end of the *Triumphus Mortis* rather than at the start of the *Triumphus Fame*. The appendix from the Giunta edition was given; however, since much of this had come from the 1514 Aldine, a note at the end cast doubt on its authenticity.

The Petrarch brought out by the Aldine press in July 1521 followed closely Aldo's edition of 1514. Now the imitation of the text by would-be Petrarchists was even more important than before: Aldo's dedication was replaced with a letter to the readers praising Petrarch as the most delightful and useful of poets whom one could read today, especially for lovers.

However, another Venetian printer still preferred to look outside his own city for alternative models to the Aldine Petrarchs. When Nicolò Zoppino printed the *Trionfi* in March 1521, he put them in an order which differed both from that of the Aldines and from that of the Fano text.[38] In December of the same year Zoppino printed both the *Canzoniere* and the *Trionfi*, and this time his model was the Giunta edition of 1515.[39]

In August 1515, some six months after Aldo's death, the Aldine press published a second edition of Dante's *Commedia*, with the title *Dante col sito et forma dell'Inferno tratta dalla istessa descrittione del poeta*. Interest in the physical and moral structure of Inferno must have been stimulated by the edition of the *Commedia* printed in Florence by the Giunta press in 1506 which contained, as the title page noted, a dialogue on the form and dimensions of Inferno ('uno Dialogo circa el sito, forma et misure dello Inferno'), illustrated by various diagrams. There is plausible evidence that a diagram of the organization of sins in Inferno contained in the second Aldine edition was provided by Bembo. It was copied in three further editions, two printed in twenty-fours by Alessandro Paganini in Venice in late 1515, the third, in octavo, printed by him together with his father Paganino in Toscolano some years afterwards. It is less likely that Bembo was involved in revising the text of the *Commedia* for any of these editions, all of which differ not only from the 1502 edition but also from each other in minor ways.[40]

As well as these editions containing the text of the *Commedia* alone, the traditional combination of text with Landino's commentary persisted. Editions based on the March 1491 text of fra Pietro da Figline were printed by Bartolomeo de' Zanni in 1507, by Bernardino Stagnino in 1512 and 1520, and by Iacob del Burgofranco for Lucantonio Giunta in 1529. Bartolomeo kept to the tradition of the previous century by using folio format and a roman typeface with woodcut illustrations; and his edition, like fra Pietro's, had *notabilia* in the margin but no index. The text of the Aldine edition, as well as the lack of commentary and the octavo format, proved too unconventional for Bartolomeo. His 1507 text was a hybrid which conflated Bembo's text with that of Landino (sometimes in its 1481 form, sometimes in one of the revised versions of the 1490s) or occasionally with that of texts from the 1470s. There was none of Bembo's new punctuation. Stagnino's two editions kept Landino's commentary, used woodcut illustrations, and still mentioned Pietro da Figline in the colophon. But he made several concessions to the Aldine innovations. As the title page advertised, he used 'littera cursiva'; moreover, the colophon added that the work had been newly revised by various 'excellent men', which meant, in effect, that the editor of the 1512 text adopted, for the most part, Bembo's text of 1502, together with its punctuation and the accented verb form *è*. He also used a smaller format, though this was quarto rather than Aldo's octavo. The 1520 and 1529 editions followed suit. In both the cases of Petrarch and Dante, then, it was the period around 1512–15 which saw the eventual dominance of Bembo's texts; but they met considerable opposition for different reasons, and both of them, especially the Petrarch, needed the support of the second Aldine editions which appeared in 1514 and 1515.

So far we have been considering Aldine editions of fourteenth-century Tuscan works. But was Aldo so scrupulous in following the intentions of authors who were still alive and who came from outside Tuscany? One can answer this question by considering two contrasting works, one non-literary, *La vita et sito de Zychi* by Giorgio Interiano of Genoa, printed in 1502, and the other by the greatest living writer from the south, the *Arcadia* of Iacopo Sannazaro, printed near the end of Aldo's career in 1514. Interiano's description of the life and customs of the Sarmatians, which Aldo dedicated to Sannazaro, was full of northern forms but seems to have been left virtually untouched in its progress through the press.[41] The *Arcadia*, on the other hand, a combination of prose and verse written in a lofty Boccaccesque style, was judged worthy of revision. This work had been written in Naples around 1485 but remained in manuscript until, after pirate editions had begun to appear in Venice from June 1502, Sannazaro authorized its printing in Naples by Sigismondo Mayr in 1504. The language of this edition was considerably revised, most probably by Pietro Summonte, so that it was brought nearer to that of Petrarch and Boccaccio and lost many of its southern dialect forms, though these could be kept especially when they coincided with Latin.[42] This text went largely unchanged when copied in another Neapolitan edition of about 1507 and in a Venetian edition of December 1512 printed by Giovanni

Rosso of Vercelli. For the Aldine of 1514 the language of the *Arcadia* was made slightly more Tuscan, generally at the expense of classicizing and southern elements.[43] But some of the latter went unaltered, and the new edition treated its source with some respect for its original character: more respect than was shown in March 1515 when, as we shall see in chapter 6, the Aldine became in turn the source of a new Florentine edition.

In 1514, when Aldo's edition of Petrarch was published, there seemed to be no loss of vigour in the best vernacular scholarship in Venice. If we move forward to 1516, however, we find a more uncertain situation. Aldo had died in February 1515. Bembo had been in Rome since 1512, and had been appointed as a secretary to Pope Leo X in 1513. Even if he had wanted to carry out any detailed editing, his opportunities were now at best limited. It remained to be seen, then, whether other printers and other scholars would be capable of continuing on the road which these two men had indicated.

In May 1516 there appeared a *Decameron* printed by Gregorio de Gregori and edited by Nicolò Delfino (or Dolfin), a Venetian patrician from Bembo's circle, though not one of his closest friends.[44] In his dedication, Delfino explained how he restored his text on the principle followed by Zeuxis when painting Helen of Troy: just as the artist compiled a portrait from 'the most excellent parts' ('le più eccellenti parti') of the most beautiful women he could find, so Delfino obtained 'many very old texts' ('molti antichissimi testi') and chose those parts which seemed 'most beautiful and most suited to the author's intention' ('più belle e più confacevoli alla intentione dello auttore'), thus restoring the work 'to its original beauty' ('alla sua prima bellezza'). For his main source he went back to the edition from which the incunabula printed in the Veneto were descended, the Mantuan edition of 1472, but he also used manuscripts, and the language of his text kept closely to Trecento Florentine norms.[45]

By modern standards this use of sources was arbitrary. But it was typical of its time: the same technique was used, for instance, in Manfredo Bonelli's 1515 edition of Cavalca's *Specchio di Croce*, where the colophon explains how a single correct text was built up by looking through many corrupt ones ('habbiamo, transcorrendo di molti [exempli] corropti, fatto uno correpto'). At least the chain of transmission from edition to edition had been broken in a way analogous to Bembo's editing of Dante. And even if Delfino had still relied principally on a much later source than Bembo's, he had ventured into new territory by applying Bembo's method to Trecento prose. He did so at the risk of attracting some criticism: as we shall see when we look at Castorio Laurario in the next chapter, Boccaccio's style still had its critics around 1516, and Delfino was brave enough to help to win respect for it. His *Decameron* won high praise from contemporaries. It was seen by an editor of the following decade, Tizzone Gaetano, as having rescued the work from being a 'tavern book' ('libro di taverne'), and it provided a text which, once its six-year book-privilege had expired, dominated Venetian printing of the work for the next decade.[46] Its text was adopted for the edition printed in 1522 by Aldo's successors, his father-in-law Andrea Torresani and Andrea's two sons. One of the latter, though, Francesco Asolano (as he signs himself) invited readers to incorporate some corrections to the spelling. For instance, he says, *bascio* should become *bacio*, *divocione* should become *divotione*, and *stratiati* should become *stracciati*, since this was the true pronunciation of Florentines (a justification which reflected the contemporary debate on how orthography should relate to speech). Francesco Asolano claimed to have studied many old manuscripts and printed texts while revising the *Decameron*, but the only visible result of this was a list of just fifteen variants, some relating merely to spelling.

These two editions, Delfino's and the Aldine, show that there was now a demand for scholarly octavo editions of the *Decameron*. From September 1525 it became even more fashionable and commercially desirable to produce such editions: in this month appeared Bembo's *Prose della volgar lingua*, in which Boccaccio's prose, particularly that of the *Decameron*, was elevated to the status of the principal literary model to be followed by contemporary writers. Two of the Venetian printing firms who had an eye on literary trends, those of Gregorio de Gregori and Giovanni Antonio Nicolini da Sabbio and his brothers, took immediate and unscrupulous advantage of the situation during the next five months by reproducing the Delfino–Aldine text with a pretence of originality, referring in the vaguest of terms to the consultation of texts, now evidently an important selling-point. But in both cases the respective 'editors' (the bookseller–publisher Nicolò Garanta and Marco Astemio of Valvasone in Friuli, who also worked in the mid 1520s for De Gregori and the printer-publisher Lorenzo Lorio) were duping their readers in order to try to profit from the new literary trend.[47]

Delfino's edition was an important advance in so far as it returned to earlier sources, including manuscripts, and created a new awareness of Boccaccio's text as worthy of serious research. In spite of its virtues, though, its importance for the text was considerably less than that of Bembo's editions of Dante and Petrarch. Nor were any of Delfino's imitators in the 1520s capable of doing for Boccaccio what Bembo and Aldo had done for Dante and Petrarch.

At the same time, and not only in Venice, there were signs of a growing interest in another still relatively unexplored area: the lyric poetry and prose written before Petrarch and Boccaccio. As we have seen, an appendix of *canzoni* by Dante was included in the *Commedia* printed in Venice in November 1491, and Aldo's 1514 Petrarch had an appendix containing *canzoni* by Cavalcanti, Dante and Cino. Another selection of early poems appeared on 27 April 1518: the *Canzoni di Dante. Madrigali del detto. Madrigali di Cino, e di messer Girardo Novello*.[48] Barbi showed that the poems were derived from a lost manuscript related to the collection which an admirer of Bembo, Antonio Isidoro Mezzabarba, copied before May 1509, adding the important note that he did so 'making no changes or additions to what I found written in very old books' ('nulla mutando overo aggiungendo di quello che io in antiquissimi libri trovai scritto'). On the basis of the apparent confusion in attributing poems, Barbi judged the 1518 collection to have been published by someone who was not very scrupulous.[49] Nevertheless, the collection represents a major step forward and was one of the sources of the Florentine *Sonetti e canzoni* of 1527, which will be discussed at the end of chapter 6.

The year 1525 saw the first dated edition of a work which Bembo's *Prose* mention frequently, the late thirteenth-century collection of short stories now known as the *Novellino*, but which was here called, with a strikingly archaizing spelling, *Le ciento novelle antike*. The place of printing was not Venice but Bologna; however, Bembo was the guiding influence behind the editor, the young Carlo Gualteruzzi of Fano (born in 1500), and it was no doubt Bembo who provided the manuscript, now lost but closely related to Vatican lat. 3214.[50] In the dedication which (as the manuscript draft shows) Bembo helped Gualteruzzi to write, there are some motifs characteristic of the Venetian scholar: the novel connoisseurship of early Tuscan, the insistence on observing its authentic forms, and not least the belief that the early literary language should be to some extent imitated as well as admired by contemporaries. And when we go on to look at the printed text of the *Novellino* and to compare it with the Vatican manuscript, we find that it has been edited in accordance with Bembo's methods. Gualteruzzi has, that is, facilitated the reading of the text by means of careful punctuation and has reproduced his source in a relatively conservative way, but he has also introduced greater regularity in spelling and morphology as well as a very few syntactic improvements suggested by his own judgement. But the principles of Bembo's two Trecento editions were not applied in a purely imitative way. For example, the Venetian had avoided the archaic letter *k*; but, since this was in keeping with Duecento usage, Gualteruzzi adopted it and even extended its use, giving it the functional role of representing velar [k] before the vowels *e* and *i*, as in *anke* or *poki*. Such respect for the orthography of one's source was most unusual, though there was a fairly recent parallel in the way in which Filippo Beroaldo had edited the sixth book of Tacitus's *Annales* in Rome in 1515, leaving untouched any passages which seemed hard to correct and keeping unusual spellings (such as *Sulla* for *Sylla*) because they were 'redolent of antiquity'.[51]

Bembo had a more indirect link with the revision of Castiglione's *Cortegiano* for its first printing by the Aldine press in 1528. When Castiglione sent him a draft of the work for comment in 1518, he said that linguistic detail ('la scrittura') would be someone else's concern.[52] Nevertheless, Bembo did see some sheets of the *Cortegiano* as they were printed, and the 'someone else' eventually chosen to 'mend and punctuate' the book ('remendarlo et apontarlo') was one of Bembo's Venetian friends, Giovan Francesco Valerio.[53] Although Castiglione did make some sporadic and inconsistent revisions himself, the bulk of the preparation of the text for printing was carried out after he had sent the final draft (now BLF MS Ashburnhamiano 409) from Spain to Venice around early April 1527.

As one would expect, given Valerio's links with Bembo, the revision brought the language of the *Cortegiano* closer to that recommended in Book 3 of the *Prose*.[54] But Valerio was not as severe as he might have been. Even if one allows for some oversights (understandable in the revision of such a long work), it looks as if he simply decided to set limits to his interventions, realizing how incongruous it would have been for the language of the *Cortegiano* to have been pure Tuscan when the author argued in the Dedication and (through Lodovico da Canossa) in Book 1 in favour of the right to use Latinizing and non-Tuscan forms. In 1.39, for example, Castiglione defends *populo* against Tuscan *popolo*, and, in accordance with this attitude, many such Latinizing spellings remained in Valerio's text.[55] Among the other forms preferred by Castiglione and respected by Valerio, even though they might seem northern rather than Tuscan, were those without the diphthongs *uo* and *ie* (*bona* and *leva* rather than *buona* and *lieva*, and so on) and those in which final *e* and *a* were lost, usually after the consonant *r*.[56] In the area of morphology, there were a number of instances where the definite articles *il* and *i* were used before *s* + consonant and after *per*, even though this was not what Bembo recommended in the *Prose* (3.9). In the present indicative of the second conjugation, first-person plural forms in *-emo* were commonly preferred to *-iamo*, and a few third-person plural forms in *-eno* were found instead of *-ono*, again in spite of Bembo's rules (*Prose*, 3.27 and 29).

Valerio's revision of the *Cortegiano* was carried out as an act of friendship. He was not a professional editor, and his only tangible reward appears to have been a gift, made at Castiglione's request, of some copies of the printed work.[57] Although the revision is characteristic of the work of editors in Venice in the 1520s in its basically Tuscanizing tendency, it can be distinguished from other, contemporary revisions by its concern not to distort the hybrid linguistic character of the original when its form was so closely bound up with its content. Valerio carried out his task with unusual tact and sensitivity, and the author would surely not have disowned the result. One could not confidently say as much of many other 'corrected' editions of this period, particularly, as we shall see in the next chapter, in the cases of those which were outside the sphere of influence of Bembo.

5 · VENETIAN EDITORS AND 'THE GRAMMATICAL NORM', 1501-1530

THE EDITIONS PRODUCED IN the first three decades of the Cinquecento by Bembo, and by editors linked with him or the Aldine press, struck a balance between a tendency to steer all texts towards a uniformity based on Trecento Tuscan and on the other hand a respect for what the author originally wrote, even if this meant allowing a few archaisms or regionalisms to survive. A hierarchy of susceptibility to editorial change was established among the different linguistic categories: interventions are found most often in orthography and phonology, then in the area of morphology, and become gradually rarer in syntax, lexis and, where relevant, questions of metre. However, as one goes further from Bembo's influence, one finds less balance in Venetian editing between the normative approach and the conservative one, so that all aspects of 'la scrittura' become subject to the editor's pen. Sometimes this was because the texts concerned were much more strongly regional in character than, for instance, the *Arcadia* of 1504, as well as being of lesser literary stature. The kind of editing which took place in these cases was of particular linguistic significance because it helped to extend a norm to a wide range of writing. But, as we shall see, even texts such as those of Boccaccio could be radically rewritten. The problem was that, once such works were given the status of models, they then had to conform with the orthographical, grammatical and metrical rules and the lexical and stylistic ideals which they were thought, rightly or wrongly, to provide. Exceptions to the rules and obscure words or awkward phrases were likely to be treated by an editor as errors, deformations inflicted by previous scribes and printers, in the apparently sincerely-held belief that he was not just conferring perfection on the text but restoring it to its original state. The editor's own linguistic principles, together with his estimate of what his readers desired, thus tended to become the basis for emending a text.

In the early years of the Cinquecento in Venice, one begins to find arbitrary alterations on the part of editors of both Trecento and later texts. The instances are as yet few in number and uneven in their execution, but they are nonetheless forerunners of the more radical reworkings carried out in the 1520s.

In 1503 Giorgio Rusconi, Milanese in origin, printed Boccaccio's letter to

Pino de' Rossi and his *Ameto*. The editor appears to have been a certain 'Zilius', author of a summary of Boccaccio's works included in the volume. He may well have been connected with the law school of Padua, since between the two works by Boccaccio there are included three vernacular poems (here unattributed) by the fifteenth-century Paduan poet and lawyer Iacopo Sanguinacci.[1] Zilius certainly came from a humanist background. He mentions Boccaccio's Latin works in some detail, but the vernacular works, other than the translation of Livy, are passed over quickly as works of entertainment. His language is very cumbersome and Latinizing (including for example 'silve obscure', 'poaeta', 'dicto'), but also shows northern features (such as 'lezenose' for 'leggonsi'). The language of his edition of the letter to Pino de' Rossi is also far from Boccaccio's original Tuscan, with Latinisms such as 'scio' for 'so' and northern forms such as 'soi', 'mazore', 'zà'.

The edition of the *Ameto* is an unusual one. On the one hand, as Quaglio has shown, the editor consulted at least two manuscript sources, was aware of the differences between their readings, and considered that some variants were important enough to be printed in the margins. However, the editor sometimes intervened on the basis of what appears to be personal taste, adding a word or a phrase or rewriting Boccaccio's text in order to elucidate what seems obscure, to add further colour, or even to add a moralizing note, as in his expansion of 'Semelè, quando divinamente cognobbe Giove' (29.57) to 'Semelè, quando divinamente Giove la mal per lei domandata grazia li concedette'. Boccaccio's works other than the *Decameron* were especially prone to editorial embellishments of this kind. An emendation in a verse passage shows the avoidance of diaeresis which became one of the major concerns of some editors in the 1520s: '*Io conosco* che li ben sovrani' (39.19, where 'io' counts as two syllables) > 'Io chiar conosco'.

One of the earliest contributions to the normalization of non-Tuscan texts is the work of Sebastiano Manilio on the *Porretane* of the Bolognese author Sabadino degli Arienti (*c.* 1445–1510) when they were printed in 1504 by Bartolomeo de Zanni.[2] The text of the first, Bolognese edition of 1483 was characteristic of the northern koinè of the late Quattrocento. It contained forms which reflect the phonology of northern dialects, such as 'quisti', 'zelosi', 'posibile', together with orthographical and lexical Latinisms such as 'epso', 'egeni' ('needy'), 'ditissimi' ('very rich'). In 1504, however, while Sabadino was still alive, Manilio considerably Tuscanized the text, so that these forms become 'questi', 'gelosi', 'possibile', 'esso', 'poveri', 'ricchissimi'. He also regularized the syntax and eliminated repetitions which he felt to be superfluous. However, he was uncertain in his use of the diphthong *uo*, which he sometimes introduced in cases where it would not be found in literary Tuscan, such as 'puoco', 'luoro', 'fuorsi'. He also allowed quite a large number of northern and Latinizing forms to survive in the text.[3] In this respect, Manilio's language had not changed a great deal since his translation of Seneca's letters printed in 1494.[4] Though Manilio described himself as Roman, the language which he adopted as a translator and editor was the sort of Tuscan prose with Veneto and Latin colouring which, as we have seen, was for the moment a norm well established in Venice.[5] In the Seneca translation he described his language as 'Tuscan vernacular' ('toscan volgare'). But in the preface he also used the phrases 'common language' ('parlar comune') and 'everyday language' ('cotidiano parlare'); and these seem more accurate descriptions of his ideal, for, while he went a long way towards using Tuscan, he was evidently no purist. His use of 'comune' seems to have similar interregional implications to those intended by certain editors and others in the 1520s and 1530s.[6] It is worth noting, too, that Manilio already displayed the Cinquecento tendency to add moral glosses to collections of short stories when he added paragraphs summarizing the lesson to be learnt from the preceding *novella*.

The next ten to twelve years were crucial in the establishment of a prose style modelled on Boccaccio's rather than on a more broadly-based language. In 1505 Aldo Manuzio printed Pietro Bembo's dialogue *Gli Asolani*, whose grammar and vocabulary imitated the manner of Boccaccio. This new direction attracted some criticism and mockery.[7] However, the principle of imitation found defenders in the Veneto even outside Bembo's circle, and the extent of support for it is shown by an edition of Boccaccio's *Corbaccio* printed in February 1516 by Alessandro Paganini in his series of texts in twenty-fours. The editor was Castorio Laurario, a law graduate from Padua.[8] One of his duties, he considered, was to defend the contents of the work, arguing for its moral usefulness as an antidote to love and distinguishing between the worth of most women and Boccaccio's attack on an exceptional case. Another of his aims was to defend Boccaccio as a stylistic model for contemporaries. There were evidently still some who ridiculed Tuscan terms which they considered to be either deformed in respect of their Latin origins, such as *stromenti* ('instruments'), or else archaic, such as *altresì* ('also'). Castorio renewed his defence of the imitation of Boccaccio in a sonnet which he added to a new edition of the *Corbaccio*, printed by Bernardino Benagli shortly after the first.

In spite of his enthusiasm for Boccaccio's elegance, it must be said that Castorio had an uncertain command of phonology and syntax. For instance, he often followed northern dialects in using single consonants where Tuscan has double ones, but he also reacted excessively against this common error; one sees both faults in the spelling 'parollete' for 'parolette'. What was still lacking for writers and editors who wished to follow Boccaccio, Dante or Petrarch was a grammar of Trecento usage. It was in response to this widely-felt need that, in September of the same year 1516, Gian Francesco Fortunio published his *Regole grammaticali della volgar lingua*. The work was printed in Ancona, where Fortunio was *podestà*, but the author was born at Pordenone in Friuli and his cultural roots were in the Veneto. Fortunio provided a series of rules, set out in a manner relatively easy to follow, which covered certain aspects of morphology, such as noun and verb endings and pronouns, and also orthography, where Fortunio was particularly anxious to do away with Latinizing spellings such as *dixi* or *scripse*. But, in giving his rules, Fortunio also offered critical comments on the texts of Petrarch and Dante (he was particularly hostile to the Aldine editions) and, to a lesser extent, on the *Decameron*. In these comments he took to its extreme the method of working from the rule to the text whose influence, we saw in the previous chapter, extended even to Bembo.[9] The search for models of style made it imperative to decide what was and what was not regular: as Fortunio wrote, citing Quintilian in his support, correct usage is based on what is normal, not on exceptions. But Fortunio could sometimes go a step further by trying to prove that what he considered irregular was a textual corruption. Thus in Book 1 he argued that *lui*, *lei* and *loro* are never subject pronouns. He was faced, however, with some cases in Dante, Petrarch and Boccaccio where they were apparently used after a verb to refer to the subject. Rather than conclude from this evidence that his rules needed to be modified, he preferred to deduce from the rarity of such examples that they were scribal or printing errors, and he corrected them with what he called 'the proper grammatical reading' ('la diritta grammaticale lettura'). Just one example: Fortunio says that where the *Decameron* reads 'elle non sanno delle sette volte le sei che elle si vogliano *loro* stesse' ('women don't know six times out of seven what they want themselves', 3.1.11), he has seen 'an old book' (he is not more specific) without 'loro'. Seizing on this reading because it happens to suit his argument, he says that the pronoun found in other texts would be 'against the grammatical norm' ('contra la grammaticale norma') and would in any case be superfluous. He supports this argument with parallel examples from Dante, concluding that, in accordance with the opinion of whoever wrote the manuscript and his own judgement, the correct reading should be 'quello che elle si vogliano istesse'; thus grammar will not be violated and the meaning will be preserved.[10] Of course, he says, if he had the autograph original or a copy derived from it, then he would follow that; but in the existing confusion the best method of re-establishing the original reading is to use good judgement, in the same way as Ermolao Barbaro emended the text of the elder Pliny in his *Castigationes*.

Fortunio's grammar provided rules and an implicit critical methodology which were potentially of considerable help for editors. However, it took a few years, until the early 1520s, for Venetian editors to try to apply his lessons systematically. The first open, and partly critical, acknowledgement of his influence came, in fact, not in Venice but in Milan, when two works by Boccaccio, the *Ameto* (1520) and the *Amorosa visione* (1521), were edited by the minor humanist Girolamo Claricio of Imola.[11] As the title pages of these editions indicated, Claricio wanted to follow in Fortunio's footsteps by using a discussion of points in the texts in order to make his own contributions to vernacular grammar. With the *Ameto* came Claricio's 'Osservationi in volgare grammatica sopra esso'; with the *Amorosa visione*, a defence of Boccaccio's poetry and further 'Osservationi di volgar grammatica del Boccaccio'. A few of the 'observations' in the *Ameto* explained obscure words in Boccaccio, such as *bozzacchioni* ('rotten plums') or *petronciani* ('aubergines'). These, and the alphabetical list of words discussed which followed them, pointed the way towards the annotations and word lists which later editors would provide for the *Decameron*. Claricio's main purpose, though, was to justify certain spellings typical of the Trecento which he had found in his sources and preserved in his edition yet which differed from the usage of his own day. He clearly respected what Fortunio had written on orthography, but felt that the *Regole* were too rigid in giving precepts which brooked no exception. For instance, when Claricio found *lui* as subject in a copy of the *Amorosa visione* and noted that this would go against Fortunio's rules, he was torn between respect for his venerable text and conjecturing that 'lui' was a corruption of 'là'; the latter alternative, he observed with some exasperation, would at least save bickering grammarians from coming to blows. In the same work Claricio explicitly agreed with Fortunio about some spellings, such as the use of *f* rather than *ph* in learned words. However, he distinguished between Boccaccio's spelling and that of Fortunio, keeping an open mind about which was preferable. He was inclined to disagree with the view of Fortunio (and Bembo) that words of Latin origin should retain initial *h*, as in *humile*; Claricio accepted that such words could be spelt with or without the *h*. Often in the *Ameto* he countered a rule of Fortunio with the evidence of the 'old text' which he was using. On the other hand, he made it clear that, although he was reproducing its archaic spellings, they were not a matter of life and death. For example, he wrote in the *Amorosa visione* that Boccaccio used *-ct-* in the Latin manner, while Fortunio said one should use *-tt-*, but that Claricio's reader should not be surprised to find him following contemporary 'common usage' rather than that of a single author, and anyone should be free to use either spelling. At times he lost patience altogether with the rules of scholars. In the *Ameto* he pointed out a misprint of 'optima' for 'ottima', as his 'old text' read, explained the rule about such spellings which Fortunio gave in Book 2 of the *Regole*, but then added, as if it were all beneath him anyway, that this was a rule known even to cobblers. At the end of the 'Osservationi' on the *Amorosa visione* he stressed that his aim both here and in the *Ameto* was not to teach but simply to point out matters worth thinking about.

Claricio's 'Osservationi' and his presentation of the text show that, as an editor, he had virtues which were most unusual for his time. First, he was concerned to reproduce what he claimed was Boccaccio's spelling, whatever other people's 'rules' said the spelling should be. Second, he went to the trouble of justifying the readings of his texts in a series of critical notes. Third, in these notes he offered guidance to his readers on their use of the language of the text as a model. He occasionally stated what he considered to be correct usage; he sometimes pointed out what he believed to be a more Tuscan usage, but without saying that this was the only valid model; and he sometimes offered alternative spellings as equally valid. Since the reading of the text was not governed by a set of rules, as it had tended to be with Fortunio, Claricio was free to preserve any unusual features of his sources. Fourth, in the *Amorosa visione* he and the printer Giovanni Castiglione used the grave accent and the apostrophe, a particularly innovative step for a work in roman type.[12]

One must give Claricio credit for approaching the role of editor in this fresh way, just as he deserves credit for his individualistic defence of Boccaccio's poetry and of the contemporary 'letteratura cortigiana'. Regrettably, though, these editorial virtues are more than outweighed by Claricio's incorrigible tendency to rewrite and even, it seems, to deceive. Apart from making exaggerated or spurious claims about his use of manuscripts and thirteenth-century grammars, he took it upon himself to revise both the *Ameto* and the *Amorosa visione* extensively in order to make them more comprehensible and to remove what he considered to be stylistic imperfections which would make them less enjoyable to read and less rewarding to imitate. He therefore replaced archaic or obscure words with more common ones, such as 'quanto' instead of 'chente'. He removed words which he found coarse or too familiar and everyday; thus, in the *Amorosa visione*, 'vacca bella' and 'schiava puttana' became respectively 'giovenca bianca e bella' and 'schiava istrana', and *mutare* was replaced by *cangiare*. He expanded Boccaccio's text, sometimes just with an adjective but also more drastically, for instance enriching the flora of the garden of Pomona (*Ameto* 26) with plants not mentioned by Boccaccio and adding reminiscences of both classical authors and vernacular ones such as Petrarch and Poliziano. Other emendations removed cases of dialoephe, both when the two vowels were unstressed, as in '*come amata* cosa, loco avesse' (*Ameto* 2.34), which became 'sì come amata', and also in more normal cases, for instance after a stressed syllable, as when he changed '*e più adentro* alquanto che la scorza' (*Ameto* 2.25) to 'et via più dentro'.

There had been before 1520 a few signs in Venice and Florence of editorial eagerness to 'improve' texts; but there had been nothing like these two editions of Claricio, either in the freedom of his treatment of the text or in the discussion of individual readings in his 'Osservationi'. Editors working in Venice began to treat the minor works of Boccaccio much more arbitrarily after 1521, as if they were following Claricio's initiative. But fortunately, although they shared some of his concerns, nobody went so far as Claricio had done in rewriting a text.

How did the majority of Venetian printers deal with the movement towards linguistic standardization in the decade and a half following the publication of

Fortunio's grammar in 1516? One can start to answer this question by tracing developments found during this period in books produced by one suitably representative candidate, Nicolò Zoppino, a printer–publisher of Ferrarese origin. His press concentrated almost entirely on vernacular texts, often ephemeral ones, and the length and success of his career in Venice (from about 1505 to 1543) can be attributed to his ability to follow what was fashionable: he was quick to imitate initiatives taken by printers in other cities, particularly Florence, but also had a flair for presenting texts in a way which caught the intellectual spirit of the moment.[13] A tiny but significant indication of his efforts to comply with standardization can be seen in the increased use of the Tuscan spelling 'Zoppino' rather than the northern 'Zopino' in his colophons from about 1525. But is there any evidence that he employed editors in order to bring his texts closer to the usage of Trecento Tuscan? And how much consistency of linguistic standards is apparent between different works printed by this one firm?

Before 1522, Zoppino's texts underwent no thorough linguistic revision, whether they were Tuscan or non-Tuscan, from the Trecento or from a more recent period. The *Vita di Cristo*, a poem in *terza rima* by Antonio Cornazzano of Piacenza (*c.* 1430–*c.* 1484) which he published or printed from 1517 onwards, was edited for him by a certain 'B. L.', probably Castorio Laurario's father Bartolomeo, a lawyer and poet.[14] The 1518 edition included forms which show the influence of Latin or northern usage or both, such as 'trahe', 'dece' (= 'dieci'), 'ardegiar', 'simplice', the preposition 'de', 'el corpo', 'gli soi', and the plural 'le rete' (in rhyme), which goes against Fortunio's rules. Most of Zoppino's editions of works of Tuscan origin simply reproduced their source edition. The translation of Petrarch's *Secretum* (1520) was derived from the Sienese edition of 1517; it contained Latinizing forms such as 'advenne', 'anxio', 'cognosciuta', 'aspecto', and the *el* which had become widespread in Quattrocento Tuscan both as a definite article ('el palazo') and as a pronoun ('io el confesso'). For Poliziano's *Stanze* in August 1521, Zoppino used the first Bolognese edition of 1494, edited by Alessandro Sarti, substituting his own name for that of the original printer in his version of Sarti's dedication.[15] Here too we find forms characteristic of Quattrocento Tuscany: 'e facti', 'sancto', 'le dolce ... cure' and so on. Zoppino's tendency, at this stage, to use non-Venetian editions as his sources and to reproduce them passively is also seen in his Petrarch of 1521, based on the Florentine edition of 1515 rather than on an Aldine edition.

A partial exception is the 1516 edition of the *Facetie* of the Piovano Arlotto, printed for Zoppino by Giorgio Rusconi. The source here is one of the Florentine editions printed a little earlier for Bernardo Pacini, and again the original dedication is reproduced. For the most part, the text is unmodified. We thus have Latinisms such as 'mi ha etiam exhortato', 'epso', 'auctore' (in the dedication) and Quattrocento Tuscan forms such as 'andorono', 'harei', 'qualunche' (in the *Facetie* themselves). However, a few changes take the text away from Florentine phonology and from Fortunio's rules, either in the direction of Latin ('popolo' > 'populo') or in that of non-Florentine dialects ('impegnerebbe' > '-arebbe'). Occasional alterations have been made to the syntax: for instance, 'che li rispose' becomes 'e lui rispose'.[16] Some slight changes are made in Zoppino's next edition of the *Facetie* in September 1518, but again they are inconsistent, and the gothic typeface used for this edition (the 1516 edition was in roman) suggests that it was intended for a popular audience which would have cared less about linguistic refinements.

A hint of an impending change of direction is seen in 1520 in the *Libro del Peregrino* of Iacopo Caviceo from Parma (1443–1511), also printed for him by Rusconi. This work, first printed in 1508, imitates Boccaccio's *Filocolo*. Zoppino's title page says that it has been 'diligently corrected into Tuscan' ('diligentemente in lingua Toscha correcto'); but the text is very similar to that of preceding editions, with only minor changes such as '*il* spirito' > 'lo'. The hesitant shift towards Tuscan is also seen in his edition, dated 20 December 1521, of the *Dialogo de Fortuna* of Antonio Fileremo Fregoso, another poem in *terza rima* by a contemporary northern author. This still has forms which are non-Tuscan or which go against Fortunio's rules of spelling and morphology.[17] However, evidence of a gradual move towards the newly-emerging norm can be seen if we compare the text with its probable source, an edition (attributable to a Milanese printer in about 1520) whose page layout is followed by Zoppino for the first nine *capitoli*.[18]

Up to the end of 1521, then, Zoppino attached at best only slight importance to the linguistic revision of his editions. In 1522, however, a certain Cassiodoro Ticinese (that is, from Pavia) provided Zoppino with an edition of the *Opere* of the Florentine Girolamo Benivieni which was based on the Giunta printing of 1519 edited by Biagio Buonaccorsi but, as the title page proclaimed, 'very recently revised and purged of many errors' ('novissimamente rivedute et da molti errori espurgate'). Just as Fortunio saw printing errors ('errori di stampa') in earlier editions of Trecento authors, so Cassiodoro considered the work of Benivieni, an author who was still alive, to have been printed less than correctly ('in stampa meno che ben corretta'). He has endeavoured, he says, to restore what he calls this jewel lying in basest dung to its original brightness ('primiera nitidezza') by observing 'rules' of the vernacular which can only be those of Fortunio. Apparently believing that he was following the author's will, but actually going against what we know of Benivieni's hostility to Fortunio, Cassiodoro removed Latinizing spellings and corrected some points of morphology.[19] He also added more punctuation and corrected hypermetric lines by making truncations. However, he left some forms which break Fortunio's rules: for example, 'e' as a masculine plural definite article, 'eccellente' as a feminine plural adjective, and, ironically, for the word 'rules' itself he even used the plural 'Regoli' rather than the correct 'Regole'.

For the rest of the 1520s there is evidence of a gradual rather than sudden increase in the use of editors by Zoppino, with results which were still inconsistent. In August 1524 the sons of Giorgio Rusconi printed for themselves and Zoppino another edition of Caviceo's *Peregrino*. Zoppino now added an important declaration to the readers in which he attributed recent improvements in printing to the adoption of a new method of spelling, rediscovered by contemporary scholars, in place of 'things very dissimilar from everyday usage' ('cose molte [sic] dissimile dall'uso cottidiano'). In order to share this spelling with others, he says, he has had the *Peregrino* diligently corrected ('correggere et emendare'). Moreover, Caviceo's obscure Latinate terminology has been translated. Zoppino judged it unsuitable in any case for prose, but his main aim was to make the work accessible to a wider public including both the learned and the unlearned ('parimente il dotto e l'indotto'); authenticity was evidently unimportant. Though he was diligent in the past, he says, he is now and will continue to be 'diligentissimo'. But the revision to the spelling was rather superficial, the most consistent change being the reduction of Latinizing *-ct-* to *-tt-* (while *-pt-* remained unaffected). Non-Tuscan features of phonology and morphology were either left untouched or revised inconsistently. The editor allowed some Latinizing words to remain (such as 'munusculi' and 'casulula' in 1.2), but he changed several other words or phrases, such as 'le excubie nocturne son emisse' > 'le guardie notturne son mandate' (1.13), 'il vagiente puerculo' > 'il piangente fanciullo' (1.17), 'rescule' > 'cose' (2.1), 'la etatula' > 'la età' (3.89).

Zoppino's *Ameto* of December 1524 took the Florentine edition, rather than Claricio's version, as its source. It removed some Latinizing spellings, such as 'extremi' > 'estremi', 'cognoscere' > 'conoscere'.[20] Zoppino's dedication simply substituted his own name for that of the Florentine printer Bernardo di Giunta; but his edition of Boccaccio's *Fiammetta* in March 1525 included a new preface, in effect another manifesto which shows how the demands of students of the vernacular ('[gli] studiosi delle volgari cose' or 'della volgare lingua'), who were reading Boccaccio not just for pleasure but also as a linguistic model, were making printers aware of the need to pay attention to the spelling of the works. Printers of the past, Zoppino says, used not to show the 'diligence about the orthography' ('diligenza cerca l'orthographia') of Latin and the vernacular which learned men show nowadays, following the 'uso cottidiano'. As for his own contribution, he has had the *Fiammetta* corrected and emended. In spite of a repeat of his claim about increased diligence, Zoppino's text followed closely that of the Giunta editions of 1517 and 1524, making only a few changes such as 'ochi' > 'occhi', 'belleza' > 'bellezza', and 'uficio' > 'ufficio'.

A superficial and inconsistent revision was made in Zoppino's 1529 edition of a treatise on fortifications by a contemporary southern author, Battista della Valle of Venafro. His work, entitled simply *Vallo*, had been printed in

Naples in 1521. But Zoppino's source was probably a later Venetian edition, such as that of 1524 attributed to Gregorio de Gregori, whose text included many Veneto forms together with some southern ones which had survived from the original version. Some did not appear in the Zoppino edition: thus for example one can contrast 'faccione' (= 'modo'), 'cana', 'zoè' in de Gregori's version with 'fattione', 'canna', 'cioè' in Zoppino's. On the other hand, very many non-Tuscan forms remained in Zoppino's text. A few seem to go back to the original southern form (such as 'tenire', 'sutto', 'ditto') but most result from the Venetian reworking of the text.[21] No thorough linguistic revision was deemed necessary, then, in a work of a highly practical nature, destined neither to be imitated as a model of good writing nor to be read as entertainment.

However, three editions of literary works printed in 1530 continued the prominent advertisement of attention to correctness, even if there was not really a great deal to boast about. In one, the *Opere d'amore* of Antonio Tebaldeo, the warrior turned man of letters Marco Guazzo included a list of errors which he had corrected. The second, Boccaccio's *Filocolo*, had a title page advertising the text as 'nuovamente revisto'. The revision was in fact that which Tizzone Gaetano had carried out three years earlier, and which will be mentioned later in this chapter. But Guazzo said in a preface that, to satisfy Zoppino, 'intimate lover of good things' ('delle cose buone intrinsico amatore'), he had expunged the title *Philopono* which Gaetano had given to the work, and he interpolated several lines in the text in support of his own etymology for 'Philocolo'. For the third of these editions, Sannazzaro's *Arcadia*, Zoppino did not simply reproduce one of his two earlier editions (1521 and 1524) but added a supplement of three poems attributed to Sannazaro.[22]

Zoppino's 1525 printing of Cornazzano's *Proverbi in facezie* still contained a number of Latinizing spellings such as 'excelsa', 'nuptiale' (feminine plural), and non-Tuscan forms such as 'doveti' (= 'dovete'), 'vedoa', 'ho cusito el zaccho', 'el meiore', 'dui basi', 'se havea olduto'. But, if we compare the 1531 edition of the same author's *Vita di Cristo* with the 1518 edition mentioned earlier, we can see that it was considerably revised according to Trecento Tuscan norms. Many Latinisms were removed, double consonants were introduced, and the Tuscan *uo* diphthong was added (as in 'figliol' > 'figliuol', 'homo' > 'huomo'). However, regularization was still rather haphazard, with forms such as 'trahe' and 'dece' surviving.

The rather inconsistent normalization found in Zoppino's books was typical of other Venetian printers and publishers of the 1520s, as a handful of examples will show. Gregorio de Gregori produced three editions of another work first printed in Naples, the translation by Paride Del Pozzo (or Paris de Puteo, 1413–93) of his own *De re militari*, now entitled *Duello: Libro de re, imperatori, principi, signori, gentil'homini, e de tutti armigeri*. These editions came

73

out in 1521, 1523 and 1525, the first and third printings being executed for Melchior Sessa and Pietro Ravani. In comparison with the Neapolitan edition of 1518, the first de Gregori text made changes which brought the language rather closer to Tuscan (as in 'incomencia' > 'incomincia', 'sarria' > 'saria', 'duncha' > 'dunque'), but it kept some Latinizing and southern forms. Both the 1523 and the 1525 texts were derived directly from this one, and their language shows very little further change.

The gradual Tuscanization of the *Porretane* of Sabadino degli Arienti, begun in 1504 by Sebastiano Manilio, was taken further by the Venetian editions of 1525 (de Gregori) and 1531 (Sessa). The former still contained a number of northern and Latinizing forms.[23] But it made a few changes to Manilio's text, such as replacing -*ct*- (in 'nocte', 'aspecto' and so on) with -*tt*- and 'città' with 'città'. The 1531 edition made many more changes of this sort, but they were still unsystematic.[24] The task of revision was not an easy one, given the strongly non-Tuscan colouring of the original, but the lack of linguistic consistency in the 1525 and 1531 editions may also be linked with the relatively unsophisticated level of readership at which a collection of short stories such as the *Porretane* was aimed.

The firm of Francesco Bindoni and Maffeo Pasini was another which brought out corrected editions of non-Tuscan authors in these years. The revision of a series of works by the friar and poet Olimpo da Sassoferrato, printed in 1524–5, was entrusted to a certain Girolamo Severo, also from Sassoferrato (in the Marche), because of the author's absence from Venice. As was by now usual, hypermetric lines were shortened, Latinizing spellings were removed and Tuscan diphthongs were added; but there was some overstepping of the mark here, so that the 'nove Muse' ('nine Muses') became the 'nuove Muse' ('new Muses').[25] A superficial revision of the text of Boiardo's *Orlando innamorato* was made in connection with an edition printed in 1527 by the same firm. When Nicolò Garanta requested a book-privilege in order to publish this work, he claimed that the text had been revised, and the editor may have been Giovanni Battista Dragoncino of Fano. In some copies of this edition, the punctuation and language of the first gathering have been revised, but the changes are uncertain.[26] A slight acquaintance with Tuscan was, it appears, deemed sufficient for an editor to set to work on a text of this nature.

In a similar case, a 1511 edition of the chivalrous epic *Danese Ugieri* was revised for a new Venetian edition of 1517, but only in the first three stanzas and the very last one. This 'topping and tailing' technique could have been encouraged by a law passed by the Venetian Senate on 1 August 1517, which cancelled all previous book-privileges granted to protect against plagiarism in publishing and stated that in future they would be granted only for works which were genuinely new and had not been printed previously. However, the lack of a precise definition of what 'new' meant in this context encouraged printers to make superficial revisions or slight additions to previous versions of a work, an abuse mentioned in another law of 4 June 1537.[27] These revisions would have tended to concentrate on the parts of the book at which a potential purchaser was most likely to glance.[28]

In Venetian books of the 1520s, one begins to find editors of literary texts being named not just in one text, as with Cassiodoro Ticinese, but in a number of texts printed over several years. Two such men, from opposite ends of the peninsula, who (like Squarzafico earlier) pursued editing assiduously for a period were Lucio Paolo Rosello of Padua and Tizzone Gaetano of Pofi.

Rosello's work as an editor embraced texts from several of the main genres of works printed in Venice. These included a dictionary, Latin legal texts (Rosello studied law at Padua), a Latin philosophical text, religious works in Latin and Italian (he was a priest by 1522), and Italian literary prose works, most notably the *Novellino* of Masuccio Salernitano and Boccaccio's *Corbaccio* together with the letter to Pino de' Rossi, printed by Gregorio de Gregori in 1522 and about 1525 respectively.[29]

In editing the *Novellino*, Rosello set himself a task similar to that of Sebastiano Manilio when he edited the *Porretane* nearly twenty years earlier, that of eliminating Latinisms and non-Tuscan forms from a collection of short stories written in the second half of the Quattrocento, though this time the author was of course from southern Italy. He claimed to be bringing the stories back to life: as with Cassiodoro Ticinese, it was assumed that printing corrupts, or had corrupted up to now. It was not conceivable to Rosello that Masuccio should be printed in a language other than Tuscan. Rosello did not specify any sources, and it is evident that the restoration of the *Novellino* was as arbitrary as it was wide-ranging, including not just the usual phonetic and morphological corrections but also changes to syntax (for instance the recasting of sentences which Rosello felt to be awkwardly constructed) and to vocabulary.[30] The editing of the *Corbaccio* volume a few years later was similar in character. Only a few of Rosello's principles had shifted since 1522: for example, he substituted *-ci-* for *-ti-* or *-zi-* in words such as 'inditio', 'malitia', 'diliziosa'; the masculine singular definite article was now predominantly *il* rather than *el*; and he was more willing to paraphrase or reinforce the original text. By the mid 1520s, then, it was considered legitimate, even welcome, for an editor to intervene in a Tuscan Trecento work, and texts which were as wide apart both geographically and chronologically as Masuccio's *Novellino* and Boccaccio's *Corbaccio* were tending to be reduced to the same linguistic norm.

Tizzone Gaetano's editorial career ran from 1524 to 1528.[31] The Venetian printers for whom he worked were Bernardino Vitali and then Iacopo and Girolamo Penzio of Lecco. Gaetano 'revised', as he consistently put it, a much wider range of vernacular texts than did Rosello: prose and verse by Boccaccio as well as a work by another Trecento author, Simone da Cascia, and Poliziano's *Stanze per la giostra*. His special interest was in Boccaccio. Like Squarzafico, Castorio Laurario and Claricio before him, he wanted to champion an author whom he considered unjustly maligned; and, like Claricio, he did this on his own terms, by 'diligently revising' the texts so that they departed even further from their original state in the direction of his own Cinquecento tastes.

In only one edition, the *Filocolo* of 1527, is a manuscript source mentioned, when Gaetano says that 'by chance' he had access to a text which could be contemporary with Boccaccio and which, together with his own interventions, helped him to restore the work. It is true that he corrects errors found in earlier editions, as one sees in two passages from Book 2 which he cites as examples of his improvements; but he never drew on manuscripts consistently.[32]

Gaetano was a grammarian as well as an editor (his grammar was published posthumously in 1539), and in some respects his rules of phonology and morphology were independent of those of Fortunio and Bembo. Gaetano calls the literary language 'common' and 'Italian'. His opinions belong, in this respect, to the same current as those of editors such as Manilio and Claricio and of writers such as Castiglione and Trissino, and were exactly of the sort which Machiavelli had attacked in the *Discorso intorno alla nostra lingua*.[33] For example, metaphonic diphthongization of the kind found in the Naples area, from where Gaetano came, may have influenced his attempt at a functional explanation of the variety of diphthongized and undiphthongized forms found in the literary language. Gaetano's curious rule is that the diphthongs *ie* and *uo* appear in some second-person singular verb forms (and in subjunctives ending in -*i*) and in masculine nouns and adjectives, but not in other verb forms or in feminine nouns and adjectives. There are some exceptions, and Gaetano does not use diphthongs in closed syllables. Nevertheless, there is a striking similarity with the Neapolitan type of conditioned diphthongization, caused by the original presence of the high final vowels -*u* and -*i* but not found in words which originally ended in -*o* and -*a*.[34] Gaetano also shows his independence from Tuscan-based grammars in his use of the third-person plural subjunctive ending of the *avesseno* type.[35]

It is probable that one consistent lexical change in the *Teseida* of 1528 can be ascribed to Gaetano: the replacement of *tornare* in the sense of 'to become' with verbs such as *divenire*. There are, though, very few cases of intervention in the vocabulary of either the *Teseida* or the *Stanze*, no doubt because such changes often entailed recasting the verse. Metre was, on the other hand, an area in which he tried to improve on earlier editions, shortening hypermetric verses and avoiding dialoephe after oxytone words.[36]

In the prose texts which he reworked, Gaetano was guided by the two principles of the application of orthographical and grammatical rules and the elucidation of what he considered to be obscurities. A comparison of the 'Prologo' of the *Fiammetta* in Gaetano's version (1524) and in his principal source, the Giunta edition of 1517, shows that he altered the text quite freely. He removed spellings which were Latinizing or which seemed to be too Florentine and insufficiently 'common' ('comuni').[37] The definite article also came under his scrutiny, with *li* being altered to *i* before a consonant or to *gli* before a vowel. But Gaetano did not stop here, as almost all editors of Boccaccio would have done before the 1520s. Some alterations were made to Boccaccio's syntax.[38] Gaetano was also quite prepared to alter Boccaccio's prose vocabulary, to delete or add words and phrases, to change the word order, or to recast sentences which he considered inelegant or unclear.[39] By using these techniques, he could claim in the dedication that he had made the work easy to understand again ('io al potere essere agevolmente bene intesa, l'ho ridotta') and that the *Fiammetta* was now 'reborn, since it had been killed by so many lacerations suffered until now' ('rinata ... per ciò che morta era da tanti insino a qui laceramenti sofferti').

While Boccaccio's works were being revised by editors such as Rosello and Gaetano, no editor in Venice had attempted to revise the text of Petrarch since Aldo's edition of 1514, nor had any new commentary been written since Squarzafico's in 1484. In August 1525 this situation changed with the appearance of *Le volgari opere* of Petrarch, edited and with a commentary by Alessandro Vellutello.[40] Originally from Lucca, Vellutello left this city in 1516 and moved to Venice not long before 1525. Once in Venice, he had some contact with Bembo's circle, but signs of a divergence between the two men were apparent in the section following the dedication, a 'Trattato' on the new order which Vellutello had given to the *Canzoniere*, where he said that, in spite of Aldo's claims, the main manuscript used for Bembo's edition was not Petrarch's 'originale'. If this was an attempt to retain Bembo's favour by disagreeing only with Aldo, who had died a decade earlier, it not surprisingly failed. Thereafter Vellutello did not trouble to hide his independence from the Venetian.

The texts of the poems in the *Canzoniere* and of the *Trionfi* were based on the Aldine edition of 1501. But in other respects Vellutello's approach to Petrarch could hardly have been more different from that of Bembo. The Venetian had let the voice of the poet speak for itself. Vellutello surrounded the text with a commentary which provided straightforward exposition of the meaning of the poems and, where relevant, their historical context. It was exceptional for Vellutello to discuss variants, as he did when rejecting the infinitive 'por' at 366.103, the Aldine reading, in favour of 'pon' which, he explained in later editions, made better sense. He provided a new life of the poet and an essay on the investigations into Laura's origins which he had conducted on the spot in the Avignon area, using local baptismal records. The edition included a map of the Vaucluse region, an unprecedented device in an edition of Petrarch.[41] Vellutello's principal innovation, though, was a radically different approach to the question of the organization of the *Canzoniere*. He believed that Petrarch's poems were put in order by someone who divided them into poems 'in vita' and 'in morte di madonna Laura'. Vellutello preserved this division but put into a third part all the poems, whenever written, that dealt with persons other than Laura and with subjects other than love. A short appendix contained only three of the sonnets addressed to Petrarch from the Aldine appendix of 1514, together with a novelty, a sonnet addressed to Petrarch by Stramazzo da Perugia.

Vellutello did not, then, alter the substance of his texts in the way that Rosello and Gaetano had done. He also undertook original historical researches which went well beyond anything which they contemplated. Yet his reorganization of the *Canzoniere* shows that, like them, he ignored or undervalued the evidence of the best available sources. As with many of the other editors mentioned in this chapter, what counted most was his own conception of what the text should be like.

6 · STANDARDIZATION AND SCHOLARSHIP: EDITING IN FLORENCE, 1501–1530

PRINTING IN FLORENCE in the first three decades of the Cinquecento was dominated by the press established by Filippo di Giunta. His output differed markedly from that of other contemporary printers in his city: while he concentrated on the standard texts of classical and Italian literature, imitating the Aldine press in Venice, others produced mainly short and relatively ephemeral works, often religious in content. In order to set the activities of the Giunta press in perspective, it will be useful to consider first a few examples of the editions produced by other printing houses.

To judge from the books printed for them, Florentine readers had sophisticated literary tastes but were at the same time linguistically provincial. Around 1510, for example, Bernardo Zucchetta undertook the first printing, without the author's knowledge or consent, of two works by Lodovico Ariosto, *La Cassaria* and *I Suppositi*, which belonged to the completely new genre of comedies on the model of those of Plautus and Terence but written in the vernacular. But the author was Ferrarese, and the language had to be adapted before it was acceptable to local readers. As a result, both these editions contained many forms typical of contemporary Florentine and foreign to Ariosto's own usage.[1]

The same combination of characteristics can be found in some of the books published by Bernardo Pacini. He was the son of the leading Florentine publisher in the last years of the Quattrocento and at the start of the Cinquecento, Piero Pacini of Pescia. Bernardo took up publishing in 1515, after his father's death in the previous year, but he was far less successful and after 1525 concentrated on an alternative career as a notary.[2] Both favoured works of a popular and often edifying nature, but at least two of the books published by Bernardo were revised for publication by an editor.

One was the first edition of the *Motti et facetie del Piovano Arlotto*, printed for him by Zucchetta at some time before the end of 1516, the year in which the text prepared for Pacini was copied in Venice by Zoppino. Pacini's dedication to Pietro Salviati (eldest son of Iacopo Salviati and Lucrezia di Lorenzo de' Medici) explained how the work was reorganized before printing. He had refused at first to have the *Motti et facetie* printed because they were collected in such a way that they could not be read with pleasure. But now, he said, a learned friend had promised to polish them, not completely (which would have been impossible in a work written in the manner of the spoken language), but enough to make them intelligible. The reworking which resulted, and which can be assessed by comparing Pacini's edition with a manuscript copied later in the century, was every bit as radical as some of those which we encountered in the previous chapter.[3]

Some minor phonological and morphological changes were made in order to raise the linguistic level of the text above the popular Florentine of the original. But in a work of this sort, and at such an early date in the century, one could not expect a Florentine editor to have been particularly interested in such details, and certainly not in a revision of the sort which would have conformed with the new principles emerging from the Veneto and enshrined in Fortunio's *Regole*.[4] Rather, the reviser focused on bringing the sophistication now expected of a written text to the syntax (there is more use of hypotaxis and indirect speech, for instance) and to the vocabulary (introducing more high-sounding or Latinate terms, so that for example 'uno bellissimo desinare di degne vivande' became 'uno splendidissimo convito e di lautissime vivande' and 'andare' became 'ire'). Sometimes he expanded the original text, strengthening the punchline or providing more explanatory detail or descriptive colouring, but he did not hesitate to abbreviate phrases or whole passages which seemed too long winded.

He also intervened in the content of the stories and in the structure of the work as a whole, which was uncertain and repetitive. He reduced the importance of the work's original context by cutting out all references to the author and to the witnessing of the *facezie* by contemporaries. He provided a neat conclusion to the work by putting at the end the *motto* about the Piovano's epitaph. Several stories were omitted altogether. The editor removed some that seemed not to belong to the humorous spirit of the work, such as the accounts of the Piovano's works of charity in numbers 124, 140 and 141, and some of the philosophical material from the end. It seems probable that other stories were left out because they implied criticism of the clergy or because they were blasphemous, obscene or otherwise potentially offensive.[5] A few *facezie* may have been omitted for political reasons.[6]

The reworking of the *Facezie* for print in about 1515–16 showed little respect for the design and detail of its manuscript original, even though this original could not have been more Florentine and though it dated from less than fifty years earlier. Another popular work published by Bernardo Pacini, but this time one from north-eastern Italy, needed a similarly thorough linguistic revision in order to be successful with a Florentine readership in this period. It was a chivalrous poem about a knight called Falconetto, one version of which had been printed in Venice in 1500 by Zuan Battista Sessa. Although the language of Sessa's edition was strongly influenced by Venetian, it had been reprinted in Florence in 1508 by someone who even gave his name in the northern form 'Zuanne Bonacorso' and with hardly any revisions for the

Florentine public.[7] Not surprisingly, Bonacorso's printing career seems to have begun and ended with this extraordinarily ill-judged enterprise. When Pacini republished the poem, he not only added a supplement of eleven stanzas but ensured that the rest of the poem was more appetising to a Florentine reader. This meant, of course, removing traces of northern phonology, morphology and syntax; but it also meant bringing some forms which reflected Trecento Tuscan more into line with contemporary spoken Florentine.[8] Some changes went further. In each of the first two editions of the *Falconetto*, the transposition of some lines in canto 1, stanzas 5 and 6, threw the rhyme scheme into disorder; this was remedied for Pacini's edition. Lines of ten or twelve syllables were adapted so that they become hendecasyllables. Near the beginning of the first canto (that is, where the potential purchaser was most likely to glance), the editor rewrote stanzas 3 and 6–10 to a considerable extent. Here is the 1508 version of stanza 3:

> E perché sei piatosa assai, ne vegno,
> pregoti alquanto de tuo aiuto
> presti al debil mio caduco ingegno,
> e il leve legno in mar quasi perduto
> tu, priego, guida al disïato segno
> che da gran scogli sia poco offenduto,
> sì ch'io raconti i facti alti e divini
> del magno Carlo e gli altri paladini.

The version printed for Pacini makes the syntax more sophisticated, eliminating the parataxis of lines 1 to 5 and replacing 'che' in line 6 with the clearer 'in modo tal che'. The metaphor of the boat reaching its destination safely is revised, and in the rewriting the rhyme-word 'offenduto' is replaced with 'tuto', a Latinism being preferred to a dialect form. In addition, two hypometric lines (2 and 3) are rectified. Here is the result:

> E perché se' pietosa, ad te ne vegno,
> pregandoti che alquanto del tuo adiuto
> tu presti al debil mio caduto ingegno
> et al mio cor ch'è in mar quasi perduto,
> e che quel guidi al disïato segno
> in modo tal che giunga in porto tuto,
> sì che io racconti e facti alti e divini
> del magno Carlo e gli altri paladini.

By around 1520 (the date suggested by the British Library catalogue), the Florentine audience clearly expected a high level of poetic craftsmanship even in a work of entertainment.

Filippo di Giunta, unlike Aldo Manuzio, was not a scholar. He had begun to print in 1497, but the success of his press was in large part a result of a decision reached by 1503 to imitate the initiatives which Aldo was taking in the production of Latin and (to a lesser extent) vernacular books in italic type and in octavo format. Filippo's main link with Venice was his younger brother Lucantonio, who had become rich there as a stationer and bookseller and then as a publisher in the 1480s and 1490s, and in 1491 had formed a partnership with Filippo which was to last until 1510.[9]

From 1503 until 1514, when Filippo's son Bernardo became involved in the production of the firm's books, Filippo allowed his learned collaborators and editors to identify themselves by signing dedicatory letters and other extratextual material. He employed several local teachers, some of them clerics, to edit classical texts for him, and this editorial tradition was doubtless one of the factors which in due course helped to foster in Florence a serious approach to the editing of vernacular texts.[10]

Like the Petrarch of 1504, mentioned in chapter 4, the Giunta Dante of 1506 did not fall completely in line with Bembo's Aldine edition. The editor who took on the task of responding to the Venetian challenge was Girolamo Benivieni. Although he followed Bembo in discarding his predecessor Landino's commentary, he added some material of his own: a long poem in praise of Dante and his work and two dialogues on the form of the Inferno accompanied by seven illustrative woodcuts. Useful though these discussions were, Benivieni's text was an unhappy compromise, following the Aldine to some extent but without offering a carefully researched alternative.[11] It seems certain that the Florentine printers were given an emended copy of the Aldine edition from which to work, since they followed its page layout from *Purgatorio* I onwards. Like other editors, Benivieni reacted against Bembo's punctuation and his use of apocope and elision. He did not use the semicolon at all, nor the apostrophe, so that the apocopated possessives *mi*, *tu* and *su* became *mio*, *tuo*, *suo*. On the other hand, the Florentines were not to be outdone on this question, and a different system of accentuation was used.[12] As one would expect in the light of Benivieni's subsequent attack on Fortunio, he frequently made the spelling more Latinizing.[13] When he introduced new readings, he returned most often to Landino's text, but also appears to have consulted more than one earlier printed source (including perhaps Berardi's edition of 1477) and possibly one or two manuscript sources.

In spite of the effort that went into this revision, Benivieni did not offer any justification for his new text. It is as if such matters were too trivial to mention. We know that later on Benivieni held strong views on linguistic usage, views which had perhaps been exacerbated by the fact that, as we have seen, his poetry had been the subject of editorial interventions both in Florence (probably within his own family) and in 1522 in Venice. But his silence on the subject of language in the 1506 edition suggests that, for him as an editor, the importance of the language of the *Commedia* ranked well below that of the content and form of the work. This approach was no match for the revolutionary editorial methods of Bembo, and it is not surprising that Benivieni's edition was never reprinted. Indeed, Florence did not produce another *Commedia* until the very different edition of 1595.

From about 1515, the year of Aldo's death, the Giunta press enjoyed a period of expansion which marked a turning point in the firm's relationship with Venetian printing. Filippo and Bernardo, who was associated in the production of editions from March 1515, began to print Greek and Latin texts which Aldo and his successors had not yet published, and they soon found themselves being imitated by their Venetian rivals. A new outlook was also apparent in the publishing of vernacular texts. There was still the traditional disdain for editions produced 'in foreign regions' ('in aliene regioni'), as the preface of the 1504 Petrarch had put it, but there was no longer a completely ostrich-like attitude. It was recognized that there had been faults of neglect on the Florentine side and that texts of the Trecento needed to be studied with care and respect.[14] Unfortunately, it is not possible to say who were those responsible for this change of direction. Unlike the vernacular works printed while Filippo was running the press on his own, those appearing from 1516 onwards contain no clues as to the identity of their editors. When there is a dedication, it appears either in the name of the (deceased) author or, more usually, in the name of Bernardo, who, unlike his father, evidently wanted to seem to be filling the same sort of role in Florence as Aldo had filled in Venice.

The year 1516 also saw the start of a new editorial initiative in the choice of texts. Filippo di Giunta had concentrated on Trecento verse; now he and Bernardo redressed the balance by beginning an ambitious series of texts by Boccaccio of which all but one were entirely or principally in prose. The first text in the series was the *Decameron*. Delfino had brought out his edition in May, two months earlier, and, though he was not named, he must have been the main target of the attack in the Giuntine preface on the presumption of outsiders who thought they knew Florentine better than the Florentines. The language of the Giuntine edition was indeed closer to contemporary Florentine than that of Delfino's text, but the sense of innate superiority apparent in the preface was not based only on a belief that Florentines had a natural instinct for the correct reading. The main source, which naturally enough was the previous Florentine edition of 1483, had in fact been corrected with manuscripts including that transcribed in 1384, in all probability from one of Boccaccio's autograph originals, by Francesco Mannelli.[15] Three Florentine *novelle* from the late Trecento and the Quattrocento were added to form an appendix which was similar in intention to that of the Aldine Petrarch of 1514 and which also pointed to the continuity of the Florentine short story tradition.

The dedication to the *Corbaccio* and the letter to Pino de' Rossi, also of 1516, did not mention any new sources but talked of how 'the lovers of the Tuscan language' ('gli amatori della lingua toscana') had revived it after a period of near-burial since Boccaccio's time: confirmation that some Florentines were now aware of the need to restore such texts and of the fact that the Tuscan language had changed since the Trecento. In the same way, the dedication of the *Fiammetta* of April 1517 acknowledged that the task of editing the work had been a difficult one, not just because of the corruptions introduced by non-Florentine printers but also because Florentines had copied it carelessly.[16] The same theme of neglect returned in the dedication of the *Ameto* (1521). The Tuscan language, said the writer, had been both written and printed carelessly from the time of Boccaccio almost up to the present day. But by now, he continued, the situation was more complex because of the differences of opinion which were becoming apparent among the admirers of the language. Well-intentioned though these people were, the result was that some, with their 'false grammatical observations' (and the allusion was evidently to Claricio's Milanese editions), had transformed Boccaccio's language out of all recognition. It was probably the spur of this rivalry which led Bernardo to claim that the sources used for his revision were 'some very old texts ... compared together' ('alcuni antichissimi testi ... conferiti insieme') and even certain things in Boccaccio's own hand which were compared with the text of the *Ameto* and found to be very similar in style. It is disappointing to find, then, that the principal source was none other than Claricio's rewritten version of 1520. Yet the Giunta editor has repaired some of Claricio's damage, using one or more manuscripts as well as what appear to be conjectures of his own, including the shrewd emendation of 'tumultuose labbra' to 'tumorose labbra' in IX.15.[17]

The sort of Florentinization found in the Giunta editions of Boccaccio was also applied if the author was not Tuscan. For Sannazaro's *Arcadia* (March 1514/15) the Aldine edition of the previous September was used but the linguistic revisions went rather further. Many southern forms remained unchanged, as in the Aldine. However, where Aldo's editor was happy to leave non-Florentine forms such as 'duono', 'boscareccia', 'agiuto', Bernardo's editor introduced 'dono', 'boscareccia', 'adiuto'. As the last of these changes and other similar ones suggests, the influence of Latin on spelling in Florence was still strong. Much more extensive editorial interventions are found in a work from outside the literary canon, the *De re militari* of Antonio Cornazzano.[18] When this long poem in *terza rima* was offered to the Florentine public in May 1520, Bernardo (or the editor writing in his name) noted briefly the usefulness of its contents but devoted the rest of his preface to a criticism of its barbarous language and to an explanation of the revision to which it had been subjected. After much effort it had been found impossible to make the work more elegant, he said, so it was decided to make it more intelligible, in so far as this could be done while preserving the original rhyme scheme.

The previous edition of the *De re militari*, printed in Ortona in 1518, had gone some way towards correcting the language along the lines suggested by Fortunio.[19] Like the 1518 editor, the Giunta editor appears to have used as his source the Pesaro edition of 1507; but his aim was a different one. He was not much concerned with the elimination of Latinisms, nor did he alter plurals such as 'tante madre'; and there are other features which give a strong contemporary Florentine flavour to the revision.[20] Rather, the editor's two main concerns were vocabulary and metre. The suggestion in the preface that there was an initial effort to make Cornazzano's work more elegant is borne out by the first thirty-three lines, where whole tercets were rewritten. Thereafter, alterations usually affected only one line at a time, though the amount of rewriting was still considerable. Individual words might be changed, but whole phrases continued to be replaced with others which appeared stylistically or metrically more elegant.[21]

The fourth Giuntine edition of Petrarch, which appeared in July 1522, was altogether different from the three editions of Francesco Alfieri.[22] The editor had studied both Bembo's Petrarch of 1501 and Aldo's version of 1514 (or its successor of 1521) and accepted some of their features. The *Canzoniere* was divided into poems 'in vita' and 'in morte', the second part starting with 'I' vo pensando' as in Bembo's text. The *capitoli* of the *Triumphus Cupidinis* and the *Triumphus Castitatis* followed the pattern of all three Aldines. Thereafter, the 1522 edition arrived at a compromise between Bembo's ordering and that of other editions.[23] The text itself followed the Aldines closely, but did not use any one of them slavishly and in the *Trionfi* sometimes preferred readings found in more than one of the incunabula.[24]

The 1522 editor also introduced some new features of his own. In the *Canzoniere*, each poem which was not a sonnet was identified as a *madriale*, a *ballata* or a *canzone*, and the appendix was the most complete to date, containing all the *canzoni* and sonnets of the 1514 Aldine, plus the *ballata* 'Nova bellezza' and four new correspondence sonnets: two addressed to Petrarch and two more from his *rime disperse*.[25] There is also an important letter to the reader which set out to justify the differences between this edition and others, particularly as regards the much-disputed ordering of the *Trionfi*, and then went on to discuss the appendix. It is made clear that these extra poems were included only in order to indulge the current fashion for such appendixes and to encourage young writers by showing that even Petrarch composed poems with which he was dissatisfied. Moreover, the supplementary poems had been separated into two sections, printed in separate gatherings, according to whether or not they were closely related to the *Canzoniere*.

It is evident from the quality of this Petrarch of 1522 that Florentine philology was now based on firmer foundations than in the previous decade. It still proclaimed its independence from Venice, but it had ceased to assume the automatic superiority of any printed edition produced in Florence and had begun to learn, from the example of Bembo and Aldo, the importance of a careful investigation of manuscript source material. Pride in what was

Florentine was now combined with an unprejudiced respect for the lessons which had been taught by Venice.

Though Venetian editing of Trecento works was passing into the hands of editors such as Rosello and Gaetano, Venice still had much to teach Florence. Those Florentines who were responsible for, or interested in, the new philological direction recently taken in their city must have been helped by the appearance of Bembo's *Prose della volgar lingua* in Venice in 1525, and in particular by Bembo's attention to the evolution of the literary language from the Duecento onwards, his careful identification of the essential grammatical characteristics of the prose and verse of the Due and Trecento, and the new attention with which he studied Boccaccio's prose. And two years later, in 1527, the Giunta press produced two remarkable editions which confirmed the re-establishment, at last, of a Florentine vernacular philology capable of both learning from and rivalling that of Venice.

The first to appear, in April, was a new edition of Boccaccio's *Decameron*. It was said later to have been produced by a team which included various young members of distinguished Florentine families, such as Bardo Segni and Piero Vettori, together with the humanist Antonio Francini, who had been a regular editor of Greek and Latin texts for the Giunta press since 1516.[26] There is a striking similarity here with the two patrician editors in Venice, Bembo and Delfino, who applied to the vernacular the methods which others had been using for the classics. The Florentine editors' relationship of dependence on Venetian precedents and yet progress beyond them is summed up in their working practice. Their starting point was a copy of the 1522 Aldine 'reprint' of Delfino's edition. As this copy (which still survives in Florence) shows, they corrected Delfino's text extensively with the help of manuscripts including, at a late stage, the authoritative Mannelli copy which had already, but to a much lesser extent, influenced the Giunta edition of 1516.[27]

The other noteworthy edition of 1527 appeared only a few months later, in July. Again, the material came from the earlier stages of Tuscan literary history, and again the importance of the study of the subject matter had been indicated by Bembo. This time, however, Bernardo di Giunta was publishing not a well-known work, printed and reprinted many times before, but a collection of 289 early poems of which the great majority had never appeared in print: the *Sonetti e canzoni di diversi antichi autori toscani*.[28] And, while the volume owed an obvious debt to Bembo, a certain element of rivalry with Venice was now apparent. The purpose of the collection, the preface stated, was to remind those who had cultivated Tuscan poetry of the need for gratitude to early poets as well as to Petrarch, and there is a deliberate reaction here to Bembo's criticism of the crudeness of early poetry and his stress on the poetic supremacy of Petrarch (*Prose*, 1.17 and 2.2).[29]

The editor of the *Sonetti e canzoni* was the same Bardo Segni who also played an important part in the 1527 *Decameron*.[30] The first four books are devoted to

Dante's lyric poetry, and the next six books work backwards (as the preface explained) towards the oldest poems. Books v–viii are dedicated respectively to Cino, Cavalcanti, Dante da Maiano, and Guittone d'Arezzo. Book ix contains one poem each by twelve earlier poets, including Guinizelli ('Al cor gentil'), Bonagiunta da Lucca, Giacomo da Lentini and Guido delle Colonne. 'Canzoni antiche di autori incerti' ('old *canzoni* of uncertain authorship') are collected in Book x. There then follows a previously unannounced eleventh book containing two appendixes, one with two anonymous *sestine*, the other with correspondence sonnets by some of the poets mentioned earlier.

The preface draws an analogy between the editor's work and the excavation and cleaning of ancient statues damaged and scattered in the course of time. Once the texts had been gathered, Segni carried out this 'cleaning' process with great patience and care. This is apparent, for instance, in his listing of variants for Dante's poems and for Cavalcanti's 'Donna mi priega', a practice which had not been followed in Florence since the Jacopone of 1490. He also paid close attention to punctuation, as did Bembo, using the apostrophe carefully and applying an unprecedentedly elaborate system of accentuation which went far beyond the practice of Aldine texts. This system is to be linked with the debates which Trissino had initiated in 1524 on the spelling and pronunciation of the vernacular: like Trissino's alphabet, it distinguished between words which would otherwise be spelt in the same way but, instead of Trissino's Greek letters, it used the grave accent for close *e* and *o* and the acute accent for the open varieties.[31]

As one would expect, Segni emended the texts where their incomprehensibility led him to believe them to be corrupt. He left the identity of his sources vague: they were described simply as 'many various and old texts'. In a few cases, however, it is possible to identify his sources, and a comparison of these with his edition shows that Segni respected some unusual and archaic forms, never rewriting extensively in the manner of a Claricio or a Gaetano, but that he sought to improve the texts in various ways, polishing their surface just as the preface describes the way in which the ancient statues unearthed in Rome before the recent Sack were cleansed of blemishes. Throughout, he modernized and regularized spellings, though within limits, so that an archaic patina was still visible in the printed text. For the first eight sonnets by Cavalcanti in Book vi, where the Giunta text derives from the Venetian tradition, Segni restored Tuscan forms altered by an earlier scribe.[32] He also contaminated different sources: in Book ii, for example, some poems incorporated readings from the Venetian tradition alongside others characteristic of the 'Raccolta aragonese', while a third tradition was used for sonnets 10–12.[33] He also appears to have used conjecture with some freedom, but he usually took the reading of his source as his point of departure and did not emend simply in order to make the text conform with different aesthetic criteria.[34]

The nature of Segni's interventions can also be judged from his editing of the first two of the three *canzoni* by Guittone in Book VIII ('Sè di voi, donna gente' and 'Tutto 'l dolor ch'eo mai portai fú gioia'), where he was faced by a source whose corruptions affected both sense and metre. Sometimes he let the reading of his manuscript stand, especially if the words are individually comprehensible, as in the phrase 'lo saver di voi canzone' (*canzone* 1.44). But more often he intervened in order to give some sense to the text. If he suspected a word or phrase of being corrupt, he might substitute something which resembled it, as in 'grastia' (for 'grassia') > 'carstia' (i.e. 'carestia', 1.31) or 'verso mectendo' (i.e. 'ver ciò mettendo') > 'pur sommettendo' (1.29). He might resort to an emendation based on the rearrangement of existing elements. An example is his treatment of 'kecio kede come divoi nasce unde somigla permia fede altrui' (where other manuscripts have the correct 'ché ciò che l'om di voi conosce e vede | semiglia per mia fede'). With some redeployments, excisions and additions this became 'chè di voi nasce ció ch'é bel frà noi; | onde simiglia altrui' (1.16–17). But his surgery might lead to the replacement of the text of his source with something more remote from it, as in 'oservare' > 'aguagliare' (1.55), or 'verso contradiosa' (i.e. 'ver ciò contraríosa') > 'chem'sia tanto gioiosa' (1.43). In II.8 and 26 he invented lines in order to fill lacunae in his source but did not indicate that he had done so. Segni was alert to possible problems of metre: he reduced the severely hypermetric 'bon fidele avoi come non truovo alcuna cosa' to 'à voi fidel, com'eo non trovo cosa' (1.42); the absence of a stressed syllable in either the fourth or the sixth position was remedied in 'certo lotardare mi pare macto' > 'certo lungo tardar mi pare matto' (1.86); and the correct rhyme scheme was restored with the change of the corrupt 'comeve lasso veo divita fiore' to 'ed eo pur vivo lasso isventurato' in II.36.[35]

A similar balance between regularization and conservatism can be seen in Segni's editing in Book XI of the two anonymous *sestine* (by an author perhaps from southern Tuscany or Umbria), for which one can be sure of the exact identity of his source manuscript.[36] Segni modernized the spelling to a greater extent than Gualteruzzi did in the *Novellino*.[37] Phonology was likewise brought into line with the type of usage of which Segni seems to have approved, neither Latinizing nor too popular.[38] He also made some cosmetic changes in the area of morphology, introducing forms preferred by Bembo in Book 3 of the *Prose*.[39] As elsewhere, Segni regularized the metre, emending hypometric and hypermetric lines, eliminating dialoephe after a monosyllable or an oxytone word in cases where he could introduce a transitional consonant (as in '*e aven una* ch'è vestita a verde' > 'ed havene una', 1.4), and avoiding synaloephe where three consecutive vowels were involved.[40] As for Segni's emendations where his manuscript presented or seemed to him to present problems of comprehension, one of his changes, in Debenedetti's judgement, was unnecessary (II.16 'risplende' for 'risponde') and others could be bettered, but several others were clearly both necessary and acceptable. It is notable, too, that Segni always tried to base his emendation as closely as possible on the reading of the manuscript, reconstructing the original rather than guessing at it. An example is at II.5, where his manuscript read 'e formo nel suo amor, chome mur petra', and the Giuntina has 'è fermo ne 'l suo Amor, come in mur pietra'. Even though Segni has smoothed and polished the surface of these relics of early poetry, there is a world of difference between his relatively conservative editing and the unrestrained reworkings of his contemporary Venetian counterparts.

Both of these 1527 editions, the *Decameron* and the *Sonetti e canzoni*, were ambitious and unprecedentedly successful undertakings. In each case the task of editing had been carried out with the help of printed editions but also by using manuscript sources of high quality, some dating back to the fourteenth century. The editors still saw it as a normal part of their work to make conjectures, so that the sense would be improved, and to modernize usage to some extent, so that the texts would be more readily legible. But these operations were carried out within limits of discretion and respect for the original that were much more narrow than had been the case with some recent examples from north of the Apennines. Between them, these two editions gave Florence the initiative in the philological rivalry between herself and Venice. The Giunta text of the *Decameron* was the basis of most of the editions printed in the Cinquecento. The *Sonetti e canzoni* were copied in Venice in 1532 by Giovanni Antonio Nicolini, and the *Vita nuova* poems were used in the first edition of this work (Florence, Sermartelli, 1576). But it also inspired further study of pre-Petrarchan poetry in manuscript form. Annotated copies of the 1527 edition bear witness to the way in which Cinquecento scholars collated manuscripts with its text. The Giunta collection also provided the basis for new manuscript collections, in particular that put together between 1527 and 1533 by the Florentine abate Lorenzo Bartolini. The 'Raccolta Bartoliniana' shows how the earlier antagonism between Venice and Florence could be replaced by more fruitful cooperation, since one of Bartolini's three main sources was a manuscript which belonged to Bembo, and the collection was most probably compiled not in Florence but in Padua in 1529.[41]

7 · TOWARDS A WIDER READERSHIP: EDITING IN VENICE, 1531-1545

THE 1530S WERE YEARS OF TRANSITION in which Venetian editing began to move away from the patterns of the first three decades of the century with gradual innovations which then became more firmly established in the 1540s. Behind these changes lay two simultaneous and interrelated developments: an expansion in the printing industry and an increase in the number of those wanting to read and use the vernacular.

Relatively few new presses had opened in the period of turmoil which followed the disastrous defeat at Agnadello in 1509, and the late 1520s brought the closure of three major printing houses, those of Giorgio Rusconi and his heirs (1527), Gregorio de Gregori (1528), and Bartolomeo and Agostino de Zanni (also 1528). But many presses successfully continued production from the 1520s into the politically calmer years that followed. The Aldine press closed temporarily at the end of 1529 with the death of Andrea Torresani but was given a new lease of life from 1533 onwards by Aldo's enterprising and combative son Paolo. Other examples of continuity are Bernardino Stagnino and Bernardino Vitali, who went on printing until 1538 and 1539 respectively; Bernardino Benagli and Nicolò Zoppino, who printed until the early 1540s; and the presses of the Sessa, Scoto, Nicolini da Sabbio and Bindoni families, which were well enough established to last until near the end of the century or beyond. And these presses soon faced stiff competition from a number of prolific newcomers, such as the printers and publishers Giovanni Andrea Valvassori (who began operations in 1530), Francesco Marcolini (1535), Giovanni Giolito (1536) and especially his son Gabriele (on his own from 1541), Vincenzo Valgrisi (1539), Michele Tramezzino (1539), Comin da Trino (1539 or 1540) and Giovanni Griffio (1544).[1]

This expansion would not have been possible without a corresponding rise in the supply of, and demand for, writings in the vernacular. The achievements of contemporary authors such as Bembo, Sannazaro and Ariosto, the increasing standardization of the vernacular through the efforts of linguists and editors alike: all this had by now created confidence in the new language, a greater interest in it, and a greater desire to use it correctly. The whetted appetite of the reading public, together with the opportunities for earning income from the press, led in this period to the rise of what the French termed *polygraphes* (a term first recorded in 1536 and from which Italian *poligrafi* was then derived): the jacks-of-all-trades who would turn their hand to practically any subject matter in their own writings and who often served printers as translators into the vernacular and anthologizers as well as editors. However, the flourishing of vernacular printing did not benefit writers alone. It also made an unprecedentedly wide range of works available to the literate: and not just to those who traditionally belonged to the ranks of consumers of books, but also to less experienced readers. Indeed, greater competition between printers meant that books tended to be aimed at an ever broader public.

In this new climate, the role of the editor was more crucial to the success of printers and publishers than ever before. Texts which did not conform with the new linguistic norms became increasingly rare, because texts were appreciated for the correctness of their usage as well as for the entertainment or improvement which they offered. Editors were therefore expected to standardize the language of all but a handful of literary classics. As for these classics, editors had to devise new ways of attracting those whose needs and expectations could not be easily satisfied by the type of book produced up to now: in particular, younger and female readers and the less well educated.

The extent and nature of this new demand can be judged from some works of practical guidance for users of the vernacular which appeared in Venice in the 1520s and early 1530s. Nicolò Liburnio's *Le vulgari elegantie*, published in 1521, was a compendium of advice on stylistic elegance, which included a repertoire of sentences listed under themes such as 'Love', 'Sorrow', 'Night', 'Sighs', comments on the usage of Dante, Petrarch and Boccaccio, a list of differences between prose and verse usage in old texts, a list of nouns and their corresponding epithets 'necessary to the composer of verses' ('necessari al componitore di versi'), and two lists of similes. Throughout, Liburnio stressed that he was writing for the young who wanted to learn to write with ease in prose or verse and for whom he did not consider Fortunio's grammar helpful enough. In February 1526 Liburnio produced a fuller guide to grammar and eloquence, based on the same three Tuscan sources; hence the title of the work, *Le tre fontane*. He added another helpful device to his repertoire: synonyms for some difficult words. Bembo's *Prose* had been printed some five months earlier, but, said Liburnio, with their Ciceronian dialogue form they would only be of use to those already expert in Latin and knowledgeable about the style of the Tuscan writers. His work, in contrast, was aimed at beginners. He envisaged his readers as young people and women, and he foresaw them going on to use what they had learned in their everyday lives, winning the hearts of lovers or making a good impression as courtiers. As for their level of preparation, all he asks is that they should have a basic grounding in the Donatus ('qualche prattica del Donato grammaticale'); this should not be a restriction, he says, for what Italian child, of however poor a father, does not learn to give the equivalent of Latin nouns and verbs in his native variety of the vernacular?[2] And the reference here is not to the *ars minor* of Donatus but to the medieval compilation (known also as *Ianua*) used in the early stages of education, perhaps at first simply as a reading text, then in order to achieve a first understanding of Latin with the help of an integrated translation.[3]

To a similar audience of young students of the vernacular was addressed a rhyming dictionary, the *Rimario de tutte le cadentie di Dante e Petrarca* of Pellegrino Moreto (or Morato), printed in 1528. The author hoped, as his dedication shows, that this would not only teach how to write verse but also provide help on a problem faced by beginners from northern Italy, when to use double or single consonants. In two months, he said, he would be producing an alphabetical dictionary of all the obscure words of Dante and Petrarch and explanations of the most difficult passages in Petrarch. His notes on Petrarch remained in manuscript (Fowler, *Catalogue*, p. 374), but a guide to the *Luoghi difficili* of Petrarch was compiled by the Florentine Giovan Battista da Castiglione and printed in 1532. Concentrating on the sense rather than the style of the poems, Castiglione gave several comments on older and contemporary Tuscan usage. His point of view was that of a Florentine exasperated with so-called 'reformers' who insisted on rules based on the Trecento, while Tuscan was still alive and in use. Unlike them, he said, he did not consider it a sin against the Holy Spirit to say 'io cantavo' rather than 'cantava'.[4]

Liburnio, Moreto and Castiglione were trying to bring the Tuscan classics within reach of a new class of aspiring readers and users of the language, something that the grammars of Fortunio and Bembo did not attempt. Nobody had yet tried a parallel operation in the context of editing. Few editors in Venice or Florence had been involved in providing any sort of extratextual material, though there had been important exceptions such as the commentaries and biographies by Landino, Squarzafico and Vellutello, the notes on Jacopone's language in the Florentine *Laude* of 1490, the defence of Boccaccio by Castorio Laurario; there was also, from Milan, the example of Claricio's 'observations' on the language and text of Boccaccio. But for those with little or no experience of older or literary Italian, for those who would have got little out of austere editions such as those of the Aldine press, what was now needed was a presentation which would complement the Aldine model by giving the sort of down-to-earth, practical help at which *Le vulgari elegantie* and so on were aiming.

The main Tuscan texts to be affected by these new pressures were the two which Bembo's *Prose* had promoted to the status of principal models for verse and prose, namely Petrarch's *Canzoniere* and Boccaccio's *Decameron*. Further impetus to the vogue for the imitation of Petrarch was given by the printing in 1530 of the lyric poems of the two greatest modern exponents of the genre, Bembo himself and Sannazaro.[5] But by the mid 1530s a third text was rapidly acquiring a similar status. The new situation is illustrated in a dialogue of Pietro Aretino first printed in 1536, where the courtesan Nanna advises her daughter Pippa that, to be fashionable, she must have on her table at all times Petrarch, the *Decameron*, and also the *Orlando furioso* of Lodovico Ariosto.[6] This epic poem, centred on the paladins of Charlemagne but with elements inspired by Arthurian romance, was a very recent work: it had been printed in Ferrara first in 1516, then with revisions in 1521, and again in 1532 with additions and after the author had excised much of the influence of northern Italian usage and of Latin in order to make it conform more closely with the doctrines of Bembo.[7] Of these three works, as we shall see, Petrarch attracted the most editorial interest at first, but it was with the works of Boccaccio and Ariosto that printers and editors subsequently introduced more radical changes of presentation.

Alessandro Vellutello had already brought out the first new Cinquecento commentary on Petrarch in 1525. It proved extremely popular. He published a second, revised, version in 1528, a third in 1532, a fourth in 1538, and there were many later reprints.[8] But Vellutello's reorganization of the *Canzoniere* did not convince everyone, and his commentary gave almost no help to readers and would-be writers on questions of style, metre or language. The first alternative new edition was printed in 1532 by the firm of Francesco Bindoni and Maffeo Pasini. The editor, Sebastiano Fausto of Longiano, near Cesena, provided not only a commentary and brief lives of Petrarch and Laura but also a *rimario* with index and an alphabetical list of nouns with their epithets: in other words, the kind of material which Liburnio and Moreto had been offering to aspiring Petrarchists in the 1520s.[9] However, the level at which Fausto pitched the rest of the edition was a relatively high one. His commentary set out to explain the circumstances and purpose of each poem; he did not see its main purpose as being to provide an exegesis of difficult words or phrases, though his comments on the *canzoni* were fuller and more helpful in this respect than those on the sonnets. He was not interested in helping readers by comparing Petrarch's usage with regional differences in the vernacular. Fausto was addressing himself, rather, to the kind of reader who was interested in the question of the texts of the *Canzoniere* and the *Trionfi*. He was not happy either with Bembo's edition or with Vellutello's. As he explained in an introductory section on the ordering of the *Canzoniere*, he had made every effort to see old texts, and, among many others, had found one written during Petrarch's lifetime and another written two years after his death; but all differed in their ordering, nor did he consider that there existed an autograph manuscript which could resolve the problem. Unlike Vellutello, he did not think it worthwhile to carry out a major reordering, but 'for greater convenience' ('per più commodo') he separated the sonnets and other compositions into two groups. In the rest of the volume he adopted a combination of the Aldine texts but managed to suggest in various ways that these did not provide the last word. His appendix to the sonnets included four which had not appeared in earlier editions. The commentary often gave alternative readings from sources vaguely identified as 'some texts' or 'old texts'.[10] As for the *Trionfi*, his commentary provided more variants from unspecified sources; he restored 'Nel cor pien' to the start of the *Triumphus Fame* (though he did explain the reasons for which some had omitted it); and he included in his commentary the fragment beginning 'Quanti già ne l'età matura', saying he could not understand why it had been removed. From a philological point of view, this sort of approach was backward-looking, while on the other hand this was by no means a Petrarch for a wide readership. Not surprisingly, it was never reprinted.

The summer of 1533 brought two new Venetian Petrarchs: one printed by the heirs of Aldo Manuzio in June, the other edited by Giovanni Andrea Gesualdo of Naples and printed in July by the firm of Nicolini.[11] The material for Gesualdo's edition had, however, been sent to Venice by 1530; Sebastiano Fausto seems to have had access to it, directly or indirectly, and the Aldine editor had certainly had a chance to read it and, consequently, to decide that he would follow a different path.

Gesualdo returned to Bembo's 1501 text, except that he divided the *Canzoniere* at 'Oimè il bel viso'. In other respects his edition bore a superficial resemblance to Vellutello's. It had a short *giunta* of five sonnets addressed to Petrarch, derived from both Aldo's and Vellutello's appendices. Gesualdo provided a long biography and a 'spositione' of the poems. But his commentary, dedicated to a woman, was longer than Vellutello's and had a different emphasis: it was the first to provide much grammatical and other linguistic guidance. Gesualdo repeated, for instance, Bembo's view that, in the phrase 'ardendo *lei*' (*Canz*. 125.11), 'lei' has the sense of *colei* (*Prose*, 3.48) and, when commenting on 'lo cui' (*Canz*. 85.7), he followed Bembo in saying that *lo* is used either before monosyllables or before *s* + consonant (*Prose*, 3.9). But he also gave some much more basic guidance. Writing on the phrase 'e humile' (*Canz*. 114.13), for instance, he explained in detail how to use *e* and *et* (or *ed*) in verse. He was the first commentator to pay particular attention to the needs of non-Tuscan readers, pointing out which of certain alternative forms he considered to be the Tuscan one.[12] He made many points which would interest readers from his own region, the south.[13] One of Vellutello's rare linguistic notes had explained the literal meaning of *guado* ('ford', *Canz*. 366.129). Gesualdo gave a succinct definition but also related the word to *vado*, a form which many non-Tuscans would recognize either because it was close to the Latin root (*vadum*) or as a word still used in some parts of Italy.

The Aldine Petrarch constituted in some respects the sort of reaction to such heavily-commentated editions which one would expect from this press. The text was presented without any introduction and was substantially the same as that which Aldo had printed in 1514, though now the different types of composition in the *Canzoniere* were numbered in separate series. Aldo had written in 1514 that he was going to provide a short 'espositione' of obscurities in Petrarch, and the dedication signed by his son Paolo, just twenty-one years old, said that he wanted to fulfil his father's promise. It is doubtful, however, whether this Petrarch was edited by Paolo, who showed no great interest in the vernacular later in his life. Furthermore, the anonymous editorial material contains references to the culture of the Neapolitan area and a few unorthodox forms, such as 'deto' ('finger'), 'aitri', 'nobbeli', 'nui', which suggest that its author may have come from central or southern Italy.[14] The editor was young, he tells us, still owing everything to an unnamed master, and he wanted to provide guidance on imitating good writers both to 'our delicate ladies' and to young people who were less well educated – studious but inexperienced in languages because of poverty or other misfortune. The help which he offered them in his edition was of two sorts. Firstly, he introduced a system based on the same three diacritic accents (grave, acute, circumflex) which Aldo had used for Latin texts.[15] Although, he said, it was more ornamental than essential (it did not show how to pronounce e and o), it would help the interpretation of the text by removing all ambiguity. Secondly, just as Aldo had occasionally given selective explanatory notes at the end of some texts, so this Aldine edition ended with an 'ispositione' of difficulties in Petrarch's poems. Introducing this to readers, the editor said that some would expect him to provide the range of information found in Servius's commentary on Virgil: studies of the poet's life, the title of the work, the quality of the verse, the writer's intention, the number of books, then the explanation of the text. But he saw no point in doing this, because so many others had done it so fully. His reference to 'la 'ntentione dello scrittore' shows that he was thinking partly of Gesualdo, who used exactly that spelling.[16] After his explanation of his spelling system and a defence of the study of Petrarch and of Tuscan, he ended his introduction by urging aspiring young authors to practise their Tuscan continually.[17] Sparse and brief though his notes were, they offered genuine assistance to the relatively inexperienced, such as synonyms for a few words 'not known to all' and paraphrases of some difficult expressions.

In 1535 there appeared an edition of Ariosto's *Orlando furioso*, printed by Bindoni and Pasini, which began the processes of presenting the poem as a worthy successor to the classical epic and of helping readers to understand and imitate Ariosto's verse. The edition included an *Apologia* against the author's detractors (which recalls Claricio's defence of Boccaccio in his *Amorosa visione*), an index to the additions made by Ariosto in 1532, a table showing the appearances of characters, and a brief explanation of some difficulties ('Dechiaratione di alcuni vocaboli e luoghi difficili', which recalls recent guides to the problems of Petrarch). The contributor of this material and the reviser of the text (he had punctuated it according to the Aldine model) was Lodovico Dolce. This Venetian had already written and published some verse works but was to achieve greater fame as a prolific editor and translator.[18] In his dedication Dolce said that he had now made the *Furioso* 'easier to use' ('più commodo'); as he explained in his *Apologia*, he had added, at Pasini's request, notes on words which were somewhat obscure because they were only rarely accepted by common usage, doing so in order to help the reader who was inexperienced in the literary vernacular and those ignorant of Latin literature ('al comodo del lettore non molto esercitato nella lingua, e di quelli che non hanno cognitione delle latine lettere'). As an added attraction to their intended audience, the printers chose to use a cheaper octavo format and a more popular semigothic (rather than roman or italic) typeface, though accented characters came from an italic fount. Dolce's 'Dechiaratione' is only two and a half pages long. Its main purpose (apart from elucidating three references to classical proper names) was to give synonyms for uncommon words or phrases. But Dolce was as concerned with justifying Ariosto's usage as he was with explaining it, for with only one exception he gave quotations of the same terms in Dante, Petrarch or Boccaccio. One can see that his purpose here was not only to assist readers but to carry on from the *Apologia* his defence of Ariosto as fundamentally a 'good observer' ('buono osservatore') of the rules of vernacular grammar who had nevertheless used with discretion the licence, which all poets should have, to deviate from past usage.

The success of Dolce's edition can be gauged by two reactions from contemporaries, even though each of them was galling in its way. The first was Niccolò Franco's satire in the third of his *Dialogi piacevoli* (Venice, Giovanni Giolito, 1539) on the way in which the character Eolophilo (a name which suggests restless changing with the times) had parasitically gained fame and future employment just by putting his name at the front of the work of a famous writer with an *Apologia* against detractors and a table of contents. Now, as soon as someone wanted to print any little work, Eolophilo would be asked to write an introduction. The second sign of the influence of Dolce's *Furioso* was the plagiarized edition which Giovanni Giolito published in 1536, carrying out his work in Turin because of the ten-year book-privilege accorded to the 1535 edition in Venice. The only similar innovation introduced by Venetian printers of the *Furioso* during the rest of the 1530s was the inclusion by Zoppino and Alvise Torti in their editions of 1536 of some 'notationi' by Zoppino's old collaborator, Marco Guazzo; but these merely listed the places where Ariosto had expanded the poem in 1532. Torti also had the idea, which was to prove influential later, of printing the work more or less simultaneously in different formats aimed at different readerships: a quarto in roman type and an octavo in gothic type which would have cost about half as much.[19]

In 1541, following the death of Giovanni Giolito, his innovative son Gabriele reopened the family's press in Venice. Gabriele must have been impressed, like his father, by Dolce's edition, for in 1542 he asked the

Venetian to prepare an augmented version of it which was to be dedicated to no less a personage than the heir to the French throne, Henri, who had married Caterina de' Medici in 1533. Dolce kept the table of contents from his 1535 edition and expanded his 'Dechiaratione' into a longer 'Espositione' of (he claimed) all difficult words and references. In explaining unusual words he used standard vernacular or Latin equivalents; however, there was also a hint of a method which was to become more widespread when he glossed 'zolle' ('clods') not only with Latin 'glebe' but also with a Venetian dialect term (used by 'our peasants'), 'zope'.[20] He now emphasized the moral benefits of the work by starting each canto with an allegorical interpretation of its events. Another way of asserting the high status of the poem was to point out its descent from a classical lineage; hence he added a comparison of some passages with classical Latin sources, especially Virgil and Ovid. Dolce also carried out a more thorough revision of Ariosto's spelling than he had done in 1535, correcting for instance the use of double consonants and diphthongs and removing some Latinisms.[21] The 1542 printing was in quarto format and italic type. In 1543 Giolito published Dolce's edition in the cheaper octavo format and in roman type, and he continued to print these two complementary types of *Furioso* until 1559–1560, reaching totals of fifteen quartos and twelve octavos.[22]

With Dolce in Giolito's camp, Bindoni and Pasini were anxious not to lose the initiative gained with their *Furioso* of 1535. In the same year 1542 they followed Giolito's edition with a new presentation of the poem, using contributions completed at least two years earlier by Domenico Tullio Fausto, brother of the Sebastiano who had edited their Petrarch of 1532. Like Torti in 1536, Bindoni and Pasini printed their *Furioso* in two formats, a quarto with roman characters and an octavo with gothic characters.[23] These two versions between them contained twelve extra items, a greater variety of material than anyone had yet provided for any vernacular work. Two brief indexes were taken by the printers from Dolce's two editions, and some of Fausto's material covered similar ground to Dolce's (explanations of difficult words and references, comparisons with Virgil and Ovid). But there were also items relating to the literary tradition and the historical subject of the poem (with emphasis on Ariosto's debts not just to Virgil but also to Homer and especially to Boiardo, and the false suggestion that both Italians were indebted to a Spanish source), and, in the quarto edition only, the two aids to composition which Sebastiano had provided in his Petrarch: a list of epithets and a *rimario*.

This plethora of extratextual matter seems only to have shown that editors could go too far. Bindoni and Pasini never printed the *Furioso* again, nor did any Venetian press reproduce Fausto's presentation. For nearly two decades the Giolito edition held sway. However, to keep competitors at bay, Dolce kept an eye on the text itself and expanded his contributions, without ever going to Fausto's extremes. In 1544 he included a letter to Giolito defending himself against some who apparently wanted to replace archaic words with better ones according to their own 'Italian' ideal. The letter precedes his 'Espositione' of difficult words, but it looks as if he was referring to a threat to the text of the poem. He also paid tribute to the corrections of Bernardino Merato, a copy-editor or proof-reader.[24] Dolce explained several new words and place-names in his 'Espositione' and, as an aid to poets, added a selection of descriptions (of day, night and the seasons) and of proverbs and observations 'which can be used easily by every skilful mind' ('delle quali ciascun destro ingegno si può commodamente servire'). As we shall see in chapter 9, this new item may have been suggested by a Florentine edition which had appeared earlier in 1544.

In 1545 the Aldine press joined the fray. Paolo Manuzio kept to family traditions by printing the naked text (in a not particularly correct version, derived perhaps from Giolito's most recent edition with some spelling revisions). As with his father's Petrarch of 1514, he managed to provide an excitingly original appendix: the *Cinque canti* which Ariosto had written as an addition to the poem but then rejected. Manuzio's source for this was Ariosto's son Virginio.

At the same time as Giolito and others were making efforts to widen access to poetic models, the same process was beginning with Boccaccio's *Decameron*. The first step was taken in March 1535 by Bernardino Vitali, who printed the work together with a wordlist which served partly as a glossary. The compiler of this list, Lucilio Minerbi, addressed the work to youths and 'amorous ladies' who wished to imitate Boccaccio in their own writing. One of the pitfalls to which prose writers were alert was that of using archaisms; but Minerbi only exceptionally warned that a word was no longer in use, as he did with *arrubinare* ('to make red'). His approach to the *Decameron* was suggested by the third book of Liburnio's *Tre fontane*, which had included lists of words from the work classified by parts of speech and with Tuscan synonyms in a few cases. What was new, though, in comparison with Liburnio, was Minerbi's frequent use of northern Italian or even specifically Veneto dialect terms in his definitions. This method would certainly have been welcome to those who had not yet mastered Tuscan sufficiently in order to approach it through the medium of Tuscan; on the other hand, it meant that his readership would be primarily a fairly local one.[25]

In 1538 Giovanni Giolito published a new type of *Decameron*. The editor was Antonio Brucioli, a Florentine who had come to Venice after being banished from his native city in 1529.[26] He already had considerable experience of working for the press, albeit in a completely different field, having produced six translated and commented editions of books of the Old Testament for Aurelio Pincio and Bartolomeo Zanetti between 1533 and 1538. Like Minerbi, he intended to explain Boccaccio's terms (as he wrote in his dedication) for the benefit of non-Tuscans. But his method was to provide annotations at the end of stories together with an index of all the terms annotated. The level of comprehension which he expected was not high, since he explained words such as *guancie* ('cheeks') and *zio* ('uncle'). He was aiming at a more broadly northern readership than Minerbi: he used not just Veneto terms in his definitions but also what he often called 'Lombard' ones, such as *lavandiera* for *lavandaia* ('washerwoman'), *paviere* for *lucignolo* ('wick').[27] Where Minerbi had used only Venetian *corlo* as a synonym for *arcolaio* ('wool-winder'), Brucioli gave both *corlo* and a term used more widely in the north, *guindolo*; for *smucciare* ('to slip') he gave local *slizzicare* but also Tuscan *sdrucciolare*.[28] He extended the field of reference still further by using his experience of France (where he had lived between 1523 and 1525) to point out words which he thought were French (such as *giamai*, *altresi*) or to give French equivalents (such as *traino* for *salmeria*, 'baggage train'). Some of his notes were relatively extensive, of the sort that one would find in a commentary, and they also occasionally elucidated proverbs and references to places. Not surprisingly, Brucioli used the Florentine 1527 text predominantly. But he seems to have started his edition with ambitions of producing a new text, since in the early part of the work he adopted some readings from both the Florentine and Venetian editions of 1516. As for questions of usage, he sometimes alerted his readers to the fact that words or forms were old-fashioned. But he was respectful of Trecento usage. In the previous year, when Bartolomeo Zanetti had printed the first edition of Giovanni Villani's *Croniche*, the letter to the readers had explained that Brucioli had been foremost in recommending that the source manuscript should be reproduced without modernization, and it had justified this unusual conservatism by saying that readers would be able to reflect on linguistic change and that it was in any case presumptuous to alter what others had written 'according to our feeling' ('secondo il nostro sentimento').[29]

In 1541 Dolce edited a *Decameron* printed by Bindoni and Pasini.[30] He wanted to appear to take the question of the text seriously, for he dedicated the edition to no less a figure than Pietro Bembo and used his letter of dedication to analyse the failings of previous editors as textual critics. Some, he wrote, had found discrepancies between texts and had selected variants according to personal preference, without further investigation. Others had corrupted good texts because they wanted to modernize Boccaccio's language. This was an accurate analysis of the faults of much contemporary editing, but Dolce then suggested that his main or only guide to the constitution of the text was the grammatical norm, for he went on to say that editors had reduced the *Decameron* to a state in which one could find in it few of the rules which had been derived from it. Even the diligent Delfino, he said, had distorted verb forms and committed other errors. Nor did Dolce have any new evidence to offer. In spite of these criticisms of his fellow-Venetian, he returned to

Delfino's 1516 text, making only some superficial changes such as the alteration of *allui* to *a lui*.[31] This patriotic gesture and the choice of Bembo as dedicatee were intended to reassert the claims of Venetian scholarship in the wake of the recent edition of the Florentine Brucioli. And although Dolce took over Brucioli's technique of giving explanatory notes at the end of stories, and indeed took over much of what Brucioli had written, his notes were fewer and briefer, and he distanced himself (at the end of 3.8) from the Florentine's explanation of proverbs, saying that he wanted only to elucidate words which many found obscure.[32]

A year after the provocative appearance of this rival edition, Brucioli revised his own version of the *Decameron*. This time the publisher was Gabriele Giolito. It was decided to cover the market as widely as possible by publishing a quarto edition in italic type and a more portable version in sixteens and in roman type. There was also a difference in contents between the editions: if one paid more for the quarto, one got Brucioli's notes, now expanded in order to give more help with the rules of Tuscan, whose indispensability to both writers and speakers of the language was pointed out on the title page and again in the dedication. The additional notes covered points such as noun plurals, definite articles and pronouns, and gave more help on the question (particularly important in speech as well as in prose writing) of which of Boccaccio's forms were no longer in use.

Although Petrarch had inspired a flurry of new presentations in Venice in 1532 and 1533, editors were at first reluctant to treat his poetry in the same way as they were treating the *Orlando furioso* and the *Decameron*. The works of Liburnio and others had suggested that there were some who wanted to read Petrarch and write Petrarchan verse but who also needed some fairly elementary linguistic guidance, preferably not buried in a commentary. Yet such help was not immediately forthcoming, for a number of reasons. Petrarch's verse was, no doubt, seen as so refined that it did not lend itself easily to annotations for beginners. There was already a strongly established tradition that the alternative to printing the text on its own was to print it with a full commentary, and this made it harder to find a middle way than in the cases of the *Decameron* (which had never had a commentary) and the very recent *Furioso*. More experienced men of letters would have despised the compilation of *rimari*, dictionaries and lists of descriptions: Speroni makes the poet Antonio Brocardo depict these as sterile exercises in his *Dialogo della retorica*, written probably in 1542. Furthermore, the principal editors involved in the presentation of Boccaccio and Ariosto in this period did not have a strong interest in Petrarch. Dolce wrote his more important verse compositions in *ottava rima*, a metric form shunned by the Tuscan poet. He shared the antipathy of his friend Pietro Aretino towards a strict imitation of older Tuscan. When he became the main defender of Ariosto, he was also going against the tastes of

Bembo, who seems never to have responded to Ariosto's public declarations of admiration.³³ As for Brucioli, he was a writer of prose rather than verse. Nor did most printers see a need to do anything other than to continue to reproduce, with slight if any variations, one of the editions which existed by 1533. Of the commented editions, by far the favourite model was that of Vellutello, but there were also imitations of the Aldine editions of 1533 and (in one case) 1521 and of the Giunta edition of 1522. The new Aldine edition of 1546 followed the 1533 edition very closely but neither used its experimental accent system nor republished its concluding notes. Giolito began a quarto series, printed from 1544 onwards, which reproduced Vellutello's text and commentary with a more comprehensive appendix derived from the Aldine edition of 1514 and the Florentine edition of 1522; this appears to have been due to Lodovico Domenichi of Piacenza, then at the start of his editorial career.³⁴

The study of Petrarch's sources and his autograph drafts continued to flourish in circles linked with Bembo. In one case it appears that this was going to lead to the publication of some new method of consulting the poems: in 1532/3 a book-privilege request was made in Venice to print Petrarch together with an 'artificio' (perhaps 'device' or 'system') by a friend of Bembo, Giulio Camillo Delminio.³⁵ As we shall see in chapter 8, however, it was not until twenty years later that Dolce edited a version of some of Camillo's notes on Petrarch. Meanwhile, a compilation of them made by someone else continued to circulate in manuscript form.³⁶ An intermediary was also responsible for the publication in the 1540s of some of the fruits of the study of Petrarch by Trifon Gabriele, one of Bembo's closest friends. Bernardino Daniello of Lucca, who had been a pupil of Gabriele in the 1520s, compiled a new commentary on Petrarch which he acknowledged was much indebted to his master. It was printed at the Nicolini press first in 1541 and then, with important revisions and additions, in 1549.³⁷ Daniello gave prominence to the drafts to which Gabriele had access: he mentioned several variants rejected by Petrarch and in the second edition grouped them in a section at the start, adding new examples. His purpose in doing so was didactic. Just as Aldo had pointed out the value which a study of Petrarch's revisions could have for poets, so Daniello emphasized that a knowledge of these variants would teach students of Petrarch to have better judgement in their own works. His text was based on the Aldines of 1514 and 1533, but he included at least one innovation of his own and he discussed, probably in the light of Gabriele's teaching, the problem of the ordering of the *Trionfi*.³⁸ Daniello's commentary was designed to give more help with Petrarch's language than some of its predecessors. As he explained in his dedication, he gave less historical information than others; instead, especially in 1549, he provided more synonyms or paraphrases of Petrarch's vocabulary and commented on a few points of grammar.

In spite of the conservatism which continued to be shown in respect of

Petrarch by the leading Venetian printers, the new approach to the presentation of literary texts pioneered with Ariosto and Boccaccio did eventually begin to have its effect on the publishing of his verse. This can be seen first in an edition printed in 1539 by Francesco Marcolini. The text of the poems followed the Aldine model (without any appendix), but it was accompanied by the *Osservationi sopra il Petrarcha* of Francesco Alunno, a Ferrarese teacher of mathematics and calligraphy with a passion for lexicography. A letter of recommendation from Aretino praised Alunno's devotion to those who become entangled in Petrarchan usage ('pietade ... inverso di coloro che s'intricano ne i modi usati dal toscano Poeta'). What Alunno provided in order to extricate readers was, in effect, a concordance, and the companion text accordingly had page and line numbers for purposes of cross-reference. Readers could also use this system to find epithets, as he explained, by looking up a word such as *cielo* ('sky'), then the adjectives such as *stellato* ('starry') which might go with it, and then seeing if the references coincided. This was a laborious process, though; and another feature of this edition which would have limited its usefulness for beginners was that definitions were normally given only to distinguish between different senses of the same word.

Just as Bembo's *Prose* encouraged the fashion for Petrarch, so their negative judgement on Dante helped to discourage interest in the *Commedia*. From 1530 to 1550 there were only four editions of the work in Italy. But one of these, printed in 1544 by Marcolini, contained a new commentary by Vellutello.[39] This Dante continued the stand against Bembo's literary and editorial principles which Vellutello had begun in his Petrarch nearly twenty years earlier. The letter to the readers called the Aldine text of 1502 the most incorrect of all and accused Bembo of having provided Aldo with badly corrected texts of both Petrarch and Dante. Vellutello said he had derived his own text 'from different and older ones, those which are recognized as being less corrupt than all the others' ('da diversi e più antichi testi, quelli che di tutti gli altri meno si conoscano esser vitiati'). He used these eclectically, as he explained with a welcome openness about his methods:

And although they are all, as I say, very incorrect, yet I found that in such a great number what one does not say, another does, and where I saw that the meaning was missing or understood that it was altered and off the point, by ruminating diligently over them I arrived, as I firmly believe, at the truth.

(E benché tutti, com'io dico, siano incorrettissimi, pur ho trovato che in tanto numero quello che non dice l'uno, dice l'altro, e dove ho veduto mancar la sentenza, o compreso esser alterata e fuori del proposito, ruminando diligentemente in quelli, ne sono venuto, secondo il fermo creder mio, su la verità.)

In practice, this meant that Vellutello used Bembo's edition as his main source but also borrowed from Landino's text and occasionally from other sources.[40] He also took the decision (a very reactionary one in 1544) to use no apostrophes. As well as his commentary, which concentrated on lengthy exegesis of the text rather than of its historical and philosophical background, Vellutello provided a life of Dante and detailed descriptions of each part of the afterworld, with several new diagrams for the Inferno.

Vellutello was the only editor of the three great Trecento writers in this period who attempted to use textual sources dating from before his century. Daniello drew on Petrarch's drafts in his commentary, but for the *Canzoniere* and *Trionfi* editors accepted, with few if any changes, one of the texts printed between 1501 and 1525. There was some experimentation with the *Decameron*, but always on the basis of the editions of 1516 or 1527. As far as printed texts are concerned, this picture shows a falling-off from the vernacular textual criticism of Bembo. The sense of decline becomes even more marked if one contrasts the efforts which Paolo Manuzio was putting into the editing of classical texts. A good example of the application of his methods is his 1540 edition of Cicero's *Epistolae familiares*, in which he carried on a polemic with the edition which the Florentine Piero Vettori had published in 1535. The dedication began with a detailed description of how he carried out emendations, first comparing manuscripts with printed books, then selecting the best readings, preferably in consultation with other scholars. The letters are followed by a list of variants in which conjectures were carefully separated from readings of 'old manuscripts' (a distinction not made in Manuzio's earlier edition of 1533), and by a long series of discussions of obscure passages and of controversial readings in which Manuzio referred to his manuscript sources (usually vaguely, but sometimes by the name of their owner) and duly acknowledged conjectures suggested by friends from Rome and France.[41] He used concluding scholia as an effective device for explaining obscurities. This edition, which Manuzio reprinted more than once with revisions, suggested strong and active links between Venice and humanists from centres to the north and south, while at the same time it offered a strong challenge to Florentine scholarship. His emphasis on exegesis of the text, which was one of the issues on which he differed from Vettori, had a parallel in the attitudes of contemporary vernacular editors in Venice; but there was no such continuity of method when it came to the establishment of texts. The great achievement of Venice in the age of Aldo was to have treated the editing of older vernacular texts with the same seriousness as that of classical ones. In the Venice of his son, though, the bridge between the two branches of scholarship was severed.

When editors were preparing for publication recent or contemporary literary works which did not conform with puristic ideals, it was now normal for them to subject the text to a thorough revision. Their priority was not authenticity but to present a polished text, assimilated to the ideals embodied by accepted models, even if the result went against the views which the author stated in the text.[42]

Among the many works which were revised in this way were two Quattrocento epics for which the language of the *Orlando furioso* of 1532 now came to act as a model: the *Orlando innamorato* by Matteo Maria Boiardo (1441–94) and the *Morgante* of the Florentine Luigi Pulci (1432–84). Up to 1544 the *Innamorato* had been printed many times (notably by Rusconi, Zoppino, Bindoni and Pasini, Nicolini, and Alvise Torti) in a form not radically different from that which had first appeared in the author's native Emilia in about 1482. In 1545, however, soon after the publication of a completely rewritten version by Francesco Berni, Girolamo Scoto printed the poem 'newly reformed' ('nuovamente riformato') by Lodovico Domenichi.[43] In imitation of Dolce's editions of the *Furioso*, each canto began with an illustration and a short summary (though not an 'allegory'), and there was an index. But the editor's main task was the revision of Boiardo's language. In his preliminary dedication, Domenichi took the important step of making it clear that the poem was not in its original state but had been brought into line with contemporary standards. Boiardo had been prevented from doing this himself, he said, not just by his death but also by the roughness ('rozzezza') of his time, in which 'this Italian language' lacked the polish ('pulitezza') of the present day. In another dedication at the end of the work, he used the metaphor of someone purging 'with the sickle of judgement' ('con la falce del giudicio') the useless weeds which had taken root in a wheatfield 'through the fog of their dark times' ('per la nebbia de i lor tempi oscuri'). The first eight stanzas were rewritten quite freely (again one sees the importance of making a good impression on the customer browsing at a bookseller's), but thereafter Domenichi kept close to the original text. Many of his interventions were caused by Boiardo's northern phonology. If the form in question was in rhyme, then the change could trigger off a more extensive revision.[44] The morphology was also subject to correction. Two 'wrong' definite article forms were eliminated, for example, in 'col specchio al scudo occise quel serpente' (2.1.27.8) > 'uccise con lo specchio quel serpente'; a pronoun form was changed in 'e disse ad esso "voi sète prigione"' (3.1.23.3) > 'gli disse, "hora voi sète mio prigione"'; and examples of corrections to verb forms are 'èi' > 'sei', 'porrai' > 'potrai', 'furno' > 'furon', and second-person plurals of the present tense in *-ati* > *-ate*.[45] Where earlier texts had something like 'onde convien ch'il mondo si commova, | e questo un pezzo, e quel un altro piglia; | il mondo tutto a guerra se scompiglia' (2.1.6.6–8), Domenichi corrected the syntax so that the last two verbs became subjunctives ('pigli' and 'scompigli') depending on 'convien' just as 'commova' did. He often raised the tone of the vocabulary by removing words which were dialectal or which he felt to be unsuitable for other reasons.[46]

Shortly afterwards, Domenichi edited Pulci's epic for Scoto. Here too he judged the poem to deserve (as he put it) smarter clothing. In Canto 1, his revisions (based probably on one of the more recent Venetian editions) were fairly free, affecting metre and choice of vocabulary.[47] But the *Morgante*'s Quattrocento Florentine garb needed much less attention than did the language of the *Innamorato*, and for the most part Domenichi was content to correct noun and verb endings and pronoun forms.[48] He also provided canto summaries, a subject index, and a 'dichiaratione' of some difficult words or proverbs in each canto. He claimed in his dedication that 'the sacred mysteries of Tuscan can be understood by men born inside the bounds of Italy and brought up in the reading of good books, without further recourse to the oracles and priests of that province' ('i sacri misteri della lingua Toscana possono essere compresi da gli huomini nati dentro i termini d'Italia e nodriti nella lettione de i buoni libri, senza altrimenti haver ricorso a gli oracoli et a i sacerdoti di quella provincia'). While Domenichi, unlike Castiglione in the *Cortegiano* (1.37), was confident that non-Tuscans need feel no terror of the 'mysteries' of Tuscan, he resented the way in which Tuscans still wanted to control the cult of their language, as if carrying on the cerimonial traditions for which their Etruscan forebears were famous. Perhaps he was alluding specifically to the spelling system which Pierfrancesco Giambullari (under the pseudonym Neri Dortelata) had used in 1544 to teach Florentine pronunciation to those not fortunate enough to have been brought up by the Arno.[49]

At about the same time another *Morgante* was printed by Comin da Trino. The colophon has the date 1545, but the title page has 1546, so it seems best to assume that it followed and reacted against Domenichi's edition with its revisions and anti-Florentine tone, rather than that Domenichi's comments were aimed at this edition. Externally, Comin's edition resembles Scoto's closely, with its quarto format and its use of woodcut illustrations, *argomenti* and linguistic notes to accompany each canto. But the similarity is only superficial. The reader is immediately confronted by a letter signed by Comin which claimed, for once justifiably, that the work was now being printed with a true understanding of its Florentine nature. It explained that errors had crept into previous editions printed by men with little knowledge 'of [Pulci's] pure Florentine language' ('del suo parlare mero fiorentino'); these errors had led in turn to false corrections. But now a Giovanni Pulci who said he was the author's nephew or grandson ('il quale, per quanto si ha da esso, è nipote dello autore': Comin did not seem too convinced) had provided Luigi's original 'corrected just as he composed it' ('corretto nel modo proprio che esso lo compose').[50] The book was now consequently 'in no part altered' ('in nessuna parte alterato'). This probably meant that Giovanni was using an early printed copy, perhaps the Florentine edition printed by Francesco di Dino in 1482/3, whose colophon says that the text is 'drawn from the true original and revised and corrected by its own author' ('ritracto dallo originale vero e riveduto e correcto dal proprio auctore'). That his source was this edition and not, for instance, Bonelli's Venetian edition of 1494 whose colophon repeats the same form of words, is suggested in the first few stanzas by his restoration of two readings, 'figlia e madre' (1.2.1) and 'Filomena' (1.3.1), which are found in the Florentine text but not in that of 1494, nor in two later Venetian editions used as controls, those of 1507 and 1537, nor in Domenichi's text, all of which have 'figlia, madre' and 'Philomena'. Domenichi's revisions were ignored, and other restorations went completely against the trends of editions from earlier in the Cinquecento both in metre and language. Lines of twelve syllables which had already been reduced to eleven in the 1537 edition were reinstated.[51] The correction of 'a Orlando' to 'ad Orlando' (2.6.6), which in the 1537 and Scoto editions eliminated a case of dialoephe, was similarly reversed. Giovanni Pulci's copy also restored authentic forms such as 'la *suo* gloria' (1.4.6), '*el* più savio' (1.8.2), '*mie* barca' (2.1.5), which had been replaced by at least the time of the 1507 edition with 'sua', 'il', 'mia'. However, Comin's text did not always follow the Florentine incunable faithfully.[52]

Giovanni Pulci, according to Comin's letter, was also the author of the linguistic notes which follow each canto. Again the need for a Florentine editor was stressed: the aim was to explain 'those places [which are] not understood, such as words, sayings, proverbs, and other particular Florentine expressions of those times, unknown to the majority of other Italians' ('que' luoghi non intesi, come vocaboli, detti, proverbi, et altri proprij parlari fiorentini di que' tempi, ignoti [*sic*] alla maggior parte de gli altri italiani'). These notes are independent of Domenichi's and often much fuller. With a local readership in mind, Giovanni still used Venetian as well as Tuscan in some of his definitions.[53] The *argomenti*, however, seem to be abbreviated reworkings of those in the Scoto edition.

Prose works were also liable to be modernized. There was, though, a tendency to exaggerate the extent of the editor's labours. The fourteenth-century Florentine translation of Pietro Crescenzi's *Libro della agricultura* was printed in 1536 by Bernardino Viani in a version which, the title page claimed, was 'newly restored with very great labour to its pristine form' ('novamente con grandissima fatica alla sua pristina forma restituita').[54] But most of the changes were minor adjustments to the spelling. Brucioli edited in 1539 and 1540 a number of sermons by the Ferrarese friar Girolamo Savonarola, whose involvement in Florentine politics had led to his death in 1494. In the dedication of one volume, the *Prediche per tutto l'anno* (Brandino and Ottaviano Scoto, 1 March 1539), Brucioli said that he had removed the work from linguistic barbarity, though he blamed this on the neglect of printers. In fact, his text is practically identical with that of earlier Venetian editions. The same is true of his edition of the translation of Savonarola's sermons on *Quam bonus* which followed from the same press on 16 March.[55]

When Brucioli prepared Landino's translation of Pliny's *Historia naturale* for Giolito in 1543, he used the 1534 edition printed by Tommaso Ballarino. The title page of this earlier volume said that the text had been corrected with the greatest diligence, and in fact a certain Giovan de Francesio had carried out a fairly consistent revision.[56] Brucioli, as he said in his dedication, made few changes. Nevertheless, Giolito's title page now credited him with the revision, no doubt in the hope of attracting purchasers with what was by now a well-known name in Venetian editing. The title page also mentioned Brucioli's main contribution, a series of explanations (given in the margins and then grouped in a concluding glossary) of many of Landino's terms, using equivalents which were often from the Veneto rather than from Tuscany.[57]

Similar operations were conducted on other fifteenth-century works which were being dusted off to help satisfy the new passion for translations. Sebastiano Fausto, for instance, corrected Landino's version of Giovanni Simonetta's *Sforziada* for publication by Curzio Navò in 1543.[58] Domenichi made a few changes to the translation of Bruni's history of the First Punic War (previously published by the Giunta press in 1526) and compiled a table of contents for the Giolito edition of 1545. Doni revised Manilio's translation of Seneca's letters for Aurelio Pincio in 1549.

The prose of Niccolò Machiavelli (1469–1527) now looked outdated too. As his works passed through the Aldine editions of 1540 and 1546 and then Giolito's edition of 1550, their language was rendered progressively more bland through the toning down of two influences which gave it colour, those of contemporary spoken Florentine and of Latin.[59]

So far two principal developments in Venetian editing during the 1530s and early 1540s have been identified: the presentation of the most admired literary texts in such a way that they could be more easily understood and imitated, and, especially in the later part of this period, the extension of linguistic normalization to an ever wider range of works. Another important task which Venetian editors took on towards the end of this period was the compilation of anthologies of prose and verse. Behind the popularity of such volumes lay a hunger on the part of readers for models of letters, poems, speeches and so on by the new wave of contemporary authors. Previously, readers had been looking back towards masterpieces of the past; these anthologies showed what could be achieved in the present.

Anthologies of letters were particularly popular. There was evidently a need for as many vernacular models of this genre as possible, in order to provide a parallel with the dominant role played by the letters of Cicero in the teaching of Latin prose writing.[60] Collections of letters will also have satisfied readers' curiosity by giving them behind-the-scenes glimpses of the private lives and emotions, and the opinions on literary, religious and other themes, of the pillars of the literary establishment as well as of its lesser members.[61] The first printed collection of letters by a single author, Pietro Aretino, went through Marcolini's press in 1538 and was rapidly followed by other such collections. Then in 1542 came the first two anthologies containing letters by several authors and thus offering a wider range of styles and subjects. The initiative for the first of these was taken, it seems, by Paolo Manuzio, and its success led to a second book of letters put together by his brother Antonio in 1545. Both volumes were republished several times with revisions. Other publishers were quick to follow suit: Curzio Navò in 1542, for instance, and Paolo Gherardo in 1544 and 1545. The Manuzio brothers, Navò and Gherardo all pointed to the role of letters in teaching good writing: the first three in their dedications, Gherardo by including in second place a long letter in which Marco Antonio Flaminio recommended a prominent place for letter-writing in education.[62]

In order to put a letter anthology together, printers and publishers would use collaborators to approach well-known writers on their behalf. Paolo Manuzio, for instance, was helped by Iacopo Nardi, another Florentine exile in Venice, and by Benedetto Ramberti, who included some letters relating to the problems of compiling such anthologies.[63] There is one in the second book of the Aldine collection, dated 14 December 1542, in which he compares the difficulty and rewards of finding letters to fill the book with gathering flowers in autumn. In the first book there is a complementary pair of letters showing the opposite reactions of two leading writers to requests for contributions: Speroni, in a splendid letter to Ramberti, pours scorn on the idea of having his private correspondence published, while the poet Francesco Maria Molza, writing to Manuzio, welcomes the idea of an anthology because of its usefulness in providing models of style. Book 2 included letters from Claudio Tolomei, another much-admired writer who was reluctant to commit his works to print. His final letter accused Bartolomeo Paganucci, evidently another Aldine agent, of making off with some of his correspondence with a view to printing it before Tolomei had had a chance to correct it. As a last resort, he was willing to let Manuzio print the letters as long as this letter accompanied them. As other evidence confirms, some letters were revised before publication, often no doubt by the authors themselves.[64] A few letters seem to be completely fictitious, mere exercises in style.

Verse anthologies had existed since classical times, and collections of vernacular verse had been printed in Venice since the early Cinquecento.[65] Manuzio had been interested in printing a collection of burlesque poetry in 1537, but, perhaps because Curzio Navò printed two collections in 1537 and 1538, his plans came to nothing.[66] In 1545 Giolito began a series of volumes of lyric poetry, the *Rime diverse di molti eccellentissimi auttori*. For the first book, Domenichi collected works by ninety-one poets, one (Lorenzo de' Medici) from the Quattrocento but the rest from his own century. This proved as successful an idea as that of letter collections, and inspired a host of imitations and variations on the same idea, such as anthologies of poets from a particular city.[67]

8 · THE EDITOR TRIUMPHANT: EDITING IN VENICE, 1546–1560

TWO ADMITTEDLY PARTIAL STATISTICAL ANALYSES provide evidence that the next fifteen years brought a continued expansion in the output of Venetian presses. A survey of the British Library's holdings suggests that there were more printers and more books printed in Venice between 1551 and 1575 than at any other time in the Cinquecento. Her overall dominance of the Italian market was still overwhelming, even though it declined somewhat from the heights of 1526–50 as the shares of some minor centres, including Florence, rose. The number of imprimaturs granted to works not previously published in the Venetian state rose steeply during the 1550s. And the output of the leading press, that of Gabriel Giolito, reached its peak around 1555.[1] The contribution which editors had made to this success story was reflected in their increasing prominence as individuals, each with his own distinctive approach to the shaping of a publication. Rivalries between them became more heated: hence the outbreak of polemics in print, similar to those to which classical editors had devoted such energy since the Quattrocento. One controversy involved Domenichi and the Florentine Anton Francesco Doni. In another, Dolce exchanged vitriolic words with Girolamo Ruscelli of Viterbo, who even published a book attacking his rival, the *Tre discorsi* of 1553.[2]

Ruscelli's editorial career, which lasted from 1551 until his death in 1566 and has recently been illuminated by Paolo Trovato, exemplifies well the positions of agent and entrepreneur which such a person could now occupy.[3] In the dedication (dated 5 September 1551) of one of the first works edited by him, the *De venatione* of Natale de' Conti, Ruscelli presented himself to the printer Paolo Manuzio as someone who had been given a copy of the poem by its author and was now graciously releasing it at the request of friends and printers alike. Between 1553 and 1555, Ruscelli belonged to a company which employed Plinio Pietrasanta to print works edited by Ruscelli. And we have already seen in chapter 1 how Bernardo Tasso turned to him for help when he wanted his works printed in Venice in 1557.

Vernacular editors in Venice in this period appear to have exerted more influence than before over the contents of books. The major texts were accompanied by more and more exegetic items. There was an ever freer use of linguistic revisions based on editors' subjective opinions on what the author should have written. In contrast, in the field of classical editing Francesco

Robortello was helping to foster a clear distinction between emendation from written sources and conjectural emendation. He did not share the preference for the former which, as we shall see in the next chapter, characterized classical studies in Florence, but his example encouraged scholars to use conjecture with restraint, according to relatively scientific principles, and honestly. When Robortello edited the plays of Aeschylus (Venice, 1552), he listed his conjectures, as Paolo Manuzio had done for Cicero; two years later, in his edition of Longinus, *De sublimi orationis genere* (printed in Basel), he used asterisks to indicate lacunae or corruptions which he could not emend, rather than creating a readable text at all costs. Then in 1557 Robortello published in Padua the first work to analyse emendation, the *De arte sive ratione corrigendi veteres authores* ('On the art or theory of correcting old authors'). Here he gave advice on the use of old manuscripts and on the correction of different kinds of corruption by means of conjecture, which was to be used when old manuscripts were of no help; and he stressed the need for scrupulous good faith ('bona fides') about the sources of an editor's text, whether these were written or conjectural.[4] As we go on to look at the editing of different kinds of vernacular works, firstly Trecento and modern classics and then those of lesser status, we shall see that it is rare to find a Venetian editor paying more than lip service to the comparison of sources or to the study of other indirect evidence which might help to establish an authentic text; their priorities usually lay elsewhere.

By 1546 Giolito had decided to present the *Decameron* in a more elaborate fashion than previously, so that it would form a diptych together with his *Furioso* of 1542. In order to stress this pairing, he dedicated the volume to Caterina de' Medici, reminding her that her husband had already received Ariosto's martial epic from him. The *Decameron*, as he pointed out, was a work from which one learned regular Tuscan ('la regolata lingua Thoscana'), and with this in mind he felt it necessary to provide yet more information and help for the reader. He turned to a young editor, Francesco Sansovino, born in Rome in 1521 and brought up as the son of the Florentine sculptor Iacopo (there is a shadow of doubt over his exact paternity). Francesco and Iacopo had moved to Venice in 1527.[5] Francesco's credentials for editing the work were principally a collection of letters published in 1542, based on the structure of Boccaccio's masterpiece, and the edition of the *Ameto* which he had produced for Giolito in 1545. This latter represented a confident debut on the part of someone so young who was offering the first really new edition since those of Claricio (1520) and the Giunta press (1521). He followed the Florentine *Ameto* with some borrowings from the Milanese version, and preceded it with a short explanation of its difficulties ('dichiaratione de i luoghi difficili') addressed to Gaspara Stampa.[6] Here he suggested possible interpretations of the work before going on to explain a few mythological and

historical references. Sansovino then corrected some of the spellings used by
Claricio and the Florentine editor and gave synonyms or explanations for a
handful of words. He justified the brevity of his treatment with the lame
excuse that Gaspara would understand the rest of Boccaccio's Latinizing
vocabulary. More important factors will have been the pressure of time under
which he said he was working and his declared distaste for the obscure and
over-elaborate style of this work (inferior to Castiglione's style, he said, while
the latter was itself inferior to that of the *Decameron*).[7]

But in the following year he and Giolito took an altogether more responsible
and enterprising attitude to the *Decameron*, best-seller as it was. The title page
announced that the text had been emended 'according to the old copies'
('secondo gli antichi esemplari') by several people and that variants were
given in the margins: the first time this had been done for the *Decameron*,
though there was a precedent in the *Ameto* of 1503. A letter signed by Giolito
underlined the importance of restoring the text not through the editor's taste
but through the comparison of 'various old books' ('diversi antichi libri'). In
practice, this meant that Sansovino adopted a procedure similar to that of his
Ameto: his text was based on that of the 1527 Giunta edition, but he borrowed
from its main rivals, the Florentine and Venetian editions of 1516, sometimes
preferring their less archaic or rare forms.[8] On the other hand (and this was
the point the letter was making), Sansovino had not invented readings as a
Claricio might have done. His notes demonstrate a good knowledge of the
usage of other Tuscan writers from the Duecento to his own times: he justified,
for instance, *boce* ('voice') and *atare* ('to help') with references to the *Novellino*
and Fazio degli Uberti. 'Avacciandosi', he said, was used by Villani and was
still alive in the Florentine *contado* as opposed to the city ('è più propio de'
nostri contadini, che nostro'). He is aware of the importance of wide reading
and of the contribution that could be made by a knowledge of rural usage, and
this links his editing with Bembo's letter to the reader in the 1501 Petrarch and
foreshadows the researches of Borghini later in the century. Sansovino's
marginal variants become fewer as the work goes on, but they too represent an
open and informative approach to the problem of the text.[9] He also contributed four new items of his own which made this by far the most accessible
and useful *Decameron* to date: a life of the author, a 'dichiaratione' of words,
proverbs and difficult places, a note on some proper names, and a list of
epithets. The 'dichiaratione' is in part a glossary, and one which takes
advantage not only of the work of Brucioli but also of the publication in 1543
of dictionaries by Alberto Acarisio and Francesco Alunno. Like Brucioli,
Sansovino defined some terms with northern equivalents.[10] But, as well as
explaining meanings, Sansovino also discussed social customs, proverbs,
idioms and alternative readings, with quotations showing similar usage in a
wide range of other writers from the Duecento to his own times. He provided
separate lists of Boccaccio's sources for some historical characters and of Guelf

and Ghibelline families. He also added a brief moral to be gleaned from each story. Manilio had already done something similar with the *Porretane* of Sabadino degli Arienti in 1504, but Sansovino's innovation will have been inspired by the *allegorie* of Dolce's Ariosto.

Dolce in turn had to respond to Sansovino's challenge when in 1552 he came to revise the *Decameron* which Bindoni and Pasini had printed in 1541, this time for publication by Giolito. In the earlier edition he had followed Delfino's text; now, to keep up to date, he had to introduce several of Sansovino's readings, even though most of these were derived from the Florentine edition of 1527.[11] Dolce also adopted the device of giving variants in the margin and the other assistance to the reader which Sansovino had provided. It was not easy for a Venetian to borrow from Florentine editors in this way, and it was doubtless embarrassing for Dolce, the doyen of Venetian editors, to have to revise the editorial policy used in his pre-Giolito days. The period in which Venetian editors felt the need to attack those of Florence was now coming to an end, but Dolce did all he could in the letter to the readers to set his edition apart from Florentine influences. He inveighed against 'the Florentine printing' ('la impressione fiorentina') as the worst of all, without making it clear that he was referring to the earlier 1516 edition. He attacked archaic or popular forms such as *stea, amenduni, atare, boce* found in Sansovino's text, arguing that, even if they survived in popular usage into Boccaccio's time, the narrators of the stories would not have used them. Here the Venetian Dolce was revealing his sympathy with the same 'courtly' current of linguistic thought to which Castiglione had belonged, and much later, in his 1568 edition of the *Orlando furioso*, he still made a point of praising Ariosto's vocabulary because it was not 'pure Tuscan' but had an interregional, 'common' quality. As for his sources, instead of acknowledging the influence of Sansovino's edition, he claimed that his own text had been composed eclectically, by consulting 'many very ancient volumes' ('molti antichissimi volumi') and choosing the parts which were best constructed and which best expressed the author's meaning. And many of his marginal variants, he said, were owed to the diligence of Delfino, whose text conformed for the most part with the oldest exemplars he had consulted.

Some defensive remarks in Dolce's letter show that he was aware that another rival *Decameron* was already being published by Vincenzo Valgrisi. Its editor was Girolamo Ruscelli, who had only just made his debut on the Venetian editorial scene but, never short of self-confidence, was already questioning the authority of Dolce. Ruscelli was proudest of all of having introduced what he considered a more rational system of spelling and punctuation, and which certainly anticipates modern usage. The title page of this first edition of his *Decameron* said that the work had been brought back to perfection 'no less in its writing than in its words' ('non meno nella scrittura che nelle parole'). As he explained in the notes at the end of Day 6, he wanted to free the vernacular from servility to Greek and Latin, for instance in its excessive use of *h*, even if this meant sacrificing spellings used in the Trecento. Ruscelli was unable to approach the problem of the text in a similarly systematic way. He declared that his text was based on 'the common printings' ('le stampe communi'). His main sources were in fact the editions of 1527 and 1546, but his notes refer to the consultation of printed and manuscript copies of acquaintances and to the judgement of others, in the manner of Paolo Manuzio in his Cicero of 1540. Where he found differences between texts, he says, he followed the majority ('li più') and noted variants in the margins. The margins also contain notes which acknowledge other criteria: correct Tuscan usage and personal preference. But in practice he was reluctant to change the printed vulgate, so that his favoured readings were (unsatisfactorily) in the margins rather than in the text. He occasionally offered emendations of his own invention, as in '*nella* gratia sua' for 'della' (2.7.32) or the more radical suggestion 'non già avvenne, che de' maggiori quasi era tenuto' for the corrupt 'non miga giovene, ma di quelli che di maggior case si era tenuto' (4.2.7). He incorporated the first emendation in a later edition; the second was prudently left among his annotations.

Even if Ruscelli's methods were haphazard, he at least chose to be much more open about his sources than Dolce. He also opposed Dolce's treatment of archaisms in the Bindoni and Pasini edition, though Ruscelli supported his case mainly with his belief that what Giolito had been printing up to then could not be wrong. He was quite clear that Boccaccio's usage should not be confused with that of Petrarch, or with the 'common usage' ('uso commune') of Italy, or with one's subjective opinions about what was right. But he used his notes in the margins and at the end of each day in order to give grammatical advice as well as for textual criticism, and he often suggested that Boccaccio's usage could be bettered. Ruscelli might say that it was too Tuscan: for example, he used *disiderare* because it was in his Giuntine source and native to Florence, but he felt that the Latinate form in *de-* was 'more beautiful and more to be used' ('più bello et più da usare').[12] Or Boccaccio might go against Ruscelli's beloved rules, as with *per lo* rather than *per il* or *davanti da* rather than *davanti a* or *di*. Or there were forms which seemed to him affected or 'hard' ('duro'), such as *amenduni* and *orrevole*. The borderline between maintaining Boccaccio's authentic usage and evaluating its shortcomings as a model was thus blurred. In any case, the limits of Ruscelli's knowledge of older Tuscan were revealed when he unhesitatingly dismissed as spurious words such as 'habituri' (1 intr. 48), 'aguale' (2 concl. 14) and 'mazzerare' (4.3.28).[13]

However, Ruscelli would have felt that his superficial work on the establishment of the text was less important than the information which his notes and glossary offered on linguistic usage. Two novelties here were his advice on the pronunciation of open and closed *e* and *o* and on where to place stress accents, and his use of southern synonyms in his definitions alongside the northern and Tuscan ones given by previous editors.[14]

Dolce replied to Ruscelli's accusations in a letter appended to the second edition of his *Osservationi nella volgar lingua* (1552) but then banned in 1553.[15] He maintained that he was right to exclude words such as *stea* and *amenduni* as spurious, though he had not wished to imply that archaic words should never be used in his own day, and he attacked Ruscelli for errors of judgement and of language in his annotations and text.

Unannotated editions of Petrarch on the Aldine model continued to be published. In 1547 Giolito decided to complement his existing series of Petrarchs in quarto with a new series in smaller format, rather as he had done with Ariosto. The duodecimo volumes which began to appear then were edited by Dolce and simply reproduced the Aldine edition of the previous year; both these types of Petrarch were, therefore, very derivative, as was the edition with Gesualdo's commentary which he added to his catalogue in 1553.

A not very effective challenge to the 1501 Aldine text came in 1549 in a Petrarch printed by Valgrisi and edited by Giovanni Antonio Clario of Eboli under the pseudonym Apollonio Campano.[16] He had thought, he wrote in his dedication, that the Aldine text needed no correction and that in any case he would have no time to carry out corrections. In the event, he found that his first supposition was partly wrong and that he had time to compile some 'Annotationi intorno la correttione di questo Petrarcha' ('Annotations on the correction of this Petrarch'). These merely made a few minor spelling changes and corrected (using as justification the sense, the rhyme, or the evidence of old and modern texts) a handful of errors which had in any case been put right in later Aldines. Clario divided the *Canzoniere* as Aldo had done in 1514.

In 1546 Francesco Sansovino began an attempt to create an edition of Petrarch modelled on Paolo Manuzio's of 1533, but with rather fuller concluding notes. As well as mentioning a few variants and pointing out some classical allusions, he helped the less experienced reader by explaining points of usage such as the different senses of *ne*, the inadmissibility of the *-emo* ending in *amemo* as opposed to that in *semo* or *havemo*, the meaning of 'per innanzi' (*Canz.* 23.68) which 'Nicolò D.' (presumably Delfino) wrongly explained as 'in future' instead of 'previously', or the difference between *chiunque* and *qualunque*. But after poem 33 Sansovino simply reproduced Manuzio's notes. This edition was printed by the heirs of Pietro Ravani in August, and it may be that Sansovino wanted to devote his time to the potentially more prestigious edition of the *Decameron* for Giolito: the dedication of the latter work dates from the last day of the same month.

Even if Sansovino had continued his annotations, they would have made Petrarch only slightly more accessible than before. It was not until 1548 that an edition was designed for the reader with little or no experience of older or more literary Italian. The editor was Antonio Brucioli, and the work came from the press which he had set up with his two brothers in 1541. Using the Aldine text and following the pattern of his own *Decameron*, he gave after each poem in the *Canzoniere* a summary of its meaning and a brief, straightforward explanation of some of its vocabulary, including words which were by no means obscure: in the opening sonnet, for instance, he glossed 'sovente' ('often') with 'spesso', 'sparse' ('scattered') with 'diffuse'. He also gave an index of the words and proverbs which he had explained.

In 1550 Alunno brought out an edition which included a much augmented version of his *Osservationi*. There were more entries than in 1539, and each entry contained a definition and a quotation showing its use in context. Alunno also gave some advice on when to use alternative spellings or forms, such as *tosco* or *thosco* (meaning 'poison' and 'Tuscan' respectively), *unquanco* or *unquanche* (the former used in verse, the latter in prose).

In 1553–4 Dolce set to work once more to provide a new type of Petrarch for Giolito. The texts were 'naked and without commentaries' ('ignudi e senza appostille'), in the Aldine arrangement and with the 1514 appendix. But Dolce also gave the reader every imaginable assistance in understanding and imitating the poems. The latter function was stressed in two letters signed by Giolito: as one said, it was better to use a living language than dead ones, and these things would help to achieve 'perfection in vernacular matters' ('perfettione nelle volgari cose'). There was a life of Petrarch and, from the edition of 1554, some notes by Giulio Camillo which explained Petrarch's intentions or pointed to parallel passages in other authors. But the bulk of the ancillary material consisted of 'indexes' ('indici') which covered many aspects of style and language, such as 'concepts' ('Allegrezza', 'Amore', 'Anima', 'Anni' and so on), comparisons, pairs of opposites (such as 'breve' and 'lungo') and metaphors, fine expressions, epithets, a concordance in which most words had a definition, and a rhyming dictionary. Dolce had nothing to say on the text here, but he gave a few comments on it in his *Osservationi nella volgar lingua*, a grammar which first appeared in 1550 and which he continued to revise for publication by Giolito. In Book 1 he claimed to have seen, incorporated in Petrarch's own manuscript of the *Canzoniere* which Bembo had used in 1501 and later owned, two corrections of 'in la' to 'ne la', similar to those proposed by Bembo in the 1538 edition of the *Prose della volgar lingua* (3.58). Dolce had introduced these readings in his text because, according to him, that was how Petrarch had left them. He was quite right, on the other hand, to correct Bembo's slip of the pen, 'habito gentile' (228.10), to 'habito celeste' both on the authority of Petrarch's manuscript and 'as necessity requires' ('come la necessità lo ricerca'), since 'celeste' was in rhyme.[17] He took the opportunity to take Vellutello to task (without naming him) for keeping 'gentile' and, in order to preserve the rhyme scheme, for reading 'sembiante humile' (a reading which Dolce quotes himself near the end of Book 1) instead of 'preghiere honeste' in line 13. For once, Dolce was completely justified in using his correction as the occasion to reflect on the need to emend printing errors 'with the authority of the correct exemplars' ('con l'autorità de' corretti essemplari') and on how rash it was 'to change words at one's whim, thinking that only what we like is right' ('il mutar le parole a sua voglia, stimando che tanto solamente stia bene, quanto piace a noi'). This, he said, was not the method of Barbaro, Poliziano, Aldo, Vettori (as editor of Cicero) and Carlo Sigonio (as editor of the Aldine Livy of 1555). At least Dolce was aware of the ideals of the best classical scholarship, and he himself never perpetrated any radical revisions of the sort of which some contemporaries were capable. However, because of his lack of awareness of the historical development of the vernacular, the texts which he edited were far from being exact reproductions of the author's intentions, so that he was in effect practising the opposite of what he preached.

In 1554 it was the turn of Ruscelli to give his attention to Petrarch. First of all, he challenged Giolito and Domenichi by revising their version of Vellutello's edition, now ten years old. This Petrarch was printed by Giovanni Griffio with a preface by Ruscelli, dated 8 March, which suggested that both the poems and their commentaries needed linguistic revision: a revision which was made, of course, according to his new principles of punctuation and spelling.[18] About seven months later (the dedication is dated 21 October) Pietrasanta printed another, more original edition by Ruscelli.[19] Again, much stress was laid on the 'perfect spelling' ('perfetta ortografia') which it used with the justification (as in the Boccaccio) that spelling was still in its infancy in the Trecento. Ruscelli explained in detail how he used the grave accent, the apostrophe, *h*, capital letters for personifications, *baciare* not *basciare*, and single *z* for a voiced consonant against *zz* for a voiceless one (as in *mezo* and *pezzo*). The text was followed by a wealth of supplementary material from various sources: Vellutello's biographies of Petrarch and Laura, Aldo's 1514 appendix and letter, a rhyming dictionary compiled by Lanfranco Parmegiano, a list of epithets, and a glossary (by Ruscelli) of names and words with digressions on spellings, etymologies and regional usage. Such notes were no doubt informative and helpful for many readers, but Ruscelli stated that his aim in recording variations such as Roman *annà* for *andate* was to show how Tuscan deserved supremacy because of the way in which Tuscan writers had cultivated their language. As for the details of the text, he showed that he was aware that a manuscript judged to be the best available was used in the preparation of the 1501 Aldine, but he doubted that this manuscript was autograph and found its ordering confused. He therefore followed, once more, Vellutello's arrangement of the *Canzoniere*, with corrections which went beyond those made in the Griffio edition. Many of these were orthographical, but some implemented Ruscelli's interpretation of the rules of Tuscan phonology. He preferred Vellutello's 'dolcior' to the Aldine 'dolzor' at 191.13; unaware that the latter was a Provençal spelling, he explained in his glossary that it reflected the pronunciation of the area north of Tuscany and the Marche. He refrained from correcting 'fia chi nol schifi' (105.41) but pointed out in his notes on spelling that it was a most manifest printing error which created a most manifest linguistic error (*il* before *s* + consonant); 'nol' should really be 'non'.

This was the only edition of a major author by Ruscelli not to be reprinted. In contrast, Dolce's more compact, less idiosyncratic new edition, which could have been occasioned by prior knowledge that Ruscelli was working on Petrarch, continued to be published by Giolito.

One classic had not so far been printed by Gabriele Giolito or edited by Dolce: Dante's *Commedia*. They jointly made good this omission in 1555 with an edition which added the adjective 'Divina' to the title for the first time.[20] A radically reformed text seemed to be promised by the title page, which claimed that the *Commedia* had been revised with the help of 'many very ancient copies' ('molti antichissimi esemplari'); Dolce's dedication said that he had used a copy of a copy written by none other than one of the poet's sons. For the most part, his text stayed close to the 1515 Aldine, but he does seem to have consulted some unpublished material, including Boccaccio's commentary on the *Inferno*, as well as the texts of Landino and Vellutello.[21] Dolce was much more original in his presentation of help for the reader. His dedication sought to show that there was a place for Dante's learning and majesty ('dottrina' and 'maestà') alongside the Petrarchan tradition of beauty ('vaghezza'), just as (in the terms of the contemporary debate on the qualities of Florentine and Venetian painting) *disegno* had its place alongside *colore*. He used his experience as an editor, and his observations of the techniques used by other editors, in order to devise an alternative to the old-fashioned commentaries of Landino and Vellutello. The canti were preceded by short summaries, as in Berardi's Dante of 1477 or (a more recent model) Domenichi's *Innamorato* of 1545. They were concluded by *allegorie* comparable with those which Dolce had written for the *Furioso* or those which Sansovino had introduced into his *Decameron*. The margins were used, as in his arch-rival Ruscelli's *Decameron*, for explanatory notes (though Dolce's were only brief) and for a few variants; and two indexes, one of words, the other of subjects, referred to these notes. In this way the *Commedia* was brought into line with recent Venetian editions of other classics.

The by now dominant trio of Dolce, Ruscelli and Sansovino also turned their attention in the 1550s to the two most imitated modern poets, Sannazaro and Ariosto. The language of the Neapolitan's *Arcadia* underwent in 1559 a revision by Sansovino (for the printer Rampazetto) which left unaffected strictly orthographical matters, such as the use of *h*, in contrast with Ruscelli's practice, but which altered points of phonology, morphology and syntax. On the other hand, some genuine errors were corrected, presumably with the use of an edition from the first half of the century. In this respect, Sansovino continued to prove himself a relatively conscientious editor; and he diligently provided the reader with ancillary material, including a glossary of Latinisms and southern forms.[22]

After Paolo Manuzio's *Orlando furioso* of 1545, Giolito continued to introduce novelties which would attract purchasers to his successive editions. First, for his 1546 edition he managed to obtain from Virginio Ariosto another rejected *giunta*, eighty-four stanzas dealing with the history of Italy. Then in 1548 he added the *Cinque canti* in a version which differed in some details from that printed by Manuzio.[23] In 1552, however, there was a change of strategy. Giolito and Dolce omitted some of the ancillary material and shifted attention onto the text. Over 500 words had been emended according to the author's original, the title page announced. More details were given in a letter to the readers, almost certainly written by Dolce. Previous Giolito editions had been revised by different people, it said (a statement which confirms that Dolce had left the close scrutiny of the text to others such as Merato). The text had therefore come to diverge from the Ferrarese edition of 1532. But now it had been restored to the state in which Ariosto left it, so that it contained once more forms such as the Latinizing *populo* or *suave* which had been Tuscanized by previous editors. However, as in the Dante of 1555, these promises of textual accuracy were hardly fulfilled. A few spellings from the 1532 text were indeed introduced, but many others were not (500 changes did not go far in a poem of this length) and, inconsistently, some changes took the Giolito text even further from the original.[24] On the other hand, the letter to the readers stressed that there was no intention of revising lines that went against Bembo's rules, such as 'che 'l sciocco vulgo non *gli* vuol dar fede' (7.1.5), where (it was claimed) *lo* and *lor* would have been grammatically correct. Others had had the audacity to change them (we shall see shortly what was meant by this), but these things were trivial and in any case poets had licence for them.

Giovanni Andrea Valvassori's first *Furioso* of 1549 had a new kind of *giunta*: a canto by Nicolò Eugenico which continued from where Ariosto's poem ended. In 1553 and again in 1554 the same printer included some old material, from Fausto and from the Florentine edition of 1544, and some new: a preface and allegories by Clemente Valvassori, who emphasized the morality of the work and the lessons to be learned from 'this great Christian poet' ('questo gran Poeta Christiano').[25]

Ruscelli completed his trilogy of editions of major classics when his *Furioso* was published by Valgrisi in 1556.[26] This edition had many of the ingredients familiar from his *Decameron* of 1552 and his second Petrarch of 1554 and took some of their tendencies to further extremes. Firstly, Ruscelli declared himself even more anxious than before to help his readers to understand the work. He has omitted nothing, he tells them, which could be wanted 'by a person without letters, by a beginner, by a person whose understanding and learning is average and even above average and high' ('da persona senza lettere, da principiante, da mezanamente, e ancor da sopra il mediocre et il molto intendente e dotta'). His own contributions included plot summaries in prose, interpretative notes after each canto and at the end of the work, and a glossary of words which might be obscure to 'those who do not know Latin or Tuscan literature' ('quei che non sanno lettere latine ò toscane'). He explained the principles on which the woodcuts for each canto were designed. As in the Boccaccio and Petrarch editions, he also borrowed contributions from other people. From the *Romanzi* of Giovan Battista Pigna he drew a biography of Ariosto and a study of a hundred revisions made by the poet; each canto began, for the first time, with a verse *argomento* by Scipione Ammirato; there was a list of allusions to classical myths compiled by Nicolò Eugenico; and some notes were borrowed from the *Spositione* by Fórnari which will be mentioned in the next chapter. Ruscelli also used and expanded Domenico Tullio Fausto's list of debts to Homer and Virgil and reproduced Dolce's list of imitations almost without change, acknowledging its usefulness but not naming its author.[27] On the other hand, Ruscelli, unlike other editors, did not consider the *Cinque canti* worthy of inclusion.

The second part of Ruscelli's task was to revise his text as far as he dared. Before the start of the poem, the reader came to an account of the 'just and regulated spelling method' ('dritto et regolato modo di scrivere') which had been used. But Ruscelli could not resist venturing beyond spelling changes: he wanted to correct grammatical 'mistakes', or at least draw attention to them, and he had to give his opinions on questions of moral propriety. This was partly because of personal inclination. But he had an added motive for wanting to regularize the *Furioso*. His *Tre discorsi* of 1553 had pilloried Dolce mercilessly for making various linguistic blunders in his translation of Ovid's *Metamorphoses* and in his grammatical *Osservationi*. Ruscelli was preparing his own grammar, the *Commentarii della lingua italiana* eventually published post-humously in 1581. But some of the forms used by Dolce and condemned by him occurred in Ariosto's poem: *il* before *s* + consonant, for instance (as Dolce had pointed out in his 1552 edition), or *messe* (rather than *mise*) in the past historic of *mettere*, which was moreover used in rhyme. Ruscelli therefore wanted to find some way of preventing Dolce from appealing, as grammarian and poet, to the authority of the text of the *Furioso*.

In the *Tre discorsi* he had not dealt with this problem convincingly. On the one hand, he had approached it from the theoretical assumption that all authors, including Ariosto, wanted to write like Petrarch. If one found errors in the *Furioso*, they must therefore be caused by copyists or printers. He said that, in the annotations on Ariosto which he was preparing, he had corrected those errors which could be emended without spoiling the verse or the meaning ('senza guastar verso né sentenza'). He had earlier given the example of 'gl'incudi'; this noun was feminine, and Ariosto could only have written 'l'incudi'. The masculine article was alleged to be a printing error due mainly to editors. Dolce had seen these notes, in the days when he and Ruscelli were still on speaking terms. Hence the remark in the letter in the 1552 edition signed by Giolito, which Ruscelli naturally attributed to Dolce: though Ruscelli pointed out that Dolce had misremembered what he had said about the line 'che 'l sciocco vulgo non gli vuol dar fede', and that Ariosto's *gli* was perfectly correct in the context. On the other hand, Ruscelli then had to admit that Ariosto did occasionally use *messe*; but this was not to be imitated because it went 'against reason, rules and all good authors' ('contra le ragioni, contra le regole, et contra tutti i buoni Autori').[28] In 1553, then, Ruscelli was trying to blame some of the 'irregular' usage of the *Furioso* on its transmission in print, but he still had to admit that some was due to the author: an uncomfortable situation when he wanted to promote Ariosto as a model.

How, then, could Ruscelli justify editorial corrections which went against the evidence of the 1532 text? In editing a work of this status, he had to appear faithful to the author's wishes. He could not change the text freely, as if he were dealing with the *Orlando innamorato*. Yet he could not claim that widespread corruptions had crept in, as if this were a Trecento work. By 1556 Ruscelli had found a way out of his dilemma. He hit on a solution which would legitimize his interventions and, as a bonus, show that Ariosto reached, before his death in 1533, a linguistic position which was closer to the Petrarchan norm. According to Ruscelli, the poet's late brother Galasso had, in 1543, shown him a copy of the 1532 edition which had been revised by the author with a view to producing a corrected edition. This story could have a basis in truth, though if so it is curious that it was not mentioned in the *Tre discorsi*.[29] Details of the alleged 'changes and improvements' ('mutationi et miglioramenti') were given by Ruscelli in a separate section of his edition which was in effect an account both of the emendations introduced by him and of his reservations about the poem. Ariosto, he said, had not altered the use of accents or apostrophes, but the poet had made a few changes to punctuation, capitalization and spelling (removing *h* from *th* of Greek origin). He had also introduced a number of linguistic corrections, concerning for instance the use of diphthongs and of verb forms of which Ruscelli disapproved as un-Petrarchan, such as *veniro* or *debbe*, and he had indicated other points which for metric reasons needed more complex revision, such as *pel* (for *per lo*) or the apocope in 'mirabil cose'. Earlier, in a note on 'chi vive amando il sa, senza ch'io 'l scriva' (36.44), Ruscelli gave an example of how this sort of rewriting might be done, when he claimed that Ariosto had written four alternative versions in the margin, all avoiding the use of *il* before *s* + consonant. Of all these corrected passages, Ruscelli said, he had changed those which were clearly indicated but had left others as they stood. And, as in the *Decameron*, his criticisms were not restricted to language and style. A note on Canto 2.52, for instance, indicated his disapproval of 'rinculò', the word he would most have liked to remove from the poem. He claimed, too, that Ariosto had crossed out the story of Anselmo from Canto 43. Not even Ruscelli, then, dared introduce any radical corrections in the *Furioso*, and he contented himself with introducing formal changes and with giving his examples of Ariosto's last-minute repentance in matters of style and content.

Giolito did not print another *Furioso* until 1559. In the edition of this year he included a new letter to the readers which offered no more than vague prospects of a response to Ruscelli's edition. He hoped, he said, to give them the poem shortly in another quarto edition but 'with new annotations and comments' ('con nuove annotationi e comenti'), and also in folio format, 'so that all kinds of men may receive pleasure and usefulness from it, and there may not be in this happy work any place or word which is not fully understood by all' ('accioché ogni qualità di huomini ne riceva diletto et utile, e non sia in questa felice opera luogo né parola alcuna che non venga pienamente da tutti intesa'). Giolito's aim of bringing understanding to the whole spectrum of readers is more explicit here than ever, though one should note that he did not forget the market for expensive folio editions as well as those who were less well educated. But he never followed up these ambitious proposals. The long reign of the Giolito–Dolce editions was now in its twilight, and it finally ended in 1560.

Given the eminence of the *Furioso* of 1532 as a linguistic model, editors felt obliged to revise the works which Ariosto had written before his views on linguistic usage had moved closer to those of Bembo. In the mid 1540s, as we saw earlier, Virginio Ariosto responded to the vogue for his father's verse by making new manuscripts available for publication. Giolito received the previously unpublished verse redactions of *La Cassaria* and *I Suppositi*, the unpublished second version of *Il Negromante*, and a version of *La Lena* which differed in some respects from that printed in 1535. The *Satire* had been printed clandestinely in 1534; now Giolito obtained a manuscript with corrections by the author. The *Cassaria* was revised anonymously for printing in 1546 according to Bembo's rules, so that, for instance, *lo* (not *il*) was used before a vowel or *s* + consonant, nouns and adjectives in *-e* had their plural in *-i*, *in* combined with the definite article to give *nel* and so on (a correction which involved some rewriting in order to keep the correct number of syllables), and *tosto* was preferred to *presto*. The other three plays appeared early in 1551 (*Il Negromante* and *La Lena* being dedicated by Dolce, as they had been in 1535): they too had been linguistically revised, though to a less consistent standard than in the case of *La Cassaria*.[30] The revision of the *Satire* was carried out in 1550 by Anton Francesco Doni, along lines very similar to those used in the *Cassaria*. But Doni also felt free to try to improve on some of the corrections which Ariosto had introduced into his own text.[31]

Ruscelli decided to have his own version of the *Satire* printed in 1554 by

Pietrasanta. Competition with Giolito's edition led him to claim that he was
using no fewer than three manuscripts which he happened to have obtained in
Rome. In fact, he created a text different from Giolito's by returning to the
earlier redaction of the *Satire* found in the pre-1550 printings and adding
corrections based on his grammatical rules. He justified these corrections with
reference sometimes to his manuscripts and sometimes to rules which had been
transgressed either by copyists or by Ariosto; but, if errors such as *in la* for *nella*
were Ariosto's, they were (he said) made at an early stage in his career and he
would have removed them from the *Satire* just as he did from the *Furioso*.[32]

In 1557 Giolito had the idea of printing Ariosto's *Rime* together with the
Satire. The editor was Dolce, though he was named only in the second edition
of 1558.[33] In the preface, signed by Giolito, he drew a distinction between, on
the one hand, Ariosto's *Furioso* of 1532 and his *Satire*, a 'work corrected by him'
('opera da lui corretta') and, on the other hand, the earlier *Rime*. In these,
Dolce had not changed anything, he told readers, 'as some do' ('come alcuni
fanno'); 'hence if you find in various places that rules are not observed which
Ariosto afterwards observed fully in the last edition of the *Furioso*, it will be
because we have not departed from the author's exemplar, avoiding presumption' ('onde se troverete in diversi luoghi non essere osservate le regole che
dipoi l'Ariosto osservò pienamente nell'ultima editione del *Furioso*, sarà
perché noi non ci siamo discostati dall'esemplare dell'Autore, fuggendo la
presontione'). All this, of course, was meant as a reply to Ruscelli. The epic,
Dolce was saying, did not need the sort of revision which Ruscelli had
dreamed up, and in any case Dolce's policy of non-intervention was superior
because it respected the author's wishes.

With authors whose status did not match that of Ariosto, Ruscelli did not
trouble to create an elaborate scenario to justify his corrections. In 1552 he
edited for Giovanni Maria Bonelli the *Compendio dell'historie del Regno di Napoli*
which Pandolfo Collenuccio of Pesaro had begun in 1498 and left unfinished
on his death in 1504: a work, then, written by a non-Tuscan and before the
spread of Bembo's influence.[34] Michele Tramezzino had first printed this
work in Venice in 1539 and had seen no need to introduce changes in his three
subsequent printings. Ruscelli, though, was horrified by errors in its language
and contents, and set about revising it in such a way that the title page could
claim that the *Compendio* had been 'newly adapted to the purity of the
vernacular and completely emended' ('nuovamente alla sincerità della lingua
volgare ridotto e tutto emendato'). Although he made a hasty gesture towards
the exegetic side of editorial work by compiling a list of noble families of the
Kingdom of Naples (he was always devoted to this part of Italy and, rather
snobbishly, to its upper class), his main efforts went into revising Collenuccio's
language. Ruscelli's concluding discourse on his revisions made no pretence
that he was restoring the work to its original state; like Domenichi in the

Innamorato of 1545, he stated explicitly that his version differed from what the author wrote. The *Compendio* had been written, he said, before the rise of Tuscan, in an age when writers still used a 'courtly language' ('lingua ... cortegiana') with a greater or lesser admixture of Tuscan. Collenuccio wrote it just as we have it ('pur così come l'habbiamo'), though many of the errors were due to printers who did not scruple to put profit before diligence. Now, however, Ruscelli wanted to remove the work from that usage, 'our most beautiful vernacular having already gained such authority that one can say that, as regards the people of the present, it is supreme' ('havendo la bellissima lingua nostra volgare già pigliato tanta autorità che si può dire che in quanto a i presenti ella sia in istato'). His pride in the modern vernacular made him wish that 'all good authors' ('tutti gli autori buoni') should be available to be read after being 'corrected and made regular' ('corretti e regolati') according to the now accepted Tuscan standard. 'I therefore wanted,' he went on, 'to adapt this author to the purity of that language as regards its rules and observations and idioms and to rational spelling' ('ho voluto ridurre questo autore alla purità di essa lingua in quanto alle regole et osservationi e modi suoi et alla ragione della scrittura'). He did make some concessions, however, preserving some military words and 'some others of the modern and common ones of Italy' ('qualch'altra delle moderne e communi d'Italia'), and he drew the line at correcting factual errors.

In the first of the six books, the result was a thorough rewriting which covered all aspects of the earlier editions from punctuation and spelling to syntax and choice of vocabulary (lexical Latinisms were one of Ruscelli's main targets), and in which words or phrases were added or deleted as he saw fit.[35] In the remaining books the free manipulation of the text ceased, but he maintained his meticulous scrutiny of the language, also correcting a few genuine errors. Accentuation and punctuation became much more helpful. Latinisms in spelling and phonology were eradicated.[36] Morphology was strictly regulated, especially for definite articles, pronouns and verbs. Typical changes were 'el' > 'il', 'de li primi' > 'de' primi', 'li beni' > 'i beni', 'li anni' > 'gli anni'; 'loro' (subject) > 'essi', 'li' ('to them') > 'loro', 'quale' (relative) > 'il quale'; in the past historic, the *-ette* type was eliminated, *-orono* became *-arono*, and in the third person plural of strong verbs *-ono* and *-eno* normally gave way to *-ero*; in the imperfect subjunctive *fusse* was changed to *fosse*. This campaign against what Ruscelli considered to be impurities, whether of classicizing or regional origin, was motivated by his passion for what his dedication called 'the purity of our most beautiful vernacular' ('[la] sincerità della bellissima lingua nostra volgare'). It was a campaign waged, of course, at the expense of the authentic text; yet it was just this sort of reworking which helped to diffuse a standardized Italian literary language.

In 1554 Ruscelli edited Machiavelli's comedy *Mandragola* as part of a collection of *Commedie elette* to be printed by Pietrasanta. Ruscelli could not allow Machiavelli's contemporary Florentine forms to stand: pronouns, verbs and certain expressions were therefore revised.[37] But, for Ruscelli, a Florentine such as Machiavelli could not have made these 'mistakes' himself. To justify his corrections, Ruscelli therefore fell back on the traditional device of saying that the work had been corrupted by earlier printers. Machiavelli's language was really, he said, pure Tuscan like that of Boccaccio (just as he had found it hard to believe that Ariosto had not wanted to write like Petrarch). But the editions of Machiavelli's *Discorsi* and of other works which Blado had printed in Rome introduced errors, probably through the ignorance of those responsible. Ruscelli added some explanatory notes on the play (on 'san Cucù', for example), but typically he also criticized it: he felt that Machiavelli should have constructed it better, with a crescendo of difficulties to be resolved building up to the end of Act 4.[38]

One of the authors whom Ruscelli criticized for using the 'lingua cortigiana' was Castiglione in the early drafts of the *Libro del cortegiano*. This work had been one of the mainstays of the Aldine press, which followed its first edition of 1528 with five more up to 1547. In the dedication of the second Aldine, an octavo printed in 1533, Francesco Torresani claimed that this edition was more correct than the first, 'according to the author's autograph manuscript' ('secondo l'essemplare iscritto di mano propia d'esso Autore'), and the manuscript used in 1528 could well have been still available to the printers. The 1541 edition Tuscanized Castiglione's phonology slightly (introducing some *uo* diphthongs and correcting some double consonants, for instance). Gabriele Giolito had begun to publish the book in 1541 and offered rival editions whose title pages, characteristically, announced some added attraction. In the five editions from 1541 to 1551 this was nothing more than a new index and the assurance that the text had been diligently revised. Paolo Manuzio responded in 1547 with an appendix giving a brief list of qualities appropriate to courtiers and the claim (advertised on the title page) that the work had been 'compared with the author's manuscript' ('rincontrato con l'originale scritto di mano de l'auttore'): but this was probably only an echo from the earlier Aldines, since by now the manuscript had probably been acquired by Jean Grolier. Then in 1552 Giolito and the ever-industrious Dolce used the same ploy as in his *Furioso* of that year: their edition too had now been revised 'according to the author's copy' ('secondo l'esemplare del proprio autore'). Dolce warned readers not to be surprised to find inconsistency in spelling or in definite article forms: he judged it presumptuous and rash to depart from the author's wishes. But he must have based his claim on the title page of the 1547 Aldine, for a comparison of some chapters (1.33–6) suggests that he used the post-1541 Aldine text, taking it slightly further from the 1528 printing with changes such as 'li quali' > 'i quali' and 'bone' > 'buone', although he did restore at least one original form, 'suco' in place of 'succo'. This version continued to be published by Giolito until 1562, with slight modifications (such as 'ingeniose' > 'ingegnose', 'ponessino' > 'ponessero') and with the addition of some marginal *notabilia* in 1556 and of summaries (*argomenti*) for each book in 1559. Dolce's notes and selection of topics for his index tended to impose on the text his own conservative views, at the expense of the balance which Castiglione's dialogue had achieved.[39]

At least two Florentine editors (not counting Sansovino) were working in Venice in this period. One, we have seen, was Doni, who had returned to the city in 1548 after his unhappy experience running a press in Florence. In 1553 he edited the comic *Rime* of Burchiello, a Florentine barber who lived in the first half of the Quattrocento. The text of the poems appears to have been taken from a Florentine edition of the previous year, but he changed their order and provided a commentary which parodied those on Petrarch.[40]

Another Florentine who often worked as a translator or editor for Giolito and other printers was Remigio Nannini, a Dominican friar.[41] In 1559 his edition of Giovanni Villani's *Cronica* was printed in Venice on behalf of the Giunta in Florence. There will be more to say on this in the next chapter: for the moment, it will suffice to mention that Nannini treated his source, the Venetian printing of 1537, as the editor of that volume (with the advice of the Florentine Brucioli) had treated his own manuscript source, in other words with the conviction that an old version of a text should be respected rather than restored. This does not appear to have been a passing whim of Nannini. Even more unusually, given the tendency of his colleagues to impose rules and rationality in works outside the literary canon, he was fairly scrupulous in following the language of the fifteenth-century northern writer Antonio Cornazzano, some of whose works (as was seen in chapters 5 and 6) had already been revised by editors in the 1520s. In Nannini's dedication of Cornazzano's verse *Vita di Pietro Avogadro*, printed for the first time by Francesco Portonari in 1560, the friar admitted that a modern writer would perhaps have written more elegantly, 'since our language has today risen to much excellence and perfection' ('essendo hoggi la lingua nostra salita in molta eccellenza e perfettione'); but Cornazzano did well to write in the vernacular at all, since in his day 'it was only crude and uncultivated' ('non era però se non roza e mal tenuta'). 'I wished, therefore,' wrote Nannini, 'not to touch [Cornazzano's language], but to leave it in that simplicity and purity [in] which it was made by its author, changing neither its manner of writing nor words nor terms, and I have had it printed according to the old exemplar' ('Non l'ho voluta adunque toccare, ma lasciarla in quella semplicità e purità, ch'ella fu fatta dal proprio autore, non mutando né modo di scrivere né parole né voci, e l'ho fatta imprimere secondo l'essemplare antico').[42]

Nannini's decision to replicate the language of his manuscript source was exceptional. The predominant approach did not set such store by authenticity and seemed to presuppose that editors were now more important than the author. In dealing with the most respected authors they could ignore Robortello's 'bona fides', creating a smokescreen of spurious claims about their sources behind which they actually went ever further from the original. With other texts, they had the power to declare openly that they were imposing the standards of correctness which were current in mid-Cinquecento Venice. They could surround their text with so many trappings that the volume could become dominated by their own views and tastes, even by their personality. There was of course an element of self-aggrandizement in this. To be fair, though, one must remember the commercial pressure under which editors were working. Given the choice, the book trade preferred to publish, and most readers would have preferred to buy, a text advertised as 'alla sincerità della lingua volgare ridotto': in the case of Collenuccio's *Compendio*, Ruscelli's version immediately supplanted Tramezzino's and was reproduced for the next two centuries. One must also remember that editors and grammarians such as Dolce and Ruscelli saw themselves as champions of the vernacular against Latin. Supporters of the classical language argued that the vernacular had, among its defects, a lack of 'regularity' and a relatively small corpus of literature. Editors of the vernacular in the Cinquecento could claim that an effective way of promoting the new language was to extend to an ever-wider range of works a new regularity based on Tuscan rather than on any other regional variety or on the heritage of Latin.

9 · IN SEARCH OF A CULTURAL IDENTITY: EDITING IN FLORENCE, 1531–1560

WHILE VENETIAN PRESSES EXPANDED AND FLOURISHED in the middle third of the Cinquecento, printing in Florence was undergoing a crisis which saw the output of books and the quality of editing fall far below that of the 1520s. The roots of this crisis lay in the turbulence of the political and intellectual life of the city. In 1527 the Medici had been expelled and a new popular government had been formed. The ensuing conflict with the Pope (Clement VII, a member of the Medici family) and the Emperor Charles V had come to a climax in the heroic but unsuccessful defence of Florence against imperial forces in 1530. There followed the establishment of the Medicean duchy in 1532, the assassination of Duke Alessandro in January 1537, and the final defeat of the anti-Medicean rebels at Montemurlo later in the same year. These upheavals caused a haemorrhage of talent, one of the beneficiaries being Venice. Among the exiles who sought refuge there either briefly or for longer periods were four notable writers, Brucioli, Iacopo Nardi, Donato Giannotti and Benedetto Varchi. The Giunta press suffered badly in these years. In 1533 it suspended operations in Florence. Bernardo went to Venice to join his uncle Lucantonio, and Antonio Francini, one of his main classical editors, also left to work for the Venetian branch of the family. Only in 1537 did the Giunta presses in Florence begin to work again, but on a much reduced scale.

The 1540s brought a revival of intellectual activity. The reopening of the Studio of Pisa in 1543 attracted leading teachers to Tuscany and reduced the departure of students to northern Italian cities. The Accademia degli Umidi, founded in 1540 as a response to the new Accademia degli Infiammati in Padua, was soon brought under the control of Duke Cosimo de' Medici. Its name was changed to the more formal Accademia fiorentina, and it became the main cultural organ of the Medicean state, fostering the study of the great Florentine writers, especially Dante and Petrarch, and the use of the Florentine vernacular. However, it was bedevilled by a problem endemic in the city: bitter factions caused by both political and intellectual differences between those involved. Ideologically close to Duke Cosimo and fiercely opposed to Bembo were Pierfrancesco Giambullari, Giambattista Gelli and Carlo Lenzoni, known as the Aramei because of their bizarre theories on the links between Aramaic and Tuscan. But a group faithful to the original ideals and spirit of the Umidi centred on the poet and playwright Anton Francesco Grazzini.[1]

The restoration of university teaching and the new academy (together with the absence of a press in Pisa) brought a corresponding revival of Florentine printing, even if the city never had entrepreneurs to compare with Giolito or Paolo Manuzio. Bernardo Giunta returned in the last few years of his life, printing in Florence from 1546 to 1550. He was succeeded by his two eldest sons, Filippo and Iacopo. This family, hitherto dominant in the field of scholarly publishing, faced competition from new presses, especially the very prolific one of Lorenzo Torrentino, a native of Brabant whom Cosimo appointed as ducal printer in 1547. However, Torrentino was dogged by financial problems which led to a decline in his fortunes and a corresponding rise in those of the Giunta from about the mid 1550s.[2]

Even though printing in Florence was emerging from the recession of the 1530s, there was still an enormous gulf between the quantity of books produced there and those produced by the booming presses of Venice. Florence was also on the periphery of the book trade. Book dealers went to Venice for most of their stock, wrote Bernardo Giunta's cousin Tommaso in 1542; it was inconvenient and uneconomical for them to obtain books printed in Florence.[3] The financial problems associated with exporting most of their output (perhaps between 960 and 975 books from a print run of 1,000) were one of the main arguments used by the Giunta brothers when they applied to Duke Cosimo for tax exemptions and other privileges in 1563.[4] Nor did Florentine books now have any advantage in editorial quality. In the 1520s a scholarly approach to the editing of pre-Petrarchan poetry and of Boccaccio's prose had been consolidated, but no editions of Due and Trecento works appeared in the following few decades to maintain this tradition. Still, some members of the Florentine Academy continued to study the text of Dante. In 1546 Varchi, his friend Luca Martini and three others collated seven manuscripts of the *Commedia*, of which two were particularly authoritative, transcribing their readings into a copy of the 1515 Aldine. Modern scholars have special reason to be grateful to Martini, for two years later he copied readings from a manuscript then in Pisa, but now lost, which dated from 1330–1 and is thus the oldest recorded copy of the work.[5] Although we do not know whether a new edition of the *Commedia* was being planned, these researches show that Florentines were keeping alive their distinctive tradition of careful collation of early vernacular sources. This continuity, parallel to and reinforced by the continuity between the work of Poliziano and that of a young scholar who was to be the outstanding Italian classicist of his century, Piero Vettori, was a vital precursor of the eventual renaissance of scholarly vernacular editing in Florence in the last forty years of the century.

We saw in chapter 6 that there were signs in the 1520s that younger Florentines were not following the older generation's attitude of intransigent hostility to Venetian views on the literary language. In the following decade there was a further rapprochement between scholars in Florence and those in Venice and the closely-linked city of Padua. A warm friendship between Bembo and a group of Florentines including Varchi and Vettori is reflected in an exchange of letters and sonnets from 1535 onwards.[6] When Varchi went to live in Venice and Padua from 1537, Vettori kept in touch with him by letter, obtaining news of the classical works which Paolo Manuzio and others were publishing.[7] Around Varchi there gathered a group of young Florentines, such as Ugolino Martelli, whom Bronzino portrayed holding a volume by Bembo. Varchi, too, played an important part in the founding in Padua of the Infiammati, and Martelli lectured there on a sonnet of Bembo. When men such as these returned to their native city, they had a formative influence on the Florentine Academy.[8]

However, in the field of classical editing there were inevitably still differences in methodology between Florence and Venice. Vettori believed (like Poliziano) that extreme caution should be used in emendation, and he took the view that Paolo Manuzio was too free in his use of conjecture. He also attached less importance than Manuzio to the explanation of the text. The controversy between them arose out of the edition of Cicero's works which Vettori completed for Lucantonio Giunta between 1535 and 1537. The first of its four volumes, printed in 1534, was an essentially Venetian product, edited by the late Andrea Navagero (who had died in 1529) and printed with the benefit of advice from Bembo on the typographical presentation.[9] The three volumes edited by Vettori were just as firmly rooted in Florence, in spite of their being printed in Venice. They were dedicated to a Florentine, and Vettori acknowledged the contribution of two other fellow-citizens: a young Latin teacher, Lodovico Bonaccorsi, who had helped with the task of collation, and Francini, who had supervised printing and proof-reading. Vettori reproduced the ninth- or tenth-century Medicean codex of Cicero's letters very conservatively, and the theme that returned insistently throughout the explanations of Vettori's corrections to the text was that one must have the utmost respect for the best sources. The textual critic's role was not to emend arbitrarily but to reconstruct carefully the original state of a text, using a wide knowledge of classical Latin vocabulary as well as of paleography and epigraphy. One had to keep archaic or unusual spellings (such as *tralatio* for *translatio*) if one was to keep the language of the author's time, not to make it conform with the rule and custom of posterity ('si auctoribus sermo, qui eorum aetate vigebat, conservandus est, et non ad posteriorum normam et consuetudinem conformandus': f. 9r).[10] He made clear his disapproval of editors who intervened too much. 'We wanted,' he said, 'to restore as far as possible what Cicero genuinely wrote. When we came across defective passages, it seemed to us safer to write them as we found them in old and approved copies rather than as they had been made good by others (not moderate enough, in my opinion, in dealing with the works of others), who took care that there should be no place from which some sense could not be elicited and thought that an editor's function was to let nothing lie still' ('Nostrum consilium fuit ... ut, quantum per nos fieri possit, sinceram Ciceronis scripturam restitueremus. Quod sicubi in locos mancos incidimus, tutius nobis visum est, eos, ut in antiquis et probatis exemplaribus invenimus scribere, quam ut a quibusdam (mea quidem sententia, parum in alienis scriptis modestis) suppleti sunt, qui operam dabant ut nullus locus esset e quo non aliquis sensus eliceretur, putabantque hoc esse officium correctoris: nihil iacere pati': f. 3v). An example is his preservation of the reading 'Non habui cui potius id negotii darem, quam * darem' (*Ep. ad fam.*, 5.20.2), with an asterisk marking a supposed lacuna in the Mediceus, where others had 'quam Tullio scribae'. 'We thought it more correct,' he commented, 'here and in many other places to leave many things which were mutilated and imperfect, as they are found in old copies, rather than to read them as they are made good and restored by some' ('nos autem et hic et aliis multis locis rectius esse putavimus mutila et imperfecta multa relinquere, ut in antiquis exemplaribus inveniuntur, quam, ut a nonnullis suppleta et restituta sunt, legere': f. 8v).

In 1540, however, as was seen in chapter 7, Manuzio's new editions of Cicero's letters launched an attack on the scholarship of Vettori, to whom Manuzio referred as 'the Florentines', so that the whole city seemed to stand accused. A letter to Varchi of June 1540 shows that Vettori knew of one of these editions and had learned that Manuzio had been in Rome collecting emendations from scholars. Some of them, he wrote, would inevitably be good, but he warned of the danger of conjecture.[11] When he learned of Manuzio's open hostility, he composed a further series of annotations on the *Familiares* which were printed in Lyons in 1541. He reiterated his criticism of Manuzio's abuse of inventive conjecture ('la licentia di emendare di fantasia') in a letter of 1548 to a younger scholar who was to play the leading role in Florentine editing later in the century: Vincenzio Borghini.[12]

In the 1550s the most important Florentine Latin editions continued to be characterized by extreme respect for a single authoritative source: respect doubtless encouraged by a measure of patriotism, since the sources were then in Florence. Lelio Torelli, Duke Cosimo's legal adviser, and his son Francesco were responsible for the editing of Justinian's *Digesta* (or *Pandectae*) from the famous Medicean codex written about the year 600. This work was eventually printed in a magnificent folio edition by Torrentino in 1553. The title of the volume said, significantly, that the books had been literally 'represented from the Florentine Pandects' ('ex Florentinis Pandectis repraesentati'). It was explained to readers that the editors had not departed by a hairbreadth from the spelling of their source in any significant matter 'so that, as far as possible, its true appearance should be represented' ('ut, quoad fieri posset, eiusdem vera species repraesentaretur'). But it did not matter if there were a few uncertainties or errors in the original, for this edition was more than a diplomatic copy: it aimed to provide guidance to readers while still informing them of the original reading with scrupulous fidelity. Very basic punctuation was introduced (comma, semicolon, full stop, question mark), as was some sparing accentuation and capitalization. Unconventional spellings of the manuscript were indicated by a capital letter within a word: this stood for what would normally be a double letter (as 'celeRimo' for 'celerrimo'), a different letter (as 'popOlus' for 'populus'), or one syllable for two through haplography (as 'sENTia' for 'sententia'). Various signs were used to mark later glosses and words which probably needed to be removed or added, and asterisks indicated probable corruptions. The editors also provided a list of the few cases where they had emended the original and of variants in other old manuscripts.

Vettori attempted to follow in the footsteps of the Torelli, as he acknowledged, when he produced a new edition of Cicero's *Epistolae ad familiares* (Torrentino, 1558) intended as a 'simulacrum' of the Medicean codex of these letters, even though more recent manuscripts sometimes had preferable readings.[13] Since the Mediceus was the ancestor of the other manuscripts which Vettori had seen, he was loath to emend it. In these circumstances, he wrote, 'it is evident that its corruptions and mutilations can only be healed by conjecture: and it is established how slippery and fallacious is this means of emendation' ('patet quae in eo corrupta lacerave sunt, non posse nisi coniectura sanari: quae ratio emendandi quam lubrica fallaxque sit, exploratum est').[14] And Vettori's edition of Aristotle's *Poetics* (Giunta, 1560) shows that, when he did not have one authoritative manuscript, he was still wary about basing emendation on conjecture rather than on the evidence of old sources or of other authors.[15]

Apart from the use of emendation, another contentious issue for Florentine editors of vernacular texts was that of Bembo's views on the vernacular and in particular his argument that, if one wanted to write Florentine well, it was no longer a great advantage to be born Florentine (*Prose*, 1.16). The narrowly patriotic resentments of some Florentines, especially the Aramei, were exacerbated in 1543 when Cosimo de' Medici allowed Varchi, full of admiration for Bembo, to return to the city in return for his involvement in the Academy.[16] There was a fear that Florence's linguistic autonomy was under threat and that Florentines were abandoning control of their literary and linguistic heritage to Venice.

By the end of the 1540s some similarities with Venetian practice could certainly be seen in vernacular editing in Florence. These were encouraged in the first place by closer contacts between the Florentine presses and those of Venice. The branches of the Giunta family in the two cities had, as we have seen, an interchange of personnel and no doubt of ideas. After spending a restless period in north-eastern Italy and then in Piacenza and Venice, Anton

Francesco Doni returned to Florence in 1546 and managed a press for a couple of years. He was assisted by one of Venice's more experienced editors, Lodovico Domenichi, who then quarrelled with Doni and went on to work for Bernardo Giunta and for Torrentino. Doni felt strongly enough about the pernicious influence of editors who were 'foreign' ('forestieri', that is non-Florentine) to attack the damage they inflicted on texts and their lack of consistency: 'one editor corrects in one way and another otherwise, some delete, some insert, some flay [the text] and others damage its hide' ('un correttore corregge in un modo e quell'altro a un altro, chi lieva, chi pone, certi scorticano e certi altri intaccano la pelle'). One had to beware that works did not fall into the hands of 'stubborn editors, because they don't follow what is written but carry on in their own way' ('correttori testericci, perché non vanno secondo gli scritti, ma fanno a modo loro').[17]

A second factor was the dearth of Florentine editions. Most of the books bought in the city must have been printed elsewhere.[18] For the Trecento classics and the *Orlando furioso*, this almost inevitably meant Venetian books. Florentines would thus have become accustomed to the linguistic norms and models of presentation which were current in Venice, and this favoured a greater conformity with Venetian editorial practices in those books which were printed in Florence.

Such conformity was seen chiefly in linguistic revisions carried out before publication. These affected even works written by Tuscans in the sixteenth century. The 1532 Giunta edition of *Il principe* and other works by Machiavelli took the Roman edition of the same year as its source but substituted Machiavelli's forms with ones approved by Bembo.[19] The revision was carried further in 1540 with the elimination of some Latinizing vocabulary ('contennendo' > 'vile' or 'disprezato', 'incursioni' > 'scorrerie', and so on) and of popular Florentine forms (as in 'sirocchia' > 'sorella').[20]

In 1548 and 1549 Bernardo Giunta printed most of the works of a neglected writer of the generation after Machiavelli's, Agnolo Firenzuola (1493–1543). These were edited by Domenichi and by Lorenzo Scala, a Florentine who was one of Grazzini's associates.[21] All but one of the works were previously unpublished and they were in some cases fragmentary. Both Scala and Domenichi used metaphors found in earlier Giuntine editions when describing their editing: Scala that of healing wounds, Domenichi that of the restoration of damaged ancient statues. But, while the editor of the *Sonetti e canzoni* of 1527 had compared his task with cleaning the surface of these statues, Domenichi portrayed himself as a restorer replacing missing limbs, for he had to insert linking passages in order to present a coherent text. He did not identify his interpolations. The nature of his linguistic revisions can be gauged from the one work which had been printed before (in 1524), the *Discacciamento de le nuove lettere*: changes to it affected spelling, phonology and morphology but not Firenzuola's syntax or vocabulary.[22]

Members of the Aramei group were involved in 1550 in preparing for printing (by Torrentino) the work of a living Tuscan: the *Vite de' più eccellenti pittori, scultori e architettori* by Giorgio Vasari of Arezzo. The author took Giovio's advice about using an editor (see above, p. 13), for he explained in a concluding note that he had entrusted a person of judgement to revise the spelling ('ortografia'); he gave as examples the use of *z* instead of *t* or the use of *h* in *ho*. His one condition was that the sense was not to be changed. It emerges from his correspondence that he was helped by a quartet of Florentines. Giambullari was most closely involved with Torrentino and the Dutchman's compatriot and corrector, Arnoldo Arlenio. Borghini helped Vasari while the work was being printed, giving advice on the dedication and the conclusion, looking for errors and compiling an index; but he was probably not involved from an early stage of the revision, since he said in a letter of 24 January that the work had not been arranged ('assettata') as he would have originally liked. Cosimo Bartoli (who was associated with the Aramei) instructed the printers on Vasari's behalf and compiled a list of errors after Arlenio had given him only three days for this task. Lenzoni also gave his advice. Whoever revised the text did more than just correct the spelling: misunderstandings on the part of Torrentino's compositors have left traces of a couple of more radical interventions. When the 1550 text has, for instance, the nonsensical 'altre tempere che col tempo le faceva fuggire il tempo' ('other temperas which, in the course of time, time made them disappear'), and the second edition of 1568 has '... che col tempo si consumavano' ('which in the course of time faded'), it is clear that in the first edition someone had been trying to replace Vasari's original 'altre tempere che le faceva fuggire il tempo' ('other temperas which time made them disappear') with a phrase more elegantly worded and constructed.[23] All four of Vasari's collaborators were, in different ways, ardent defenders of their language and were doubtless anxious for a work glorifying Tuscan achievements, and dedicated to Duke Cosimo, to have a style worthy of its subject matter.

In contrast, Varchi probably adopted a conservative approach to editing the texts of his fellow Tuscans. In 1548 Bernardo Giunta published a posthumous collection of the poetic works of a contemporary of Firenzuola, Lodovico Martelli. Prefacing Martelli's verse translation of *Aeneid* IV, Bernardo wrote that this was a youthful work which the author would no doubt have revised, but that he had consulted with many learned men, especially Varchi, and they had decided to leave their manuscript (the best available to them) untouched. In the next year Varchi seems to have left a letter by Giorgio Vasari unrevised (in his own *Due lezzioni*, printed by Torrentino), whereas another version of this letter may be the result of later rewriting by Vasari's nephew.[24]

The influence of Venice extended to the only Florentine *Orlando furioso* of the century, printed by Benedetto Giunta early in 1544. The editor, Pietro

Ulivi, dedicated the work to Varchi, who had witnessed at first hand the proliferation of Venetian editions of the epic. Ulivi thus had good reason to write that Varchi would recognize how far this *Furioso* differed from others. The edition was certainly original in its scope, but Ulivi's own contributions were heavily outweighed by what he had borrowed. He was trying to combine in one volume most of what Dolce and Domenico Fausto had put into their respective editions for Giolito and for Bindoni and Pasini. The text followed Giolito's layout (even reproducing his woodcuts), and hence could use Dolce's subject index with only minor corrections. Into the volume Ulivi also crammed most of the extratextual material which Fausto and Dolce had provided, with the exception of Fausto's items on Charlemagne, his *rimario*, and Dolce's 'Espositione' (to which Fausto's much longer 'Dichiaratione' was preferred). These debts were acknowledged obliquely when Ulivi said in a letter to readers that Fausto and Dolce were among those who had made the way easier for 'every kind of man' to get maximum pleasure and benefit from the poem, Dolce being the most meritorious. However, Ulivi did make some additions to what they had written, in the short time which he said was available to him. His list of comparisons with classical authors was similar in conception to Dolce's. But his selection of passages was different and more wide ranging. Ulivi pointed to parallels with Statius and with contemporary Latin poets. Fausto had mentioned Ariosto's debt to Greek epic; Ulivi accordingly added parallels with Homer, as well as with Aesop (both quoted in Latin). He ended his list with a short separate selection of passages concerning the seasons, times of day, reflections on human nature, and so on, again with comparative Latin quotations. This section seems to have suggested to Dolce the much longer list of descriptions, proverbs and so on (without any sources) which he added to his own 1544 edition of the poem.[25]

The Venetian style of presenting major authors had no other imitators in Florence. However, there was one work which to some extent continued Ulivi's approach: the *Spositione sopra l'Orlando furioso* by Simon Fórnari of Reggio, printed by Torrentino in 1549–50. This was directed (as the dedication of the first part stated) at readers who were only 'fairly learned' ('mediocremente dotti'), and may have been intended originally as a commentary to be published with the poem: in a letter at the end of the second part, Fórnari said that the text could not be printed together with its 'explanations' ('espositioni'), as if he had hoped that this would be possible. He also recounted that printing of the first part had been inadequately supervised. He was commanded by someone authoritative (perhaps the dedicatee, Duke Cosimo?) to print the comments composed in his youth and not since revised; but, trying to correct his own and the printers' errors, he had overtired himself and fallen seriously ill, while Arlenio had not looked after Fórnari's work with his usual care.

The appearance of a cluster of anthologies in Florence in the late 1540s reflects the influence of another aspect of Venetian publishing practice. The contents of many of them, however, show that Florence was still just about able to maintain a distinctive cultural identity even though her printers were still operating in the shadow of Venice. In 1548 Bernardo Giunta printed the first book of a collection of *Opere burlesche* by Francesco Berni and other authors, including non-Florentines. Five similar collections had been published in Venice since 1537, when Curzio Navò first printed some *capitoli* by Berni and Giovanni Mauro. The Giunta collection was edited by Grazzini. The previous poor editions of Berni's works, he wrote in his dedication to Lorenzo Scala, dishonoured Florence and especially the Accademia degli Umidi, a group which he still considered to be in existence. The main calling of this Academy was the burlesque style, since everyone was now pretty sick and tired of Petrarchan and Bembesque refinements ('principalmente fa professione ... dello stil burlesco ... avendo le petrarcherie, le squisitezze e le bemberie, anzi che no, mezzo ristucco e 'nfastidito il mondo'). He had now collected these works and with the help of others had restored them 'to their pristine form', so that Berni himself (who had died in 1535) could hardly have improved them.[26] This claim, as so often, presupposed that the author's tastes would have evolved had he still been alive in the time of the editor. The hostility towards 'bemberie' was directed above all against Petrarchist content rather than the recommendations of Book 3 of the *Prose*. In order to legitimize the comic and burlesque tradition which Grazzini considered to be more genuinely Florentine than the culture of the pro-Medicean Aramei group, he was paradoxically using the norms laid down by the Venetian Bembo, sacrificing what was characteristic of contemporary Florence.[27] A number of changes were made for metrical reasons; thus the awkward synaloephe in 'se non che ch*i* al lor giogo si summise' was eliminated (together with the repetition of *ch*) in the new line 'e chi al giogo lor si sottomise' (15.7 in Chiòrboli's edition). Some revisions went in the direction of contemporary Florentine, however; there were cases when the third-person singular imperfect subjunctive ending -*sse* became -*ssi*, or when initial *schi*- (as in 'schiava') became *sti*-.[28] Grazzini also had to bear in mind the sensitive moral scruples of the age and the reactions of the Medici; prudently, then, he censored obscene or blasphemous allusions (as in poems 3 and 59) and removed the name of a political exile, Bartolomeo Cavalcanti, from the title of poem 60.

Also predominantly Florentine in content was an anthology of *Facetie e motti arguti* edited by Domenichi and printed by Torrentino in 1548. Some of these witty remarks were translated from contemporary Latin sources, but the majority consisted of an unpublished collection by Poliziano. Although it seems that they were published with the usual haste, Domenichi still had time both to alter points of phonology and morphology and to carry out a revision similar to that which the *Motti et facetie del Piovano Arlotto* had undergone before their printing some thirty years earlier: that is, blasphemous or obscene terms were toned down and stylistic traits typical of the late Florentine Quattrocento were removed.[29]

Later in his life, Doni scorned those who compiled anthologies as the lowest form of author, 'patchers and marquetry-masters' ('de' rappezzatori, de' maestri di tarsie').[30] But in 1547 he printed a collection of speeches (*Orationi diverse*) which was similar in conception to Clario's *Orationi* of 1546, and indeed had one item in common with it. He showed more originality when in the same year 1547 he edited and printed, under the title *Prose antiche*, a miscellany of medieval Tuscan letters, speeches and witty stories. About half of his material was copied from a Trecento manuscript, now BLF 42.38, with minor adjustments to the language.[31] Among the items were translations of letters by Dante and Boccaccio, the first time that this correspondence had been printed. Doni also inserted two letters which he attributed to the same two authors but which he had most probably forged.[32] In the apocryphal letter by Dante, Doni makes him question the ability of Venetians to understand Latin or Tuscan. But the publication of the texts which were genuine provided a more effective reminder that the ancestry of the literary language was medieval and Florentine, and the volume formed a doubtless intentional contrast with the mainly ephemeral letters of contemporary Italians on which so much Venetian printer's ink was being lavished.

In the 1550s the firms of Torrentino and of the Giunta brothers produced, among other works, some editions of Florentine works dating from the fourteenth century to their own day, which were either previously unpublished or which were presented in a revised form. Grazzini continued to foster the Florentine tradition of colloquial and comic poetry. In 1552 he edited for the Giunta press a collection of satirical sonnets by Burchiello. He claimed that he had found both manuscript and printed texts of the poems in need of much revision, and particularly blamed printers in an introductory sonnet which also made it clear that this edition was part of his campaign against the Aramei. Grazzini revised the arrangement of the text found in previous Florentine editions, including Lorenzo Peri's very old-fashioned printing of 1546. He divided the sonnets into those which were truly 'burchielleschi' and those composed normally, and he excluded any judged too weak or dirty and dishonourable ('troppo deboli o troppo sporchi e disonesti'). As in the *Opere burlesche*, there was also a purist revision of the language.[33]

Grazzini's next edition was an anthology, printed by Torrentino in 1559, of carnival songs and similar compositions (*Tutti i trionfi, carri, mascherate ò canti carnascialeschi*) performed in Florence since the time of Lorenzo de' Medici. This was a more difficult task than the editions of Berni and Burchiello, he said in his dedication. His sources were few, full of errors, gaps and strange abbreviations, so that, in editing the texts, his knowing and being familiar with verse and rhyme came in useful ('mi è giovato il conoscere e l'esser pratico coi versi e colle rime'). When he was accused of errors, he admitted privately to Luca Martini that there might be some unimportant ones, but insisted that he had improved the poems, 'corrected lines and false rhymes and made a thousand disagreements agree, and rewritten them all in my hand' ('racconcio versi e rime false e accordato mille discordanze, e riscrittigli di mia mano tutti quanti'), as well as improving the spelling.[34] An example of his corrections is provided by Machiavelli's *Canto di diavoli* (or *De' diavoli iscacciati di cielo*). There are several differences from the text of his manuscript (BRF 2731): not just minor alterations such as 'diaccio' > 'ghiaccio', 'vegniàno' > 'vegniamo', but ones intended to make the style more elegant. Thus Grazzini changes, for instance, '*siàno stati dal ciel* tutti schacciati' to the tautologous 'dall'alto e sommo ciel', 'chonfusion, dolor' to 'confusione e duol', '*fatti siàno e staren* principio noi' to 'stati siamo e sarem'. He removes a relative pronoun ('ch'abiàn' > 'Habbiam', line 9) so that the second stanza begins neatly with a new sentence. The semantic symmetry of the line 'e 'l pianto e riso e 'l diletto e 'l dolore' is destroyed when 'diletto' is replaced with 'canto', perhaps for the sake of the assonance with 'pianto'. In the *Canto di uomini che vendono le pine*, arbitrary changes include the transposition of two lines in the fifth stanza, the rewriting of 'bisognia tener forte' as 'bisogna haverlo stretto' in the sixth stanza, and the omission of the seventh.

This lack of respect for the original form of lesser poetic genres was not unusual. It is also seen in the *ottava rima* version of the story of Ippolito Buondelmonti and Dianora de' Bardi, attributed to Alberti, which was printed 'presso al Vescovado' in 1560. This had been printed four times in Florence in the fifteenth century.[35] The new edition introduced changes which ranged from standardizations of spelling to recastings of whole lines either because their syntax was clumsy (as in 'vaga, pulita e d'oro avie suo chiome' > 'e proprio d'oro parevon sue chiome') or for metric reasons, for instance if they had a syllable too many ('armato andava con infinita gente' > 'e giva armato con una gran gente') or had a stress on the fifth syllable ('tenevan la terra in gram differenza' > 'e tenean la terra in differenza'). Not all the changes conformed with Bembo's rules, though, as one sees from an example such as 'furon disposti di volere amarsi' > 'e tutt'a dua deliberorno amarsi'. Strangely, the rhyme scheme was broken by changes such as 'qualunche al mondo è di tal amor pieno' > 'chi gusta questo amor alto e superno' in the last stanza.[36]

The work which Bembo had quoted most often after the *Decameron* in the third book of the *Prose della volgar lingua* was the *Cronica* of Giovanni Villani (*c*. 1276–1348). Until the middle of the Cinquecento, the *Cronica* had appeared in print only once, in Venice in 1537, when Books 1–10 were published. In the second half of the century, Florence made a determined effort to regain the initiative in the diffusion and study of this work and of its continuation by Giovanni's brother Matteo. The second part (Books 11–12) of Giovanni's section and the first part of Matteo's *Cronica* were printed, for the first time, by Torrentino in 1554. The dedication of the first of these volumes was signed by the printer, that of the second by Domenichi, who was in all probability the editor of both. He stated his editorial principles in familiar fashion. The books of Matteo's *Cronica* were very incorrect through the fault of scribes. He had striven to improve them as far as possible by doing two things: collating them with old copies and taking the advice of men of judgement, but without changing the author's meaning at all ('conferendogli con essemplari antichi, e pigliando il parere d'huomini giudiciosi ... senza punto alterare i sensi dell'Auttore'). Since the late fourteenth-century manuscript used as printer's copy for Torrentino's edition of Giovanni Villani survives (BNF Palatino 1081), one can see that the editor made a few genuine corrections, possibly with the help of one or other of the two sources of emendation which Domenichi mentioned, and that he added punctuation and regularized the spelling. But he also added, removed and changed words arbitrarily. In one chapter he toned down or removed references to Venetian treachery against the Florentines: Domenichi, if he was the editor, would have wanted to keep on good terms with Venice, where works of his were still being published, as would Torrentino, in view of Venice's position as the leading trade outlet for books printed in Florence.[37] However, these interventions were superficial compared with those which Domenichi inflicted on *Il Pecorone*, a late Trecento collection of short stories by a Florentine called Ser Giovanni. This was published in Milan in 1558, while Domenichi was temporarily in his native Lombardy; perhaps the extent of his alterations would have attracted too much criticism in Florence. Three stories concerning the clergy were replaced with others taken from works edited by Domenichi, Villani's *Cronica* and Firenzuola's version of Apuleius, and the rest of the text was not just corrected but thoroughly rewritten.[38]

In 1559 Filippo and Iacopo Giunta published the whole of Giovanni Villani's *Cronica* (Books 1–12). They had this printed in Venice, because Torrentino's book-privilege for Books 11–12 had not yet expired. Their editor was Remigio Nannini, who presumably carried out his work in Venice. There was apparently some lack of communication between the two cities. The dedication of the Giunta brothers to Duke Cosimo suggested that those responsible for the previous Venetian edition (1537) had made a poor job of it because they had been baffled and confused by 'the old and rough words of this [Florentine] language' ('[i] vocaboli antichi e rozi di questa lingua'). Instead of starting again with fresh sources, they had decided to have the Venetian printing corrected and compared with old texts, thus bringing it back 'to the true and natural quality of old Florentine' ('alla vera e natural proprietà dell'antica fiorentina favella').[39] However, Nannini announced to his readers immediately afterwards that he had decided to leave the 1537 text alone, treating it like a beautiful antique ('[una] bellissima anticaglia') which should not be spoiled. But an old medal might, he said, be given a new frame, or an old statue might have a new plinth; so, in the Venetian fashion, he had added some 'dichiarationi' in the margins to explain points of vocabulary and history which might be obscure. As for spelling and expression, he said that he had changed hardly anything, firstly so as to show the extent of the difference between old and modern writers, and secondly because he knew that the slightest thing might ruin 'that antiquity and age that one desires in a beautiful old medal' ('quell'antichità e vecchiezza che si desidera in una bella medaglia antica'). This claim was substantially true: Nannini added punctuation, accentuation and capital letters to the 1537 text but changed only a few spellings (such as 'fugitivi' > 'fuggitivi', 'comune' > 'commune', 'grandeza' > 'grandezza' (but 'mezo' was unchanged, as in Ruscelli's system), 'segnoregiava' > 'signoreggiava', 'fecciono' > 'feciono', 'per li' > 'per gli') while leaving forms such as 'vulgare', 'virtudiosi', 'advento' and 'fue'. The same applied to the 1554 text, used for Books 11 and 12.[40]

Nannini's editorial skills fell a long way short of winning the approval of Florentine scholars later in the century, as we shall see in chapter 11. Yet, if one compares the editing of this Florentine working in Venice with that of his counterparts working in Florence, it must be said that Nannini was the least inclined to force his text to conform with the *norma et consuetudo* of posterity. In the case of other vernacular editors, such as Domenichi and Grazzini, there was still a gulf between their methods and those of the school of classical editing led by Piero Vettori.

10 · PIETY AND ELEGANCE: EDITING IN VENICE, 1561–1600

RUSCELLI DIED IN 1566, after a painful illness, and Dolce followed him to the grave two years later.[1] In their last years neither of them kept up the level of activity which they had achieved in the 1550s. They were growing older, of course, and they had made up their quarrels, so that they no longer had the spur of rivalry. However, their personal situations do not explain everything. The work of all editors was being affected by changes in the character of the Venetian book trade.

In the first place, the cultural climate of the Counter-Reformation was less favourable to secular vernacular and classical literature (including translations), the two areas in which editorial activity had been liveliest. Grendler's analysis of the imprimaturs granted to new titles in Venice shows that the printing of works in these categories declined in the 1560s as the book trade began to deal more in religious works.[2] The same trend appears in the output of the Giolito press, run by Gabriele until his death in 1578 and then, less successfully, by his sons Giovanni and Giovanni Paolo.[3]

The work available for Venetian editors was also reduced by a decline in the dominance of the city's printing houses within Italy. The imprimaturs and the holdings of the British Library suggest that the Serenissima's share of books printed in the peninsula fell from about 60 or 70 per cent in the third quarter of the century to about 40 or 50 per cent in the last quarter, and that there was a particularly sharp drop in output in the decade following the plague of 1575–7. At the same time, the number of presses fell.[4]

Not only was there less work for editors: the nature of their work was being influenced by the increased controls on book production and the book trade which were being imposed by Church and state in an age of increasing intolerance of the unorthodox. Since the 1540s there had been efforts to create more effective mechanisms of control, with the aim above all of repressing heretical doctrine, though these efforts had little practical effect before about 1560.[5] Pope Paul III had created in 1542 a tribunal, known as the Holy Office, to defend Catholic doctrine, and one of the tasks of this 'Roman' Inquisition was to prevent the diffusion of heretical books. The Venetian Inquisition, dormant in the Quattrocento, was active again by 1540, and the state then tightened its constitutional grip on the book trade, especially by creating in 1547 a new magistracy of laymen, the 'Tre Savii sopra eresia', which was to assist the ecclesiastical Inquisition.[6] The next step was to draw up a list of prohibited works. The Venetian government ordered one to be compiled and printed in 1549. But at this stage the problems of reconciling the interests of state and Church became apparent. The powerful lobby of Venetian bookmen managed to persuade the government that the list was too vague. They were also strong enough to force the withdrawal of another 'Catalogue of heretical books' compiled by two Dominicans, one the Master of the Sacred Palace (the Pope's theologian), and sent from Rome in 1554 or 1555. It looked as if the Church would finally get the upper hand in 1559 when the Index of Pope Paul IV was accepted in Venice after long negotiations. A significant feature of this Index was the extension of its concerns from heresy to a second category of books, those considered to be anticlerical or immoral. All the works of Machiavelli were now banned, and Boccaccio's *Decameron* could no longer be printed, as in the past, 'with intolerable errors' ('cum intollerabilibus erroribus'). Very soon, however, the Pope died and his hated Index was no longer enforced. The first Index to have a long-term effect was the *Index librorum prohibitorum* authorized by the Council of Trent and published in 1564. The *Cento novelle* or *Decameron* were still banned 'until they appeared purged by those to whom the Fathers [appointed to draft the Index] entrusted the matter' ('quamdiu expurgatae ab iis, quibus res Patres commiserunt, non prodierint').[7] And already in March 1562 the Venetian state had decreed that the Holy Office in Venice should have all new manuscripts approved before printing by a cleric and two laymen, who had to confirm that works contained nothing 'contrary to religion, princes or public morality'. This made explicit the threefold concern, including offences against ruling powers as well as against doctrine and morality, which was being expressed by the Venetian prelate Daniele Barbaro at the Council of Trent.[8]

This process of pre-publication censorship was expensive and time-consuming. It must have acted as a check on the quantity of books published and also on their nature, since printers were now more likely to submit only those works which would easily pass scrutiny. A small example of the depressing effect which censorship could have on the opportunities for editorial work is provided by a reference to Paul IV's Index in a letter of Dionigi Atanagi, who had recently arrived in Venice and hoped to find work as an editor. A manuscript of the Index was sent to Venice from Rome on 31 December 1558. Its publication was eventually authorized on 8 July 1559 and printing was completed on 21 July. Only days later, on 12 August, Atanagi was writing to a friend that it was harder than ever to find work on 'the correction of any of these books' ('la correttion d'alcuna di queste stampe') because of 'this blessed Index' ('questo benedetto Indice').[9] A note at the end of Valvassori's 1566 edition of Ariosto's *Orlando furioso* confirmed that its new editorial material had been approved by the Inquisitor.[10] The liability to censorship of editors' contributions was confirmed in the rules on expurgation appended to the last

Index of the century, published during the papacy of Clement VIII in 1596: attention was to be paid 'not only to the text, but also to the notes, summaries, marginalia, indices, prefaces, and dedications'.[11]

In the 1560s one soon finds Venetian editors removing phrases which might be morally or politically offensive. Domenichi exercised a mild kind of censorship on Giovio's *Lettere volgari* (Sessa, 1560), as well as making the usual kind of linguistic 'improvements'.[12] It was important, too, that works should not be construed as anticlerical or blasphemous. When Grazzini's comedy *La Spiritata* was printed by Rampazetto in 1561–2, the anonymous editor made one of those responsible for the girl's possession by a spirit into 'a friend' instead of her father confessor. In Bernardo Giunta's Venetian edition of 1582, this character became a doctor, and references to a saint and to Christ were removed.[13]

These changes were perhaps merely precautionary. It was a more serious matter when editors had to work with the Roman Inquisition looking over their shoulders. One such case concerned Paolo Manuzio and his son Aldo. In June 1561 Paolo left Venice for Rome in order to set up a press at the request of Paul IV. Just over a decade later, Aldo wanted to republish the collection of *Lettere volgari* which he had brought out in 1567. But in 1571 the Congregation of the Index was founded in Rome in order to oversee the expurgation of books, thus ensuring that the Church kept a continual check on book publishing. In 1573 Paolo wrote from Rome to tell Aldo that the Master of the Sacred Palace had ordered the deletion of letters by authors whose works appeared in the Index and of a love letter. When Aldo showed signs of wanting to disobey this decree, Paolo became extremely anxious about the possible consequences.[14] A few years later the Master of the Sacred Palace was scrutinizing Castiglione's *Cortegiano*, a work now at risk for its unsycophantic view of the clergy and its concluding eulogy of Platonic love.[15] He was reported in 1575 to have decided not to impose any cuts, though the pages on love would have to be censored if the book were to be printed in Rome. Nevertheless, Camillo Castiglione, the author's son, was anxious that a revised edition should be published which would protect his father from any suspicion of disrespect for religion or the clergy. The result was a *Cortegiano* printed in Venice by Bernardo Basa in 1584, in which Dolce's 1556 edition was 'revised and corrected' ('riveduto e corretto') by Antonio Ciccarelli of Foligno. This pair of participles on the title page was perhaps intended to suggest no more than a conventional piece of editing, but readers would have noted that Ciccarelli was a doctor of theology (indeed, he was probably a member of the Jesuit order).[16] His dedication made it clear that he had removed things which might give rise to scandal, though he claimed he had not strayed from Castiglione's words and expressions and had taken care that his revisions would not be noticeable. Many changes were made in the sequence of sometimes Boccaccesque witty stories in the latter part of Book 2: Ciccarelli's techniques included the turning of some clergymen into laymen, the suppression of anecdotes, and their replacement with alternatives. Almost all references to *fortuna* as a force affecting human destiny were removed, and the word was often changed to *sorte* ('fate') or *disgrazia* ('misfortune'). Words such as *divino* and *angelico* were replaced when used out of a religious context. In Book 4, Bembo's speech on love was amended: for instance, allusions to the sacred nature of human love or beauty were qualified; new marginal notes, with the function of a commentary, presented Neoplatonism as an obsolete philosophy; and a passage was added on the need for religion if man was to achieve true happiness in the contemplation of God. On the other hand, Ciccarelli allowed risqué references to remain; his concern was to protect the Church rather than morality.[17]

The *Decameron* appeared in Venice in 1588 in a version which had been censored, or rather 'reformed' ('riformato'), by the blind poet and playwright Luigi Groto, who died in 1585.[18] He prepared his edition, as a letter to the readers explained, because the Florentine edition of 1573 (which will be discussed in the next chapter) had failed to purge the work completely of 'those things which could offend the pious ears of Catholics' ('quelle cose che poteano offendere le pie orecchie de' cattolici'). Since this was a work mentioned in the Tridentine Index, the revision had to be carried out in consultation with the Inquisition. A letter written by Groto in January 1579 tells us that he had obtained permission from the commissary of the Venetian Inquisition, who had passed on to him instructions from the Roman Inquisition. Groto had been a friend of Ruscelli, as some of his *Lettere famigliari* show, and he used the text from the Valgrisi edition with its ancillary material. However, he severely censored Boccaccio's references to the clergy and to religious practices; he had been tried on charges of heresy in 1567 and will have been keen to impress the Inquisition with his zeal. As in Ciccarelli's *Cortegiano*, there were no indications of where the author's prose ended and the editor's began. The first story, for instance, centres on the false deathbed confession of the evil ser Ciappelletto: his apparent sanctity dupes a friar and Ciappelletto is subsequently venerated as a saint. Groto gives the beginning of the story, up to the point where Ciappelletto is on his deathbed, but then, without warning, continues the story in his own gruesome way. The protagonist is locked in a box and burned alive, meeting 'the end he deserved' ('quella fine che meritava'), a foretaste of the eternal flames awaiting him in Hell. The deceitful and lustful Frate Alberto (4.2) becomes Maestro Alberto, a layman and a poet (which gives Groto the opportunity to counter Boccaccio's hostility to Venice with a flattering reference to the Serenissima's welcoming attitude to poets), and he disguises himself as the god of love rather than as the Angel Gabriel. But Boccaccio's metaphors for Alberto's sexual activities (flying without wings, riding) are kept; and in other stories Ricciardo still tricks

Catella into sleeping with him (3.6), Rustico still seduces Alibech (3.10, though they put 'the dragon into the snake' rather than 'the devil into hell'), and Caterina still sleeps holding her 'nightingale' (5.4). Like Ciccarelli, Groto did not set out to bowdlerize.

Soon after the banning of the *Decameron*, Sansovino had attempted to assuage the public's demand for the work by providing an alternative collection of short stories. In 1561 he put together and printed an anthology, enterprisingly dedicated to Queen Elizabeth of England, of *Cento novelle scelte*, one hundred tales chosen from various authors. These included ser Giovanni from the Trecento, Masuccio Salernitano from the Quattrocento, and various Cinquecento figures such as Firenzuola, Parabosco and Sansovino himself. These stories were only like copper in comparison with the gold of the *Decameron*, Sansovino apologized to readers, but they were something to read until Boccaccio's work came out 'with new annotations' ('con nuove annotationi'). The stories had, he said, been 'linguistically put in order and repaired' ('rassettate e racconcie nella lingua') by him in so far as the printers' haste had allowed. He also provided a frame story, similar to that of the *Decameron*, in which noblemen and women fled Venice in 1556 for fear of the plague and sought refuge in the region of Treviso. This collection proved very successful. It was reprinted in 1562 and 1563 with revisions to the selection, then in 1566 with thirty-one stories from the *Decameron*, though Sansovino was of course careful to choose ones which did not offend against religion. A fifth edition in 1571, printed by the heirs of Melchior Sessa, added two items: firstly, a discourse on the *Decameron*, relevant (Sansovino claimed) because it showed what a novella was and because it explained the structure of his own collection, similar to that of the *Decameron*, and secondly the *Ciento novelle antike*, copied from Gualteruzzi's edition of 1525.[19]

The same combination of Sansovino and the Sessa family brought out, in 1564, a Dante which combined the editorial traditions of an earlier period with those of the mid Cinquecento.[20] The format was folio, for the first time since 1529. The text was based on that of Bembo, with some readings taken from Vellutello and some of Bembo's truncated forms written in full (as in *mio* and *cui* rather than *mi* and *cu*); and it was surrounded by two commentaries, those of Landino and Vellutello, a technique which had been common in earlier Petrarchs but had never been applied to the *Commedia*. To these backward-looking elements Sansovino added material typical of the age of Dolce: prose summaries (*argomenti*) and allegories for each canto and a glossary of difficult words. He also updated Landino's lists of famous Florentines. His glossary demonstrated the same awareness of both Trecento usage and northern regional forms which had been apparent in his *Decameron* of 1546.[21] He tried to inform readers about the Latin or French origins of some terms and to point out Dante's neologisms. Only exceptionally did he offer advice on which forms to use.[22]

Sansovino's edition was popular enough to be republished in 1578 and 1596. There was no such success, however, for another Dante based on past scholarship which was published in 1568 by the bookseller and printer Pietro da Fino, who hoped no doubt to benefit from the revival of popularity which Dante was enjoying. On this occasion the Aldine text was accompanied by the commentary of Bernardino Daniello, left unpublished on its author's death in 1565.[23] As in his Petrarchan commentary, discussed in chapter 7, Daniello owed a great debt to his teacher Trifon Gabriele, on whose unpublished commentary on Dante he drew heavily. It is rare to find Daniello mentioning textual variants as he does with '*Questi* non ciberà', which he read 'in an old text', rather than 'Questo' (*Inf.* 1.103), or 'lerze', also found 'in some old text' ('in alcun testo antico') for 'berze' (*Inf.* 18.37). His explanations of Dante's language occasionally refer to the usage of his adopted north-eastern Italy rather than to his native Tuscan. For instance, Dante's 'zanzara' ('mosquito', *Inf.* 26.28) is compared with *moscione*, 'maciulla' ('flax grinder', *Inf.* 34.56) with *gramola*.[24] 'Brulla' ('flayed', *Inf.* 34.60) is explained as *nuda* and synonymous with Paduan *sbroià*, feminine *sbroiata*. 'Biscazza' (*Inf.* 11.44), says Daniello, as a noun means a place where one gambles ('si tien giuoco'), *bettola* in Venice.

The pattern of the editing of Dante in Venice after 1568 reflects a complete loss of impetus in vernacular scholarship and severe lack of originality on the part of printers. The Sessa company, as we have seen, reprinted Sansovino's edition twice. The only other printer to take an interest in Dante was Domenico Farri. He followed two paths, both of them trodden by others before him: in 1569 and 1578 he copied Giolito's duodecimo edition of 1555, and in 1572 and 1575 he copied the *Commedia* printed in Lyons by Jean de Tournes in 1547, with canto summaries, marginal notes and *dichiarazioni* taken from Landino's commentary.[25]

With the Giolito company abandoning Petrarch and Ariosto after 1560, just as it had abandoned Dante, other Venetian printers could now contend for the leadership of the market for these two poets. But, as in the case of Dante, there were none of the original initiatives in presentation which the period from 1530 to 1550 had produced. In the 1560s Nicolò Bevilacqua, once apprenticed to Paolo Manuzio, tried three different types of editions of Petrarch, all of them derivative. Two were based on Italian models, the 1514 Aldine and Vellutello's edition. If the third type, first printed in 1562, was more innovative, this was because it reproduced an edition printed not in Italy but in Lyons, four years earlier, by Guillaume Rouillé and edited by Lucantonio Ridolfi.[26] Appended to the volume was a rhyming dictionary by Ridolfi, and each poem was followed by linguistic notes aimed at relative beginners, often giving grammatical help based on Bembo's *Prose della volgar lingua*.[27] Easily the most popular of the Italian models was the Aldine, which continued to be imitated up to the end of the century, although, from the 1573 edition of Giovanni Griffio onwards, printers omitted the three 'Babylonian' sonnets (*Canzoniere* 136-8) which attacked the vices of the papal court in Avignon. Two other types of Petrarchan edition from the first half of the Cinquecento printed in this period were the Giuntina of 1522 (once, in 1562) and Gesualdo's edition (twice, in 1574 and 1581-2). After 1580, the only signs of originality were to be found in some editions which included essays on Petrarch's poetry. The edition printed by Giorgio Angelieri in 1585 contained a 'Discourse on the nature of Petrarch's love' ('Discorso sopra la qualità dell'amore del Petrarca') composed in that year by Pietro Cresci, a member of the Accademia dei Fantastici of Ancona, who set the poems in a stern moral context by suggesting that Petrarch's intentions were far from chaste. For an edition of the following year, the work of Ridolfi in Lyons provided Angelieri with more supplementary material: from Rouillé's 1558 edition came the text with Ridolfi's annotations, while from one of the same printer's 1564 editions Angelieri took some correspondence of 1562-4 between Alfonso Cambi Importuni and Ridolfi and from Francesco Giuntini to Domenichi on the question of the time when Petrarch fell in love.[28] When Barezzo Barezzi copied the first of Angelieri's editions in 1592, he added another 'Discorso': the author this time was Tomaso Costo, and the subject was the purpose of Petrarch's poetry and the heroic nature of the *Trionfi*.

With Ariosto's *Orlando furioso*, the editorial trends of the last four decades of the Cinquecento resembled those of the *Commedia* and of Petrarch's verse. Some editions of the epic were merely imitations of earlier models. The Valgrisi press was content to keep on reprinting Ruscelli's edition up to 1587, adding in 1565 the *Cinque canti* with verse *argomenti* and prose allegories by Luigi Groto. A concluding note to the readers by Groto, however, strove to maintain Ruscelli's line strictly: the *Cinque canti*, it was argued, had been rejected by Ariosto and did not belong to the final version of the *Furioso*, as was shown by their 'infinite errors of language' ('infiniti errori di lingua').[29] This presentation of the epic proved so popular that it replaced the Giolito edition as the model for Rouillé in Lyons: for the publication of Ariosto, at least, Venice continued to export ideas to France rather than import them as with Dante and Petrarch. Comin da Trino copied Dolce's edition in 1567 and 1571.

The only alternative to such repetition seemed to be to entice readers with new supplementary material; textual accuracy was even less important than before. The period of greatest innovation came in the 1560s, as it became clear that Giolito had abandoned Ariosto's epic. One of the printers in contention for a greater market share was Francesco Rampazetto, whose first two editions of the *Furioso* had come out in 1549. The novelties of his editions of 1562 and 1564 included a biographical sketch by Sansovino, using extracts from the *Satire*, and new verse canto summaries by Livio Coraldo. Some of the canti (up to the thirty-fourth) were followed by notes in which Sansovino explained allusions to places, people, literary works and so on but had very little to say on the language. In 1563 a member of a new generation of printers, Giovanni Varisco, brought out an edition with summaries by Giovanni Andrea Anguillara, whose translation of Ovid's *Metamorphoses* had successfully rivalled Dolce's, and allegories by Giuseppe Orologgi of Vicenza.[30] This was reprinted in 1566, but Varisco's 1568 edition was, according to its title page, 'corrected and explained' ('corretto e dichiarato') by Dolce, who was returning to the poem in the last months of his life. The Venetian's first contribution was a 'Discorso' on Ariosto's poetic and linguistic achievement, insisting that the language of the poem was not pure Tuscan but contained words common to most of Italy. In a letter to the readers, Dolce reviewed his record of editing the text faithfully (he claimed), according to the 1532 text, unlike the late Ruscelli (whose attacks on Dolce were, however, here generously forgiven). Each canto of the *Furioso* had a new set of notes by Dolce, almost completely different from those in his editions for Giolito and much more detailed.

Another of Giolito's old rivals in the battle for readers of the *Furioso*, Giovanni Andrea Valvassori, brought out in 1566 a revised edition with a wealth of supplements. It began with Fórnari's life of the author and other material from Valvassori's 1553 edition. Before the canti were verse summaries by Giovanni Mario Verdezotti and allegories which had been printed before, Clemente Valvassori's on the *Furioso* and Dolce's on the *Cinque canti*. The canti were followed by a varying combination of notes on the text and on Ariosto's 'imitations', observations on the justification and conduct of the duels, and the use of history and fables. The volume was completed by indexes, a glossary arranged in alphabetical order like Ruscelli's but with definitions drawn mainly from Dolce's, and a rhyming dictionary by Giovanni Giacomo Paruta which expanded that of Fausto. The notes on history and fables discussed some of the same points covered by Eugenico for Valgrisi in 1556 and 1558, but was often fuller, particularly on the fables. This material was explicitly attributed to Tomaso Porcacchi. Born in Castiglion Fiorentino around 1530, he had provided a life of Virgil and a translation of *Aeneid* v for a collection of the works of the Latin poet translated by different writers, edited by Domenichi and printed by the Giunta press in Florence in 1556. However, he had soon left Tuscany and began to work for Giolito and then for other Venetian printers.[31]

The title page of this *Furioso* stated that the opinions on duels were of uncertain authorship ('d'incerto auttore') while the annotations and lists of imitations were by Dolce and 'by others' ('d'altri'). The notes were based, in fact, on those of Fausto and Dolce, with some being omitted, others expanded, and new ones added. The *imitationi* combined Dolce's list, in Ruscelli's slightly expanded form of 1556, with that of Ulivi, but again there was some new material, for example on the use of Statius's *Thebaid* in Canto 14.[32] Which editor or editors were responsible for these compilations? It is perhaps unlikely that Dolce would have taken Ruscelli's plagiarism as his starting point. One of the alternative candidates is Porcacchi. The 'Giudicio' in the *Furioso* printed by Gobbi in 1580 says that the Tuscan augmented Dolce's list of imitations twice, though no dates are given. The annotations to Canto 17 in the 1566 edition show that their author had friendly relations with Domenichi, as did Porcacchi. And one of the few vernacular works mentioned in the notes (twice) and in the *imitationi* (once) is Boccaccio's *Corbaccio*, which Porcacchi had edited in 1563. It is true that the notes refer to Porcacchi in the third person, but perhaps he did not want such derivative work to be explicitly attributed to him. Several flattering references in the *imitationi* to Erasmo da Valvasone, translator of Statius, might point to Erasmo's friend Atanagi, but the latter's name does not appear in the volume. On balance, then, there seems to be a slightly stronger case for the involvement of Porcacchi.

Valvassori's edition proved again that it could be counter-productive to include too much editorial material. It was another of the newer printing firms, that of the brothers Domenico and Giovanni Battista Guerra, which finally devised a *Furioso* sufficiently popular to dominate the last third of the century. Their first editions of the epic, in quarto and duodecimo formats, came out in 1568. These combined a limited amount of older material – Fórnari's biography, the glossary based on Dolce's notes, and his list of comparisons – with new canto summaries in *ottava rima* by the indefatigable Dolce and (in the quarto only) new allegories by Porcacchi. Two years later Fórnari's biography was removed, and Porcacchi added new *annotationi* on most canti. These notes, like Sansovino's, gave hardly any linguistic help. They differed from Sansovino's, though, in their emphasis on moral interpretation of the poem. This 1570 edition (in octavo) provided the model which was imitated by most Venetian printers up to 1600. Alternative, shortened forms of it were also created, for instance in 1580 by Orazio De Gobbi, who omitted Porcacchi's notes and Dolce's concluding material but added an anonymous 'judgement' ('giudicio') which praised the poem for its ability to appeal to a wide audience and defended its right to depart from some of Aristotle's precepts, creating new rules for future writers.

The only *Furioso* to stand out from the normal run of editions after 1566 was one printed in 1584 by Francesco De Franceschi. Its appeal to the reader was based, typically for its time, on the literary and artistic presentation of the poem rather than on the accuracy of the text. The basis for the edition was Ruscelli's text and annotations. New engravings by Girolamo Porro were considered important enough to be the first attraction listed after the author's name on the title page, but there were also several new ancillary texts: a life of Ariosto by Giacomo Garofalo of Ferrara, an allegorical interpretation of the whole work by Giuseppe Bononome, a list of epithets by Camillo Camilli, a list of the first lines of each stanza by Giovan Battista Rota of Padua, and, as an appendix, some important 'Osservationi' by Alberto Lavezuola on the question of imitations of classical and vernacular sources in the *Furioso* and the *Cinque canti*, a contribution which, as Javitch had shown, was far more sophisticated than the previous ones by Dolce and Ulivi.[33]

In 1561 Francesco Sansovino edited and printed a new edition of Ariosto's *Rime*, contributing a brief note on each poem. His presentation was reproduced twice by other printers, in 1564 and 1567, after which the *Rime* were no longer published on their own.[34] They continued to accompany the *Satire*, however. Giolito had printed Dolce's edition of these works for the last time in 1560, but in 1567 he brought out a new edition by the Carmelite friar Francesco Turchi of Treviso.[35] Like Sansovino, he prefaced each poem in the *Rime* with notes, though his were much fuller, and he kept Sansovino's summaries (*argomenti*) of the *Satire*. Whereas Dolce had attributed the *Satire* to Ariosto's mature years on linguistic grounds, Turchi's title page pointed out that both these works were written in Ariosto's youth. The friar's intention was perhaps to excuse aspects of their content rather than their less than perfectly Tuscan language. Not long after, in 1580, the *Satire* appeared on the supplementary Index drawn up by the Inquisition of Parma. The time was coming when Venice, too, would be unable to print the work as it stood. Turchi's edition was printed once more in the following year but it gave way from 1583 to editions, based on Dolce's, in which the *Satire* were 'newly purged' ('nuovamente purgate').[36] Ariosto's sometimes earthy language was replaced with harmless terms or periphrases, and opinions considered offensive to religion or morality were removed. In the fifth *Satira*, for instance, changes affected the advice that a man needs a wife in order to be 'perfect in goodness' ('in bontade ... perfetto') (lines 13–18), the attack on priests (22–7), and the warning against taking a wife who attends mass or confession too frequently (196–8).[37]

Domenichi's 'reformed' version of Boiardo's *Orlando innamorato* was copied several times up to 1574, when the printer Michel Bonelli decided to replace Domenichi's prose *argomenti* with ones in verse and to add allegories for each canto, thus bringing the epic more into line with the *Orlando furioso*. Two years later, though, Bonelli brought out a new revision of the *Innamorato*.[38] It drew heavily on Berni's reworking, first published in 1542. The Florentine had introduced many stylistic corrections and had also added many stanzas.[39] Bonelli was apparently referring to both of these aspects when he told readers that he too was remedying the omissions of earlier editions of the original text as well as their other errors of 'false names, corrupted senses, and transposed stanzas' ('nomi falsi, sensi corrotti, e stanze traportate'). He must have been aware that Berni alone was responsible for the differences between the two versions, since the title page of the 1542 edition stated that the poem had been 'remade' ('rifatto') by Berni. Yet Bonelli still claimed that in his printing the poem had been 'brought, with this putting in order, to its just and healthy reading' ('ridotto con questa acconciatura alla sua diritta e sana lettura').

Whoever carried out the revision did not use Berni's version alone but also made several changes of his own. Some were intended to add further elegance to the text 'reformed' by Domenichi. Others removed or watered down offensive references to clerics: thus a 'hypocritical treacherous friar' ('ipocrito frate traditore', in Berni's 19.65) became a 'fraudulent hermit' ('romito fraudolente'), and a gluttonous friar who praised fasting when his own belly was full 'and worshipped only geese' ('e sol ne l'oche avea devotïone') now merely exhorted others to prayer ('gli altri essortando a star' in oratione', a change which also avoided the diaeresis). Boiardo's laments on the woes of Italy (2.31.49, 3.9.26) were omitted, so that the poem was removed from its original political context, just as its style and content were being falsified in deference to the tastes of the 1570s. But another important function of the revision was to make the poem less licentious. In 1576, it seems, the Inquisition became stricter in its attitude to stories of love. The commissary of the Master of the Sacred Palace wrote in this year to the Inquisitor of Bologna to forbid the printing of 'histories, comedies and other vernacular books of love affairs' ('storie commedie et altri libri volgari d'innamoramenti'), which were already banned and being destroyed in Rome.[40] Some features of Bonelli's new *Innamorato* resulted perhaps from a similar order being sent to Venice. For example, two stanzas referring to Orlando's sexual desires (1.29.48, 2.4.11) were respectively omitted and rewritten, and the vigorous, repeated love-making of Brandimarte and Fiordelisa (1.19.61-3) became a relatively innocent 'amusement' ('solazzare').

Boiardo's lyric poems were treated with no more respect when Dionigi Atanagi included some of them in his anthology of *Rime di diversi nobili poeti toscani* (Venice, Avanzi, 1565) and removed their original dialectal and Latinizing colouring.[41] The tendency to standardize phonology and punctuation was applied to the lyric poems of a living author, Torquato Tasso, by Aldo Manuzio when he revised his own edition of Tasso's *Rime*, printed in 1581, for a corrected edition printed later in the same year.[42]

While less emphasis than before was placed on innovative editorial work on verse writing in this period, editors were increasingly required to help satisfy public demand for prose, and in particular historical works. Sansovino and Nannini were both active in this area, as we shall see; but one of the most interesting initiatives came from Gabriele Giolito and one of the younger generation of editors working for him, Tomaso Porcacchi. By 1563 these two had conceived the idea of joining two series of works together in what they called a 'necklace' ('collana', which became the modern Italian word for a planned series of books). A number of translations of ancient Greek historians were to be the 'rings' of the necklace linking ten modern works on military topics which were called the 'jewels'.[43] A similar concatenation including Latin historians was to follow. Even the Greek one was never completed, however, and its plan was not set out to the public by Porcacchi until 1570. Other series hardly got beyond the planning stage: a 'spiritual garland' ('ghirlanda spirituale') whose first 'flower' ('fiore') appeared in 1568, and a 'spiritual tree' ('albero spirituale') whose only 'fruit' ('frutto') was printed in 1574. Gabriele's heirs were no more successful with a 'staircase' ('scala'), to be made up of 'steps' ('gradi'), begun in 1583.[44]

Francesco Guicciardini's *Storia d'Italia* (Books 1–16), printed in Florence in 1561, aroused much interest in Venice. As with Petrarch and Ariosto, rivalry between printers led to a snowball-like accumulation of editorial matter. Sansovino brought out the *Storia* in 1562 with a brief life of the author. But Giovanni Maria Bonelli's printing of the same year had summaries, marginal *notabilia* and a subject index by Remigio Nannini. Sebastiano Bonelli, the printer's son, also claimed that its quarto format was more convenient than the folio or octavo of its predecessors. This edition proved the more successful, being copied by Bevilacqua in 1563, 1565 and 1568 with a new dedication and a brief life of the author by Sansovino.[45] The same editor got wind of the information that the Guicciardini family had left the four last books of the *Storia* unpublished because the historian had left this section in need of completion and revision when he died in 1540. Sansovino approached Lodovico Guicciardini, one of Francesco's nephews, in May 1563 with a request for these remaining books, alleging that someone else already intended to print them without permission.[46] However, another nephew, Agnolo di Girolamo, was in control of the publication of the *Storia*, and it was he who had Books 17–20 printed in 1564, choosing (for reasons which are not clear) to have this done by Giolito in Venice. The whole work was then printed by Giolito in 1567/8, with annotations on all twenty books including comparisons ('riscontri') with other historians, a life of the author, and indexes of subjects, of maxims ('sentenze') drawn from the history, and of historians used in compiling the notes, all by fra Remigio, and a sonnet by Porcacchi in praise of the dedicatee, Duke Cosimo.[47] Porcacchi edited the work himself in 1574 for Angelieri in the same style and with the addition of his *Giudicio* ('judgement') in praise of the beauty, moderation ('misura') and truth of the *Storia*. For once, Porcacchi did not interfere with the text, but he could not resist a somewhat critical comment in the *Giudicio* on what he considered to be Guicciardini's too Florentine language.

Both Sansovino and Porcacchi, with their Tuscan backgrounds, corrected histories written or translated in a manner considered to be insufficiently elegant. Sansovino printed in 1561 his own revision of the translation of Bruni's history of Florence made by the Florentine Donato Acciaiuoli in the early 1470s. In 1562 he corrected, again for his own press, Nicetas Choniates's chronicle of the twelfth-century Greek emperors in a hitherto unpublished translation by Sebastiano Fausto. We saw in chapter 7 that Fausto had revised Landino's translation of Simonetta's *Sforziada* in the 1540s, but now Fausto's prose was itself considered to be insufficiently polished. As Sansovino explained in his preface, he had revised the work because Fausto 'did not have ... as regards style, that beauty and those ornaments which usually give pleasure in writings, but, writing concisely, he did not observe rules in any elegant and pure way' ('non aveva ... quanto allo stile, quella bellezza e quegli ornamenti che sogliono piacere altrui nelle scritture, ma, scrivendo concisamente, non osservava regole in modo veruno elegante e purgato').[48]

Porcacchi edited the *Historia di Milano* of Bernardino Corio for Giorgio Cavalli in 1565. The history had been published in a splendid folio edition in Milan in 1503. It then languished in obscurity until revived in 1554 by Giovanni Maria Bonelli. On that occasion someone (Ruscelli, perhaps, following his work on Collenuccio in 1552?) had begun a revision of Corio's language but had then stopped because he was anxious, Bonelli claimed to readers, that he was changing the nature of the work. Bonelli said that he had therefore printed the work as it stood, emending only some Milanese forms. Whatever the real reason for the abandonment of the full revision, then, Bonelli's edition had ended up by being relatively sympathetic to Corio's original. Porcacchi, though, set about the work high-handedly. He cleansed it, as he put it, of 'certain errors of language, according to what everyone seemed to want' ('certi errori di lingua, secondo che 'l mondo pareva di desiderare'); if he had gone too far, he said, he had done so not to acquire honour or profit 'but only for the benefit of the work and to exalt its author' ('solo per benificio dell'opera, e per essaltation dell'Auttor d'essa'). This defensive tone is characteristic of his dedications, and it was not adopted without reason. Taking the 1554 version as his starting point, he subjected the work to a further linguistic and stylistic revision, freely rewriting sentences or omitting phrases which seemed to him redundant. Here, for example, are two sentences from the opening of Part VII in the three editions:

Doppo che tra il Duca e li Venetiani fu extincta la guerra e succedute le cosse dimonstrate ne la parte antecedente, parve ad ognuno il tutto esser constituto in pace, e non ad altro se attendeva che cumular richeze, circa dil che ogni via era concessa ... Nondimeno in essa tempestate per ogni canto le virtute per sì fato modo ribombavano che una tanta emulatione era suscitata tra Minerva e Venere, che ciaschuna di loro quanto al più poteva circhava de ornar la sua scola. (1503)

Dappoi che tra il Duca e i Venetiani fu estinta la guerra e succedute le cose dimostrate nella parte antecedente, parve ad ogn'uno il tutto esser costituito in pace, e non si attendeva ad altro che accumular ricchezze, cerca delche ogni via era concessa ... Nondimeno in esso tempo per ogni canto le virtù per sì fato modo ribombavano che una tanta emulatione era suscitata tra Minerva e Venere, che ogn'una di loro quanto più poteva cercava di ornare la sua scola. (1554)

Poi che fra 'l Duca e i Venetiani fu estinta la guerra, parve ad ogn'uno ch'ogni cosa fosse in pace, e non si attendeva ad altro che ad accumular ricchezze ... E in questo tempo da ogni canto le virtù per sì fato modo erano in pregio che concorrevano a gara co' diletti; a' quali senz'alcun riguardo molti vi trahevano, in guisa che cosa stupendissima era riputata da qualunque l'intendeva. (1565)

Porcacchi also cut out whole passages which must have seemed to him more appropriate for a diary or a chronicle than a history: an example is the colourful description of Pope Alexander VI's coronation near the start of Part VII.[49]

Since a polished surface was now everything, even a contemporary work by a Tuscan who was not primarily a man of letters might be given to Porcacchi for stylistic revision. In 1565 Varisco printed the *Libri quattro dell'antichità della città di Roma* by Bernardo Gamucci of San Gimignano, an architect and antiquarian who, said Varisco, had compiled the work at his request. Only four years later the same printer brought out a second edition of these *Antichità*, 'emended and corrected of infinite errors' ('da infiniti errori emendate e corrette') by Porcacchi. The editor explained in his dedication that he had undertaken to mend the style of the work as best he could and to transform it to an easier and more flowing arrangement of its parts ('m'ho preso carico d'accomodar, quanto ho saputo il meglio, questo stile, e di ridurlo a più facile e più corrente testura'). But on this occasion Porcacchi seems to have been boasting without justification, for the text which he claimed to have transformed was very close to that of 1565.[50]

Non-historical works also came within Porcacchi's province. In 1561 he revised, for Giolito, the *Libro de natura de Amore* by Mario Equicola, a strong supporter of an interregional 'courtly' language who died in 1525. A measure of the ruthlessness with which Porcacchi reworked texts in non-standard Italian is provided by a comparison with Dolce's much gentler treatment of Equicola's text for the same printer in 1554. On that occasion the preface noted that the work, though learned, was not in keeping with the rules and beauty ('vaghezza') of Tuscan, yet Dolce limited himself to superficial corrections to Latinizing and dialectal spellings and even left forms such as *in la* and *credeno*. The title page of Porcacchi's edition, on the other hand, ominously said the work had been 'riformato'. Like Dolce, he accused Equicola not only of breaking the rules of Tuscan but also of lacking the polish ('politezza') required nowadays. But he then set about improving Equicola's style and grammar. In Dolce's version, for example, the author's dedication began: 'Nella mia più fervida gioventù ritrovandomi in lacci d'amor involto, de miei verdi anni il tempo miglior in amor dispensai'. Porcacchi rearranged the word order and added his own embellishments to produce a less Latinizing opening: 'Ritrovandomi io ne' lacci d'amore involto, quando più il sangue in gioventù mi ribolliva, dispensai de' miei verdi anni il miglior tempo in amore'.

Porcacchi had no need of such interventions when, in the summer of 1570, he was asked by Giolito to add 'some labour of mine', as he put it ('alcuna fatica mia'), to Bembo's *Asolani*. To the edition which came out in 1571 Porcacchi contributed brief *argomenti* (provided, he said, to please Giolito rather than because Bembo's writings needed such things) and two sorts of marginal annotations. In italic type he pointed out important topics. In roman type he gave explanations of difficult words, adding some comments on whether the word was used by earlier writers or was unique to Bembo and once even daring to suggest that Bembo's form *succhio* should more correctly be *sugo* ('*sugo* pare che più propriamente dovrebbe dirsi'). In a few cases he referred the reader to the *Vocabolario* which he had already begun in 1570 and which he intended as a supplement to Alunno's *Fabrica del mondo*, covering minor approved early writers and also the best modern writers from Bembo to his own day.[51] Only on a couple of occasions did Porcacchi give supplementary explanations using Venetian dialect.[52]

Two of the members of the clergy who assisted Giolito, Remigio Nannini and Francesco Turchi, edited in 1565 and 1568 respectively the *Specchio di Croce* written around 1330 by the Tuscan friar Domenico Cavalca. Nannini's text was revised stylistically from that printed in Venice in 1515 or 1545: unlike the other texts which he had edited conservatively, this was aimed at a wide audience, and Giolito's dedication pointed out that the relatively large roman type of this quarto edition (his previous editions of 1543 and 1550 were in sixteens) made it especially suitable for old people. As for Turchi, Bongi drily remarks that he boasted with such insistence about his improvements to this treatise, born when books were still written on parchment rather than printed, that, 'without doing him an injustice, one can believe that this edition is hardly to be relied on as regards the faithfulness and integrity of the text' ('senza fargli torto può credersi che poco sia da fidare di questa edizione rispetto alla fedeltà ed integrità del testo').[53] This is a judgement which can be extended to the labours of Turchi's fellow-editors in Venice during the last decades of the Cinquecento.

11 · A 'TRUE AND LIVING IMAGE': EDITING IN FLORENCE, 1561-1600

IN CHAPTER 9 WE SAW A CONTRAST IN MID-CINQUECENTO FLORENCE between the scrupulousness with which classical texts were being edited by scholars such as Piero Vettori and, on the other hand, the dearth of scholarly editions of vernacular texts. This disparity disappeared in the last thirty or forty years of the century, thanks to a group of Florentines who studied in detail the language of the works which had laid the foundations of the literary dominance of Florence and who also gave some thought to the way in which their texts had been transmitted. Such studies were a potential source of great prestige to the Florentine state, and they were actively encouraged by the Medici family, in particular by Cosimo (who took the title of Grand Duke of Tuscany in 1569) and, after his death in 1574, by his son Francesco. Cosimo and his secretary Lelio Torelli also did their best to protect both Florentine literature and the Florentine book trade from the full force of the Inquisition. Archive material for Florence on this subject is not as rich as for Venice; but we know for instance that, when Paul IV's Index appeared in 1559, the Duke managed to arrange for two bonfires of prohibited books including a token selection of non-heretical ones instead of the complete destruction of all copies which the Church had requested, and that Torelli defended the booksellers when they complained of the effects of the Inquisition in 1570.[1]

The outstanding figure in this flourishing of vernacular studies was Vincenzio Borghini. Born in 1515, he joined the Benedictine order in 1531. His career as a scholar began in the field of Greek and Latin. From the 1540s onwards correspondence between him and Vettori show them to have been on terms of close friendship.[2] Significantly, Borghini was soon acting as an intermediary between the great classical scholar and Florentine presses. He could thus see Vettori's editorial methods at first hand. But he was already trusted to act on Vettori's behalf. In 1546 Vettori sent some emendations to Sophocles for Borghini to include, if he wished, in the edition being printed by Bernardo Giunta. He told Borghini that he had used an old manuscript from the library of San Marco (now BLF 32.9), which 'is worth a treasure for its age and excellence' ('vale un thesoro per l'antichità et bontà sua').[3] In subsequent letters, Vettori asked Borghini's opinions on textual problems in other Greek and Latin authors which he was editing for the Giunta or Torrentino presses and discussed a reading in the *Digest* of 1553. By 1550 Borghini was helping in the publication of a vernacular work, the first edition of his friend Vasari's *Vite*.[4] He was also in demand when contemporary texts needed to be censored before being submitted to the Inquisition for a licence for printing. We shall see later that he was one of those responsible for the revision of Guicciardini's *Storia d'Italia* before its publication in 1561. He may, suggests Plaisance, have been involved in the censorship of Grazzini's comedy *La Strega* for a Florentine edition planned, it seems, around 1572-4.[5]

Borghini's special expertise, though, lay in Tuscan texts of the Due and Trecento. He displayed it first in two letters addressed to Filippo Giunta in 1561 and 1562 on the preparatory work being carried out by Dionigi Atanagi for the edition of the chronicle of Matteo Villani which was printed for the Giunta firm in Venice in 1562.[6] The letters set out two important principles underlying Borghini's textual editing. On the one hand, an editor must not be afraid (as was Atanagi) to emend manifest errors which could be confidently attributed to a careless or ignorant copyist: faults in punctuation, superfluous or missing letters, or confusions in spelling such as 'astettare' for 'aspettare'. Manuscripts of this period (even Petrarch's) were 'very incorrect' ('scorrettissimi') in punctuation, and, when compositors were working from an old manuscript, as they were evidently doing here, they should remove the letter *h* when it was superfluous and add accents, apostrophes, punctuation ('punti') and capitals.[7] What an editor must avoid, on the other hand, was 'changing, according to his fancy, words and expressions which someone does not understand and takes to be incorrect, while they are not so' ('il mutare di fantasia voci e modi di dire che uno non intende, e gli paiono scorrette, e non sono'). Even in the case of clumsy syntax and archaisms, 'words should always be kept just as they are' ('le voci si dovrebbono mantenere sempre tali quali elle sono'). An editor's own taste, as Borghini made clear elsewhere, should not interfere with his choice.[8] It was, of course, potentially perilous to try to draw the line between a copyist's error and an unusual but genuine form. It was therefore essential to be thoroughly versed in the usage of the period. Such knowledge, Borghini realized, came not only from literary texts but also from practical ones and even from the spoken word: many archaic forms had been preserved in the speech of sections of the community outside the dominant Florentine male middle and upper classes, that is among women or country people or the urban lower classes. For example, Matteo Villani's form *mora* ('heap of stones', 3.57) had been changed by some to *monte*, and yet *mora* was still used in parts of Tuscany. The example to be avoided above all was that of Ruscelli, justly described by John Woodhouse as Borghini's *bête noire*.[9] Like Vettori, Borghini was instinctively conservative. 'It is a much less grave error,' he wrote, 'to leave an incorrect passage in an author than to poultice it so that it seems to be healthy' ('è molto minore errore lasciare un luogo scorretto in uno autore che impiastrarlo che paia ch'egli stia bene'); in this way the blemish was hidden and could not be healed later on. A further similarity with

Vettori which is revealed by these letters was that Borghini did not consider the explanation of the text to readers to be one of the major tasks of an editor. However, Borghini did not object to marginal glosses of good quality: if there were to be any *chiose* for Matteo Villani, he said, they should not be like the poor ones provided (by Nannini) for Giovanni Villani in 1559.[10]

Borghini was also aware by this time of another danger about which he did not feel it necessary to warn Atanagi: that of applying grammatical rules too strictly. In some notes made on Bembo's *Prose* in 1558 he discussed the old problem of whether *lui* could be used as a subject pronoun, as in Dante's 'latrando lui' (*Inf.* 32.105; *Prose* 3.48). For Borghini, it was irrelevant whether Dante used it 'either through error or through [normal] usage' ('o sia per errore o sia per uso'). Returning to this point in further notes possibly of the same period, he argued against the practice of taking rules formulated from Latin and applying them to the vernacular, as if the latter were not an independent language. 'This is what I complain about so often,' he wrote, 'that the printers' desire to squeeze our writers into Roman laws and rules has ruined an endless number of passages and has killed off many words and expressions which have to be recovered with old books' ('ecco qui quello che io mi dolgo sì spesso, che volendo far gli stampatori stregnere i nostri scrittori alle leggi e regole romane, ha guasto infiniti luoghi e spente molte voci e modi di parlare, dei quali bisogna ritrovare con i libri antichi').[11]

While the basic tenets of Borghini's method were clearly established by 1561, his own historical knowledge of the vernacular was then still at a relatively early stage. Around 1557–8, for example, Borghini used two manuscripts of *Paradiso*, one completed in 1338, to question several readings of the Aldine edition of 1515. He was seduced by the *lectio facilior* 'corrente' for 'torrente' at 17.42. But in several cases he quite rightly spotted as erroneous some readings which Bembo had derived from his father's manuscript (Vat) or from elsewhere.[12] He corrected, for example, 'fertile monte d'alta costa pende' (11.45, from Vat) to 'fertile costa d'alto monte pende' because it made better sense; 'come del *corpo* il cibo che s'appone' (16.69, from Vat) to 'vostro' because 'it seems better put and more appropriate' ('par me' detto e più proprio'); and 'sopra la *poppa* ch'al presente è carca' (16.94, perhaps Bembo's conjecture) to 'porta', arguing correctly that someone had tried here to extend the ship metaphor from 16.96. In view of the accuracy of Borghini's perceptions, Pozzi's comment that Borghini's choices were sometimes made on subjective grounds of taste may seem rather harsh. But it is true that Borghini did not back up his case with other references to Dante or, as Pozzi points out, to other texts of Dante's time. It was probably not until around 1565, when Borghini undertook his researches into the early history of Florence, that he began his systematic reading of vernacular texts from the thirteenth and fourteenth centuries.[13]

One of the Trecento authors whom Borghini studied in detail was Giovanni

Villani. He undertook the immense task of noting the variants of several manuscripts, often identified by a particular sign, in the margins of his copy of the 1559 edition.[14] He took a critical attitude to these other sources as well as to the printed text: next to one emendation he noted that 'even manuscripts are sometimes wrong' ('anche talvolta gli scritti a mano errano'). His corrections to Nannini's edition led to important methodological reflections in a series of *castigationes*, now in BNF 2.10.66, the first such vernacular work in a genre much used by classical scholars. His main point, often reiterated, was that editors should beware of the tendency of scribes and printers to substitute rare words with a *lectio facilior*; examples are 'Cipri' for 'Citri' (1.13) and 'mollume' for 'melume' in Pietro Crescenzi. He also noted that a *lectio difficilior* such as 'scipì' was unlikely to have been created by chance or by a drowsy scribe. For editors, he said, a little knowledge was a dangerous thing: they should be either ignorant, in which case they would not interfere with the text, or well informed, so that any changes were justified. The sure way was to leave ancient authors as one found them ('la sicura sarà sempre lasciar stare gli antichi come huom gli truova': p. 37). All these arguments were backed up with evidence from other fourteenth-century authors, from the speech of women and countryfolk, and from historical knowledge.

Another important source for our knowledge of Borghini's method is an unfinished 'Lettera intorno a' manoscritti antichi', written after 1569 and before 1573 in response to the queries of a friend.[15] To the question of how one was to distinguish between good and bad texts, Borghini answered that age and reliability could be judged by the handwriting and the presence or absence of authentic rather than modernized forms; comparison with other works was also important. The friend further asked why there were such differences between texts, and how one was to distinguish between better and less good readings. Borghini's response was to contrast the practices of professional scribes, prone to misinterpret abbreviations and to replace the unfamiliar with the familiar lest it be 'less saleable' ('manco vendereccio'), and the procedure of amateurs, who copied what they found faithfully, as in the case of Mannelli's *Decameron*, a manuscript which he heard to be excellent. Some scribes did not scruple to paraphrase the original. Others changed it so as to show that they knew better, sometimes because they believed that old language had to be modernized. Borghini advised editors to try to make sense of what was there, not to change it so that it said what one thought it should say. He acknowledged too the pressure of the expectations of readers, who wanted to understand everything first time. Editors should use conjectures and 'probabilities' ('verisimili') as little as possible. Contemporaries were condemned for contravening this principle as well as for making old authors 'speak in today's manner' ('parlare all'uso d'oggi'). His own edition of the *Decameron* would, he said, have readings which seemed strange, but this was part of the process of restoring Boccaccio's pure and natural language.

However, Borghini did not advocate uncritical reproduction of a manuscript. He reproached Iacopo Corbinelli (to whose ultra-conservative editing we shall return) for including even errors. One should not have blind faith in any manuscript; on the other hand, if one did change anything, one should do so only with the 'security' ('malleveria') of the best texts backed up by reason. If one was in doubt, it was better to leave the text alone so that readers could make up their own minds.

The first editorial fruit of these studies and reflections was the *Novellino* printed by the Giunta press in 1572.[16] In 1571, as we saw in chapter 10, Sansovino in Venice had simply reproduced Gualteruzzi's 1525 text. Borghini did not wish to do this, in spite of his admiration for Gualteruzzi's mentor, Bembo. One reason was the anti-clerical content of some of the stories: not only was Borghini a monk, but he would not have wanted to antagonize the Inquisition at a time when Florence was fighting to save from censorship as much as possible of a far more important work, the *Decameron*. Another reason, as we shall see shortly, was his disagreement with Gualteruzzi's editorial principles.

Borghini's work both as a censor of the *Novellino* and as its linguistic editor was helped by the availability of a manuscript, now part of BNF Panciatichiano 32, which also provided him with a new title, *Libro di novelle e di bel parlar gentile*, for what was in effect a new collection. He made Gualteruzzi's first story into a separate prologue and deleted seventeen stories from the 1525 version. Of these, two (57 and 86) were obscene. All the others were connected in some way with Christianity, for instance because of a Biblical source or because members of the clergy appeared in them. Borghini made up the total of 100 with stories from other sources, though only six and part of a seventh came from his manuscript and could thus be considered part of the *Novellino*. Four stories from the Quattrocento were added as an appendix. All the stories were subject to censorship.[17]

The philological principles of the edition were set out in the preface, signed by the brothers Filippo and Iacopo Giunta but composed by Borghini.[18] Of Gualteruzzi's editing, Borghini made the same criticism that he had made of Atanagi a decade earlier. Too much reverence had been shown for the errors of an ignorant scribe. Borghini, in contrast, had removed the idiosyncrasies of this man's spelling. In a draft of the preface he compared this process to polishing a rough gemstone to give it the true colour which the copyists had dimmed, and the final redaction used the image of a snake shedding an 'old and rough skin' ('vecchio e rozzo scoglio'). His example was the treatment of the word 'moaddo' (story 27), which he considered had never existed. When a story was found in just one of his two sources, he corrected 'only those errors which were seen to derive from the pen' ('quelli errori soli che dalla penna derivar si vedevano'). Otherwise, since Borghini's manuscript was thought to have been written by a person who understood his material much better, this was used to correct the stories which it shared with Gualteruzzi's source. The danger of making such judgements on scribes is illustrated, though, by the case of 'moaddo': Borghini preferred the *lectio facilior* 'uomo' found in his manuscript, but 'moaddo' could well be a word of oriental origin.[19] Unlike Gualteruzzi, who assumed that the *Novellino* was written by one person, Borghini argued that the variety of its style showed that it had more than one author. This belief that the work was not conceived as a unity no doubt helped to make him willing to alter the contents of the collection.

The preface also pointed out the usefulness of this work in making students of Tuscan aware of words used by past writers, though these words were not necessarily to be used. Here Borghini was perhaps correcting Sansovino's claim (on his 1571 title page) that the stories would help 'those who wanted to write according to the rules in our language' ('chi vuol regolatamente scrivere nella nostra lingua'). But Borghini also made use of the models provided by editors such as Sansovino. His text was preceded by a Venetian-style glossary of some old words in the text ('Dichiarationi d'alcune voci antiche'). The contents, though, were typical of Borghini, reference being made frequently to the usage of Trecento Florentine authors and also to contemporary rural pronunciation.[20]

When a story was found both in the 1525 text and in his manuscript, Borghini tended to follow the former, and he also followed Gualteruzzi in using the comma and full stop as his only punctuation. He did, however, use the apostrophe, if sparingly, and he removed Gualteruzzi's spellings with *k* and *cie*. In some cases he preferred forms or expressions from his manuscript.[21] He could combine the two readings, as in these cases from stories 1 and 68 respectively:

1525:	E ki avrà quore
MS:	Et quale aree [*sic*] cor
Borghini:	E quale havrà quore
1525:	E Maestro, io ò veduto cosa ke molto mi dispiacie all'animo mio
MS:	Maestro, io hoe veduta una cosa la quale mi dispiace e ingiuria l'anima mio [*sic*] molto
Borghini:	Maestro, io ho veduto cosa che mi dispiace et ingiuria l'anima mia molto

He might also omit words found in both sources or make a conjecture when he suspected that both were corrupt, as in his 'la virtù ragionevolmente manca' (story 68), where Gualteruzzi had 'la virtù ragionevole e manca' and the manuscript had 'la vertù lagiovenile mente mancha'.

In stories for which the manuscript was his only source, Borghini made his usual improvements to spelling and to article, pronoun and verb forms. Words could be omitted, altered, or inserted in order to improve the style or occasionally the sense.²² In his story 6 he inserted a passage, translated from the Latin source, without which the beginning of the tale would not have been comprehensible.

In a later unpublished essay closely related to the letter on manuscripts, Borghini went into more details on the criteria used in this edition.²³ He gave further examples of the scribal errors perpetuated by Gualteruzzi who, he concluded, could be excused for not wanting to interpose his own judgement but had nevertheless represented 'the ignorance and negligence of an ancient' ('l'ignoranza e la trascutaggine d'uno antico') rather than, as he had intended, 'the pure image of antiquity' ('[la] pura imagine della antichità'). Borghini believed it possible and necessary to distinguish between 'errors' ('errori') and genuine archaic forms which were 'characteristic qualities of the time' ('proprietà del tempo'), between what belonged to the copyist ('quello che è propio de l'uomo che scrisse') and what belonged to the time when the text was written ('quel che è propio del tempo quando fu scritto').

In this essay, however, the case of the *Novellino* was being used to justify the methods which had been followed in editing Boccaccio's *Decameron*. Cosimo de' Medici, concerned at rumours that the first expurgated *Decameron* was to be printed in Rome, had successfully asked the Vatican in 1570 to allow this edition to be prepared and printed in Florence.²⁴ In March 1571 the Spaniard fra Tommaso Manrique, Master of the Sacred Palace, despatched to Filippo Giunta both a copy of the 1522 edition containing indications of the cuts to be made and permission for Filippo to pass this copy on to the Florentine Academicians or 'others skilled in the Tuscan language' ('altri periti della lingua toscana') for them to link up ('continuar') the remaining text.²⁵ Cosimo had, in fact, appointed a number of 'Deputati' for this purpose. The surviving correspondence concerning the preparation of the edition shows that Antonio Benivieni, Bastiano Antinori, Braccio Ricasoli and Giovambattista Adriani were among those involved, that Vettori was consulted, but that the foremost was Borghini.²⁶

As Borghini's essay on their editorial criteria explained, the Deputati had applied the same distinction used in his *Novellino* between 'writing' and 'words' ('scrittura' and 'voci'). The former had been somewhat modernized: they preferred to concentrate on the 'truth' or 'substance' ('vero', 'sustanzia') rather than 'appearance' or 'accidents' ('aparenza', 'accidenti'). It was partly a question of habit, he said. One became accustomed to one's own spellings and pronunciation. But there was also a hint of patriotic pride in contemporary Florentine in the argument that, since the fourteenth century, the Tuscan language had become in some respects much 'sweetened' ('indolcita'), losing 'some of that rough hardness and uncultivated simplicity of our ancients' ('non poco di quella rozza durezza e inculta simplicità de' nostri antichi'). Borghini therefore rejected the conservative edition of Boccaccio's *Corbaccio*, published in Paris in 1569, in which Corbinelli had closely followed a source derived from the manuscript copied by Francesco Mannelli. Yet, he wrote, one must avoid making Boccaccio speak 'with the tongue of those who were born many years later' ('colla lingua di quei che nacquono molti anni poi'). Unlike other editors (and he was thinking of such as Ruscelli), the Deputati had to keep archaic words attested by the best texts. Borghini's editing, unlike that of his Venetian counterparts, was not affected by a need to see older writers as a model for one's own usage and hence to adapt them to one's expectations: 'one must leave every writer's language just as it is, and if it doesn't suit you to use it, leave it alone' ('a ogni scrittore si ha da lasciar la lingua sua tal quale ella è, e se poi non ti vien bene d'adoperarla, lasciala stare'). Portraits of old writers, he pointed out, did not show them dressed in modern clothes. On the other hand, a portrait could concentrate too much on external details rather than on the essence of the subject. Later in the essay Borghini used another metaphor from portraiture: the Deputati, he said, hoped that their method was 'a true representation of the true and living image' ('un veramente rappresentare la vera e viva immagine') of antiquity and not merely the outer garments, 'the colours of the clothes and the buttons and the folds' ('i colori delle vesti e i bottoni e le pieghe'), as might have happened if they had used spellings such as *ke* for *che* or *io to* for *io t'ho*.[27] Putting these two analogies together, we see that Borghini is saying that the editor must not portray the past anachronistically yet must not be so obsessed with extrinsic and superficial details that his work turns out to be untrue to the original.

Borghini's use of Corbinelli's *Corbaccio* as an example was particularly relevant because his main source for the *Decameron* was Mannelli's copy of this work in the same manuscript (BLF 42.1, known as Mn), of which Florentine editors had made some use in 1516 and 1527. He termed this manuscript the 'Ottimo' ('best'), and it was this source above all that justified his confidence in the authenticity of certain forms in his text. Borghini's reverence for it, in spite of its acknowledged faults, will have been increased by the fact that it was in the possession of Grand Duke Cosimo, the patron of the whole enterprise, and probably by the precedent of Vettori's confidence in the Codex Mediceus of Cicero. Borghini told Manrique in 1571 that the Deputati would be helped above all by this 'most ancient and faithful text' ('testo antichissimo e fedelissimo').[28] To Bernardo Canigiani in 1573 he wrote that they had worked not by fancy, conjecture or probability, but with the certain and sure guidance of old texts and above all the 'Ottimo', a pole star ('tramontana') to guide them on their way; all they had to do was to look for examples to confirm readings or to explain words and expressions which this text had in their pure form but which in other texts had been 'spoiled or modernized and very often Latinized by certain Ruscellis, the real plague of this language' ('guaste o ridotte all'uso moderno e bene spesso all'uso latino da certi Ruscelli, che son propriamente la peste di questa lingua').[29]

The Deputati also had much respect for the Giuntina of 1527. They used a copy of it (now BNF 22.A.5.18) to record variants and preferred some of its readings to those of Mannelli. They also consulted the copy-text (BNF 22.A.4.2) which the 1527 editors had given to the printers. Much less use was made of other sources mentioned by the Deputati: an early printed edition, probably the 'Deo Gratias', and some other manuscripts, of which the best (now lost) belonged to Lodovico Beccadelli of Bologna. They even considered the interrelationship of Mn and the 'Deo Gratias', judging that the latter was not derived from the former or from the same original source; but their main aim here was to promote the worth of Mn rather than to establish a *stemma codicum*.[30]

It was felt necessary to accompany this new text of the *Decameron* with notes which would defend its innovations and explain correct readings which had been rejected by some recent editors. The 1527 edition had had limited success, Borghini believed, because of the lack of such justification.[31] A volume of *Annotationi et discorsi sopra alcuni luoghi del Decameron* was duly printed in 1574. An important 'Proemio' described the sources used by the Deputati and the other early texts consulted and insisted that editors must avoid 'improving' things which they do not understand. The *Annotationi* themselves were based on the principles and methods which we have already seen to be characteristic of Borghini. Uncommon words were confirmed on the authority of Due or Trecento texts, of contemporary women speakers (in the case of 'iscretio'), or of labourers in the Florentine area (in the case of 'atare'). Out of the justification of 'habituri' rather than 'habitari' arose the general consideration that rules always came from natural usage, not vice versa ('le regole furon sempre cavate dall'uso naturale, e non l'uso da quelle').[32] The discussion of 'gli tenesse' rather than 'gli narrasse' (2.1.30) was used to show how more recent copyists tended to replace words 'characteristic of old books' ('[voci] propie de' libri antichi') with 'a clearer or more used word' ('una voce più chiara o più usata'). Also typical of the notes is a faith in the 'good text' and a reluctance to depart from it even when it was apparently wrong. One catches an echo of Vettori in the comment on a reading of Mn, 'poteva' in place of 'vedea' in the idiomatic phrase 'più qua né più là non vedea' (8.4.6): the former was still recorded in case a reader preferred it, for this manuscript, it was explained, rarely erred and someone might find in another text a comparison ('riscontro') to confirm it. These *Annotationi* had a completely different aim, then, from the kind of notes which Venetian editors had provided and which Borghini considered unnecessary, at least for Florentine readers.[33] He could devote himself to explaining the vocabulary and usage of a difficult early text such as the *Novellino*, but one cannot imagine him undertaking routine notes on a well-known text such as the *Decameron*. To have done so might be seen as suggesting that Florentines were not linguistically privileged by birth in the way they often proclaimed. It may seem surprising that this was acceptable to Florentine printers: they were not catering only for local readers, if we are to believe what they said about the proportion of their output which they had to export in order to survive. It is clear, though, that the home market had a high priority. This was the case, after all, even for Venetian printers, as one can see from the use of local dialect in some of their glossaries.

Concurrently with their philological operation, the Deputati were obliged to carry out the censorship ordained by Rome.[34] Manrique had indicated certain words or passages which had to be either removed or replaced in order to protect the clergy and 'other sacred things' ('altre cose sacre') from any hint of scandal, leaving to the Deputati the freedom to make proposals about what replacements, if any, were to be made. At an early stage Borghini wrote a paper to his fellow editors ('Discorso agli amici'), typically full of good sense, on the method to be followed if the wounds inflicted in Rome were to be as little apparent as possible.[35] For Borghini, the *Decameron* was both the major monument of Tuscan prose and a work of fiction: 'the maximum of the language' ('il più della lingua') had to be retained, then, but so too did verisimilitude ('il verisimile'). He was concerned on the one hand about the attitude of those who saw the text as untouchable and who wanted to remove stories entirely rather than change the language, and on the other hand about the less extremist attitude of those who thought it possible to publish the work with 'many asterisks and gaps' ('assai stelloline e vani') wherever Rome had indicated a cut. Both of these approaches would be like cutting off one's nose to spite one's face. They would ruin the work for new readers. It was far better to save everything that could be saved in the best way possible ('tutto quello che si può salvare si salvi nel miglior modo che si può'), while making changes with the greatest respect and only when necessary.

In the end, one story alone was considered so hard to censor that it was omitted: that of how a witty reply confounded the hypocrisy of a Franciscan inquisitor (1.6). In the rest of the work, the Deputati used additions, excisions and substitutions in order to satisfy Manrique. A different typeface (italic instead of roman in the titles, roman instead of italic in the stories) was used for many of the words which were not Boccaccio's own, especially in the early stories, 'so that they should deceive nobody' ('perché non ingannino alcuno').[36] Asterisks indicated cuts. Sometimes these left a sentence hanging in mid-air, but the Deputati preferred in general to delete whole passages rather than to save isolated fragments of sentences.[37] The main technique used in order to satisfy Rome was to turn members of the clergy into lay people. Once this was done, the Inquisition did not care about the characters' behaviour. At one point, for instance, the correspondence between the Deputati and Manrique touched on two cases of extramarital sex, those of Ricciardo and Catella (3.6) and Ghismonda (4.1).[38] Manrique wanted to censor the phrases 'non possendo' or 'non potendo' ('not being able'), referring to the inability of Ricciardo and Ghismonda to resist the power of love. What concerned him was that this phrase seemed to deny free will, rather than that it seemed to condone free love. The Deputati argued in reply that fiction could legitimately depict human frailties (one did not go to Boccaccio to learn how to live) and that there was a distinction between the literary depiction of the helplessness of the lover (as exemplified from Guittone to Petrarch) and the way in which philosophers or theologians would write. Borghini and the Deputati also insisted doggedly that Manrique's censorship of some words or phrases was based simply on a misunderstanding of Boccaccio's times and of his language. For example, they argued that, once they had turned fra Cipolla into a layman, 'cappuccio' ('hood') could remain in the text (6.10.34) since in those times it was worn by all. 'Santoccio' (7.3.33) had nothing to do with saints, as Manrique appeared to think, and simply meant the father of a child being baptized.[39]

Borghini, then, managed to convince his colleagues to publish a *Decameron* which, though censored, was as full and as seamless as possible. Once this internal battle was won, he and the Deputati succeeded in engaging Manrique in a dialogue, in spite of the Spaniard's exasperation over the prolonged negotiations. Through persistence and diplomacy (they even made suggestions for some additional cuts), they were able to obtain some concessions and to limit the number of 'wounds' which they had to heal. In August 1572 Manrique and the Florentine Inquisitor issued their licences for the *Decameron* to be printed.

The efforts of the Deputati were almost in vain. Manrique died in 1573, and his ecclesiastical colleagues found the censorship far too lenient for their tastes. Foreseeing the danger of further cuts, the Deputati hurried to have the work printed even before they had completed the *Annotationi* which were intended to accompany it.[40] Printing must have been carried out in the period leading up to 15 May, the date of the dedication. But in June the Congregation of the Index and the new Master of the Sacred Palace, fra Paolo Costabili, ordered the Inquisitor of Bologna to ban sales of the edition, a prohibition presumably applied in other cities, and the Giunta were ordered not to sell it.[41] A new list of cuts was despatched to Florence. Given their extreme savagery (thirty-five stories were to be removed entirely) and the despair which Borghini must have now felt, it is not surprising that Florence was unable, for the moment, to produce another *Decameron* to satisfy the Inquisition.[42]

One of the Trecento prose works cited by the Deputati in the *Annotationi* was an unpublished *Istoria delle cose avvenute in Toscana dall'anno 1300 al 1348* (History of events in Tuscany, 1300–48), better known, because the author was apparently from Pistoia, as the *Istorie pistolesi* or *pistoresi*. This work was prepared by Borghini for an edition printed by the Giunta press in 1578. From a manuscript of 1396 a copy was made for the printers.[43] The scribe introduced various modernizations, but some of the original forms were restored at a later stage.[44] The letter to the readers contained a theme dear to the great linguist's heart: that the Florentine language had been infected since the Trecento by outside ('forestiero') influence.⁴⁵ These histories represented Florentine in its uncontaminated state, like 'a virgin beautiful, intact, and ours' ('una vergine bella, intatta, e nostra'). The letter also reiterated his principle that careless scribal errors should be treated as 'pure and obvious mistakes' ('iscorrezioni mere et manifeste'), while caution should be exercised over doubtful terms which might be correct but whose meaning was now obscure. One could tolerate slight variations in spelling, such as 'uficiali' versus 'oficiali' or 'sagace' versus 'segace', the letter went on, but sometimes an editorial preference had to be exercised: 'Pistolesi', for instance, was used more than 'Pistoresi'. Once printing was over, Borghini used the convenient fiction of typographical errors to give a last-minute list of yet more improvements. One of these restored the archaic feminine gender to 'ordine' (*preso l'ordine* > 'presa'). But many others imposed Bembo's norms: for example, 'le chiave' > 'le chiavi', 'erono' > 'erano', 'deliberorono' > '-arono', 'fecenlo' > 'fecionlo'. Although Borghini believed that the author probably came from Pistoia, he still changed 'dolsoro', with the unstressed *-or-* characteristic of that area, to 'dolsono' or 'dolsero', and did the same with other such forms.⁴⁶ The original forms were dismissed as mere scribal 'habits' ('vezzi'), and so the traditional claim could be made that the book had been printed 'in the same form in which it was left to us by the author' ('nella stessa forma nella quale dal proprio autore ci fu lasciato'). Faced with non-standard noun and verb endings, then, Borghini followed editorial convention in allowing the 'golden' Florentine of Boccaccio to exercise a powerful normative influence.

Another Trecento work which Borghini wanted to see in print for the first time was the *Trecentonovelle* of Franco Sacchetti. He was discussing the possibility of an edition in November 1571, though at that point he preferred to delegate the task to someone else.⁴⁷ If the text had been printed, it would have been modernized, reorganized and, unavoidably, censored in the same way as his *Novellino*. Borghini had a copy made from the only surviving manuscript, which was already incomplete and divided into two parts midway through story 140. At first he intended to select from the first part one hundred stories of which some were to be revised ('da racconciarsi') because of their content. Later, after he had obtained the second part, a selection was made of 134 of the 222 stories which survived in whole or in part, and these stories were censored and polished linguistically.⁴⁸

In 1579 a Florentine wrote a letter which commented scathingly on Borghini's subterfuge in the errata of the *Istorie pistolesi* and which accused him of doing what he had deplored in others, that is reducing everything to the rules as if non-standard language was unworthy. Don Vincenzio, the letter went,

is showing partiality in certain things that he does not understand or does not want to understand, so as not to alter the rules of the rule-givers or of the Academies,

pretending to classify certain words as printing errors. The important thing is that the author's language went like that. And they think that what is not found in Boccaccio or Petrarch is not language. It's ignorance or cunning.
(è partiale in certe cose che non intende o non vuole intendere per non alterar le regole de' regolanti o dell'Accademie, mostrando di metter certe parole sotto scorrettione di stampa. L'importanza è che la lingua dell'autore correva a quel modo. Et par loro che ciò che non si truova nel Boccaccio o nel Petrarca non sia lingua. Ell'è ignoranza o malitia.)

These words were written, however, not on the banks of the Arno but in France, where their author had sought refuge from the Medicean regime. His name was Iacopo Corbinelli.[49] He had been strongly influenced by the teaching of the humanist Pietro Angeli da Barga, whom he described later as the 'only light' ('solo ... lume') of his early studies.[50] The foundations of his scholarship were thus laid in Florence. But he left the city by 1562 following his brother's involvement in a conspiracy. Not yet thirty at this point (he was born in 1535), Corbinelli was able to develop a highly individual method when he turned to editing vernacular texts first in Lyons and then in Paris. A general characteristic of this method was that he concentrated either on works which had been ignored by other editors or at least on lesser-known works for which he almost always had new manuscript evidence. A second characteristic, which applied to his editing of pre-Quattrocento vernacular works, was his close adherence to the spelling of his source. He would not insert anything more than basic punctuation and refused to add the accents and apostrophes which he nevertheless did not exclude from texts of the Cinquecento. Corbinelli clearly had a taste for the archaic; but he also had a sensitive ear. He argued that the original spellings had more euphony and that, by using apostrophes (as Bembo had done) to show readers which vowels were to be elided or apocopated if a line of verse was to have the correct number of syllables, one robbed poetry of grace and dignity.[51]

Corbinelli adhered steadfastly to these two principles during an editorial career which lasted for over two decades. The first edition which he brought out, in 1568, was a collection of Duecento prose works including some by Brunetto Latini, another Florentine exile in France. Corbinelli and the Lyonnais printer Jean de Tournes followed faithfully the manuscript supplied by Giovanni Francesco Pusterla.[52] One thus had spellings such as 'siae' (= 'si hae') or 'impercioke', sentences beginning with lower-case letters, two words undivided, and even *v* rather than *u* for a vowel within a word when the manuscript used it (thus 'qval', 'svo', pp. 127–8). Possible corruptions were marked with a cross; Corbinelli said he had corrected only a very few obviously wrong words and had left others 'in their incorrectness' ('nella loro scorrettione'). In the following year the *Corbaccio* already mentioned was printed in Paris. 'As I found [the work], so I give it', Corbinelli wrote in his dedication ('Quale io l'ho ritrovata, tal la do'). He explained that he intervened in Mannelli's text only to separate words, because he found the original more musical and because it was rash to try to keep to definite rules all the time. Here, as in later editions, suspect readings were marked with an asterisk. After a few years he edited some of Guicciardini's *Ricordi* (*Più consigli et avvertimenti*, Paris, Morel, 1576), the first edition of the original Latin text of the *De vulgari eloquentia*, a work full of the anti-Florentine bitterness of the recently exiled Dante (Paris, Corbon, 1577), and a poem by a contemporary Florentine who had also suffered exile, *La fisica* of Paolo del Rosso (Paris, Le Voirrier, 1578).[53] His last edition (1589, with further issues up to 1595) was a collection of verse, the *Bellamano* by Giusto de' Conti, a Roman contemporary of Petrarch, together with a collection of early poetry (*Raccolto di antiche rime*) intended as a supplement to the *Sonetti e canzoni* of 1527.[54]

There is no doubt that Corbinelli's editing was in large measure an expression of his resentment towards contemporary Florence. It was evidently influenced by a non-Florentine model, Gualteruzzi's *Novellino* of 1525. That is not to say that his methods did not have recognizable roots in at least one Florentine edition from before his exile: Torelli's edition of Justinian. But Corbinelli's work on the vernacular could not have developed as it did unless he had been free from the pressures of having to work for Italian printers and the Italian public, and free too from the conformism which would have been expected of him in Florence. Equally, his remoteness and above all his alienation from the intellectual life of Medicean Florence meant that he could never have any influence in his native city.

The Florentine scholar whom Corbinelli detested above all was a slightly younger contemporary, Lionardo Salviati (1539–89).[55] He had begun to direct his venom at Salviati soon after his exile, but his hostility rose to a peak in the 1580s, when Salviati brought out a newly revised *Decameron* (1582) and then two volumes of *Avvertimenti* on its language (1584 and 1586). The Vatican had granted permission to Filippo and Iacopo Giunta for a fresh expurgation of the *Decameron* in 1580. At first the Master of the Sacred Palace wanted thirty stories removed but, after discussions with various Florentine men of letters, Salviati offered to seek a better solution. In August the Grand Duke Francesco appointed him to 'correct and expurgate' ('correggere et purgare') the work, and the task was completed remarkably quickly, probably by May 1581 at the latest.[56] It was another year, though, before Salviati's edition was published by the Giunta brothers, first in Venice in July and then in Florence in October–November.

For his 'correction', in other words for the philological aspect of his task, Salviati followed to a great extent in the footsteps of the late Vincenzio Borghini. He came prepared, like Borghini, with a wide methodical study of published and unpublished Trecento texts. He named the same principal sources for the *Decameron*, beginning of course with the Mannelli manuscript (Mn). Like Borghini, he believed that one had to be very cautious in correcting what appeared to be errors in Mn. 'We have preferred to leave its difficulties,' he told readers, 'rather than remove them, as the phrase goes, on a whim' ('più tosto c'è piaciuto di lasciarci le difficultà che di torle via, come si dice, a capriccio'). And, like Borghini, Salviati believed that the spelling of the Trecento needed to be modernized and standardized in some respects. Nevertheless, Salviati's editing of the *Decameron* had characteristics which distinguish it from Borghini's, because he applied two of Borghini's principles with even greater rigour.

Firstly, Salviati felt even more reverence for Mn. Boccaccio's original, he said, would be hardly more reliable.[57] As a result, he was even more wary about changing the substance of its text. When he criticized in the *Avvertimenti* (Volume I, 1.11) some readings corrected conjecturally ('corretti di fantasia') by other editors, one might expect him to have been attacking those working in Venice; but his targets included the Florentine editors of 1527 and 1573. They, he considered, had wrongly abandoned readings of Mn such as 'che ne *faresti* voi più' for 'fareste' (10.8.82) or 'essendo arche grandi' for 'essendo quelle arche grandi' (6.9.10). The next chapter listed readings which Salviati had left 'lacking or imperfect' ('difettosi o imperfetti') rather than fall into the trap, as he saw it, of conjectural emendation. He even held the same two editions guilty of changing archaisms because they felt them to be wrong: they both had 'i patti *raffermati*' (3.5.10), for example, following modern usage, whereas Mn and the 'Deo Gratias' had the older form 'rifermati'.[58]

Secondly, Salviati shared Borghini's conviction that an editor should distinguish between the 'substance' and the 'accidents' of his source but, unlike Borghini, he developed and explained the rationale for his method. This he did in the letter to the readers in 1582 and then at length in the *Avvertimenti*. The letter claimed that he had followed the 'words' ('parole') of the text exactly but had not kept its spelling ('scrittura') when this did not follow 'reason' ('ragione'), which demanded that spelling should follow pronunciation. He therefore rejected spellings, many of them Latinizing, which did not represent Florentine pronunciation, such as *t* rather than *z* in *letizia* and so on, etymological *h* (though he kept this in some cases, for reasons of tradition), *optimo, Idio, magnifice*. There is a superficial resemblance here with Ruscelli's 'rational' spelling, mentioned in chapter 8, but Salviati was the first editor to argue that spelling should follow sound. Corbinelli, indeed, had taken the opposite line, saying that it was wrong to try to teach pronunciation through spelling and that, if anything, one should perhaps follow the pronunciation indicated by old spellings.[59] Salviati, in contrast, claimed in the *Avvertimenti* (Volume I, Book I) that Mannelli's spelling, probably little different from Boccaccio's, was harsh, inconsistent, irrational, at some times inadequate but at others superfluous. He followed it where he could, so that the work should remain as far as possible 'in its original simplicity' ('nella sua prima semplicità', 1.7). But some words were written 'badly'. This could be because the spelling did not follow any pronunciation, as in *apto* or *decto*, or, as in the case of *etterno*, because it followed a pronunciation which might have existed in the Trecento but which had less 'sweetness' than the modern *eterno*. Mannelli's spelling could also be irrational, using letters not needed to indicate pronunciation (as in *conosciere, ad voi, basciare*) or ones which did not represent pronunciation (as in *inpose, vitio*), or omitting letters which were necessary (as in *meglo, magnifice*).

These solutions were explained in more detail in Book 3, chapter 2 of the *Avvertimenti*. First of all, Salviati stressed the importance of spelling if the editor was to give the correct sense. Word division and the apostrophe had to be used with great care: for instance, he followed Mn in reading 'vedendolo huomo attempato' at 5.5.37, whereas other editions had 'vedendo l'huomo attempato'. Sometimes writing ('scrittura') must be improved in order to clarify the sense. Thus, in one instance where Mn had '*mai* recato' (8 concl. 10), Salviati had 'm'hai'; and elsewhere editors had to decide whether 'alloro' meant 'a loro' ('to them', with written initial doubling) or 'al loro' ('to theirs'). In general, though, and in opposition to Corbinelli, Salviati took it as axiomatic that *scrittura* had to follow pronunciation. But which pronunciation, old or modern? His answer was that fourteenth-century authors were 'in writing their words, and even perhaps in pronouncing them, less perfect in some details than the moderns are' ('in iscrivendo le voci loro, e anche forse in pronunziandole, in alcune spezietà men perfetti che i moderni non sono'). One should follow modern usage, for example, in using certain double consonants (as in *femmine* not *femine*, *sovvenire* not *sovenire*, *dicemmo* not *dicemo*) because here 'it seems easier for the tongue and far more pleasing to the ears' ('più agevole pare alla lingua e all'orecchie più dilettevole assai'), and one should avoid spellings such as *bascio* or *conscienzia* or *ad una ora* which did not correspond with the way Tuscans spoke. The rules for good spelling which Salviati went on to evolve were based on the principle that 'Tuscan pronunciation avoids effort and harshness' ('la toscana pronunzia fugge la fatica e l'asprezza').

Salviati's justification of the use of sixteenth-century Florentine pronunciation in an edition of a fourteenth-century text may have been provoked in part by Corbinelli's insistence on the primacy of old spelling. But his main motive was clearly his intensely patriotic belief that the living usage of Cinquecento Florentines had inherited and even improved on Trecento phonology, so that to be born Florentine was (contrary to Bembo's claim) an advantage. In 1825 Foscolo condemned Salviati and his contemporaries for wanting to prescribe 'old books and the new Florentine dialect' ('i libri antichi e il nuovo dialetto Fiorentino') and choosing to adapt spelling 'to modern popular pronunciation' ('alla moderna pronunzia del popolo') rather than base their laws on the uncertain orthography of the Trecento.[60] It is true that the editing of Salviati, like that of almost all Cinquecento editors, was strongly influenced by his views on the *questione della lingua*. But, in practice, the language of his *Decameron* was not so much more narrowly Florentine than that of the Deputati. If we compare, for example, the 1573 and 1582 texts of the story of Monna Belcolore (8.2), we find the most significant changes to be the ten cases in which Salviati declares he has followed Mn rather than the Deputati (even though he personally disliked one reading, 'isforzandosi' rather than 'sforzandosi'). Of the other changes, most are merely orthographic: 'e' for 'et', 'ora' ('now') for 'hora' and 'uopo' for 'huopo', 'cembalo' for 'ciembalo', 'zazzeato' for 'zazeato', and so on. There are others which would certainly affect pronunciation, such as 'fosse' for 'fusse', 'un suo orto' for 'uno suo horto', 'un asino' for 'uno asino', 'far racconciare' for 'fare racconciare', 'gonnella' for 'gonella'; but there is little here or in other corrections described in the *Avvertimenti* to justify Foscolo's allegation of popularization.

Salviati also differed from Borghini in his approach to the question of censorship. To a large extent he did so out of necessity. The Deputati had managed to persuade Manrique in 1571–2 that nobody saw the *Decameron* as a text which taught one how to live. This line was no longer tenable in 1580–1. If Salviati was to save as much as possible of the stories which Rome had wished to see deleted, he had to turn the *Decameron* into a work whose contents could, indeed should, be taken seriously. So it now became what Peter Brown has called 'a textbook of morality'.[61] Like the Deputati, Salviati replaced unacceptable phrases with acceptable ones (usually indicated by a change in typeface or with asterisks), deleted references to the Church and transformed members of the clergy into lay people, so that the priest of Varlungo, for instance, became a schoolteacher (8.2). But he went further than the Deputati by making it clear that the examples of those who went against Christian morality were to be shunned. One technique used here was to move a story to a pagan context. In the 1573 edition, Masetto da Lamporecchio (3.1), originally a gardener in a convent of nuns, was working for a sort of prematrimonial finishing school overseen by a countess. In 1582 he became a Jew whose name had originally been Massèt and who went to work in a seraglio near Alexandria. Elsewhere, even if the context remained Christian, Salviati could still stress the heinous nature of what was being described. At the hands of the Deputati, Frate Cipolla (6.10) had simply become a layman; now Salviati turned him into 'one of those who go around pretending to be friars' ('uno di quelli che vanno intorno fingendosi frati'). But, whereas the Deputati had excised whole passages in order to keep the syntactic coherence of what remained, Salviati used asterisks much more freely in order to link together a tissue of disconnected phrases. In the story of Rustico and Alibech (3.10), he explained in a marginal note that he had done this 'to save as many words and expressions as possible' ('per salvare più parole e più modi di favellare che si può').

Salviati added two new techniques in order to bring out the morality of the work. One was to change the ending of the story so that readers saw lovers marrying and immoral characters getting their just deserts or at least repenting of their actions. Thus the adulterous Catella and Ricciardo (3.6) die of remorse. Another original feature was Salviati's use of marginal notes. We have seen that editors such as Ruscelli used the margins to comment on the language or the readings of their text. Salviati, on the other hand, used them to direct the reader towards an interpretation of the text which would be acceptable to the Inquisition. His notes might refer to a single word which could be read in a way contrary to Christian doctrine. In the description of the plague, for instance, 'the cruelty of heaven' ('la crudeltà del cielo', 1 intr. 47) was glossed with 'that is, of the air' ('cioè dell'aria'). The word *fortuna* was fraught with danger, as we have already seen in the case of the *Cortegiano*. If Salviati allowed it to survive at all, he sometimes glossed it to suggest that Boccaccio was using the word in the sense of 'chance events' ('gli accidenti') or in the conventional manner of the non-Christian poetic tradition.[62] The main function of the very few notes which explained individual words was to deflect the disapproval of the Inquisition. If Ghismonda, later to take a lover, was 'wise' ('savia', 4.1.5), Salviati hastened to say that here the word meant only 'shrewd' ('accorta'). 'Culattario' (8.9.73) was probably glossed as a foolish distortion of *catalogo* so as to play down the coarse double meaning. Most of Salviati's notes, though, directed the reader towards a morally correct reading of a whole passage, pointing out the reprehensible or admirable nature of the characters' actions or the sense in which certain comments were to be taken.

The effect of all this was, naturally, to travesty Boccaccio's intentions. But Salviati saw the *Decameron* above all as a linguistic text; the contents were secondary. One can draw a parallel with his approach to the only other Trecento text which he is known to have edited, *Lo specchio di vera penitenza* by Iacopo Passavanti, printed by Sermartelli in 1585. Presenting another edition of this treatise in 1580, Francesco Diacceto, Bishop of Fiesole, had stressed first its spiritual and then its linguistic benefits. Salviati's dedication concentrated only on the latter, pointing in particular to its fine vocabulary (not too archaic, nor yet contaminated by modern accretions) and its noble yet simple style. He was publishing it 'for the glory of our vernacular' ('per gloria del volgar nostro') and as an example for modern writers. In editing the *Decameron*, too, he had tried to preserve as much of its precious language as possible, even at the expense of distorting the contents. A good example of the success of his method, but also of the price to be paid for it, is provided by the diatribe against friars in the story of Tedaldo (3.7.33–47). As was seen above, the Deputati had cut this out entirely. Salviati managed to save most of it by using the techniques of substitution and annotation. The friars became pilgrims ('peregrini') and a marginal gloss stressed that this was an attack not on pilgrims in general but on those who used pilgrim status as a pretext to cheat people. Salviati was cheating his readers, but the benefit of the deception was that he had to delete only two short sections from this part of the story.

The revival of interest in Tuscan prose of the Due and Trecento led in 1576 to the first complete edition, printed by Sermartelli, of Dante's *Vita nuova*. The poems were derived from the *Sonetti e canzoni* of 1527; now the prose parts were added from a manuscript supplied by Niccolò Carducci and identified by Barbi as BLF 40.42.[63] The religious allusions had to be censored, however. 'Beatitudine' ('beatitude') was replaced with words such as 'felicità' ('happiness'). Among the passages omitted (almost always without any indication) were Biblical quotations, the comparison of Giovanna with John the Baptist (24.4), and the passage on pilgrims (40.6-7). In the original, Beatrice's father dies 'as it pleased the glorious master who did not spare himself from death' (22.1); this was now secularized to become 'as it pleased that lively love which stamped this affection on me' ('sì come piacque a quel vivace amore, il quale impresse questo affetto in me'). Linguistically, too, as Barbi has shown, the edition was not faithful to the source manuscript: there were errors and arbitrary changes in wording, and modern spellings and forms were introduced.[64]

Although Florentine Academicians lectured on the *Commedia* and its text had been studied by Varchi, Borghini and others, Florentine men of letters managed to survive without any locally-printed edition for many years after Benivieni's of 1506. They also had to use Venetian editions of Petrarch, since the last Florentine edition had appeared in 1522. But, while there could be no serious rival to Bembo's text of Petrarch, the need for a new Florentine *Commedia*, to replace the Aldine text, began to be felt acutely towards the end of the Cinquecento. Between 1582 and 1584 Salviati had played the leading role in the creation of the new Accademia della Crusca, whose members devoted themselves to furthering his linguistic ideals. In his last years he worked with Bastiano de' Rossi, secretary of the Crusca, on a new dictionary based on approved Trecento usage, and after his death in 1589 the completion of the dictionary became the Academicians' main task. Working through the *Commedia*, however, they found the standard Aldine text too corrupt and set about researches for a new edition. Here there was no *codex optimus*, as there had been for the *Decameron*. Undaunted, the 'Crusconi' mounted a remarkable operation on a hitherto unparalleled scale, checking the readings of about one hundred manuscripts of which some forty belonged to the Laurenziana and the rest to private individuals. The result was *La Divina commedia . . . ridotta a miglior lezione dagli Accademici della Crusca* ('with the text improved by the Crusca Academicians'), printed by Domenico Manzani in 1595.[65] Rossi explained to readers the background to the edition and then the way in which the text was presented. The starting point had been the Aldine text. If the Crusca edition departed from it (other than in spelling, which followed

Salviati's model), the margins gave Bembo's original reading and a justification of the change, and the manuscript or manuscripts which contained the preferred reading were identified in a list at the end of the text. The margins also gave a few other variants, some judged to be good but inferior to those of the text, others considered to be equally good. A few of the marginal notes, especially in the *Paradiso*, were intended to explain the text.

Credit must be given to Rossi and his fellow Academicians for their feat of coordinating such a thorough survey of so many manuscripts. All their corrections were based on the evidence of at least one manuscript, not on conjecture. They corrected some of the readings peculiar to Bembo's main source (Vat) and other errors in his edition, such as 'parte' for 'gente' at *Inferno* 7.82. But their labours did not create a text radically different from the Aldine text. Their choice of this as their starting point, and their reluctance to modify it, hindered innovation. The fact was, too, that the editors did not have a method which would enable them to judge objectively the value of their variants. There was no attempt to classify their manuscripts by age or in any other way. The Academicians rightly resisted a crude, quantitative assessment of the worth of a variant according to the number of manuscripts in which it was found: quite frequently their preferred reading had only one witness. Their only guide, then, was their opinion of what fitted best with Trecento usage, the context, the sense, or good style. For example, they preferred 'Da che tu vuoi saper' (*Inf.* 2.85) to 'Poi che' because this expression was 'more used in those times' ('più usato in quei tempi'). 'Sanza infamia' (*Inf.* 3.36) seemed better than 'sanza fama' to describe the pusillanimous spirits 'in view of the company [of angels who supported neither God nor Lucifer] assigned to them by the poet' ('rispetto alla compagnia assegnata lor dal poeta'). Both these corrections are supported by Petrocchi's critical text, but there is no such agreement in cases where the Academicians used the more subjective criteria of what they thought Dante meant to say and how he should have said it. They found, for instance, that 'dal sommo' explained Dante's concept ('concetto') better than 'dal sonno' (*Inf.* 4.68). The variant 'Pietà mi vinse' (*Inf.* 5.72) found favour because 'it seems to infer greater pity' ('pare che argomenti maggior pietà') than 'giunse'; but in fact this verb, in the form 'iunse', had given rise to the *lectio facilior* 'vinse'. In the description of those looking at Aristotle (*Inf.* 4.133), 'Tutti l'ammiran' was thought better than 'Tutti lo miran' because 'it seems to amplify this concept and fit better with the character being introduced' ('pare che ... agrandisca questo concetto e convenga al personaggio che s'introduce'). Sound rather than sense was the consideration at *Inferno* 5.83, where the repetition of *al* in 'con l'ali alzare' seemed to spoil the line; 'aperte' therefore replaced 'alzate'.

One of the major publishing events of the early 1560s in Florence was the appearance of Francesco Guicciardini's *Storia d'Italia*. The ducal printer,

Torrentino, brought out two editions in 1561, one in folio format, the other in octavo. The work was not, however, in the form in which the historian had left it. For one thing, it was pruned of passages which might give offence for political and religious reasons. Two of those responsible for this aspect of the revision were Borghini and Bartolomeo Concini, secretary of Duke Cosimo. Guicciardini's nephew Agnolo revised the manuscript stylistically too, filling in gaps and giving the work greater polish.[66] In spite of this, the folio edition had a long list of errata which invited readers to make certain changes to spelling and to verb forms throughout the *Storia*, such as 'mezo' > 'mezzo', 'co gli' > 'con gli', 'cercorono' > 'cercarono', 'venneno' > 'vennero', 'furno' > 'furono', 'fussi' (third-person singular) > 'fusse', 'faccendo' > 'facendo'.

Other histories published in Florence in this period were of a parochial nature, like most products of the Florentine presses. Histories of the city satisfied both Florentine nationalism and, when they came from before the Quattrocento, the growing appetite for examples of earlier prose to set alongside that of the *Decameron*. The works on which most attention was focused were those of the Villani family. The editions of their chronicles printed by the Giunta press after 1560 reflected the revival of Florentine scholarship in this period by using new sources and by attacking previous printings, all of which had been connected directly or indirectly with non-Florentine editors.

Giovanni Villani's *Cronica* had been conservatively edited by fra Remigio on behalf of the Giunta family in 1559, we saw in chapter 8, but his sources were an edition printed in Venice and another edited by Domenichi. This edition was not well received either by Borghini or by Salviati.[67] There was evidently need of a new edition of a work so important for the study of Trecento prose, let alone for Florentine history, and in 1587 Filippo and Iacopo Giunta brought out one which had been prepared by Baccio Valori, a close friend of Salviati. Valori's dedication was more elegant than informative. He treated the Grand Duke Francesco to some puns on the 'villanous' treatment of Villani in previous printings, as he had discovered in the leisure of his villa ('in questo ozio villesco'), and on how 'civil' ('civile') was this 'Villano' with his pure Florentine style. But only from a note at the end of the volume would one have gleaned the fact that one of Valori's sources was a manuscript written before 1370 which belonged to Iacopo Contarini (now BMV It. Z.34) and that he had consulted a text belonging to Benedetto Tornaquinci.[68] The Contarini manuscript, which went up to 7.31 only, had previously been owned by Speroni and at that time had been used by Agnolo Guicciardini as the source of corrections in his copy of the 1537 edition.[69] Valori adopted many of the variants which Guicciardini had noted. But, as well as using manuscript sources, he still relied on Nannini's text, and he still used conjecture. A study of a sample from the later part of the chronicle (11.51–2) led Giuseppe Porta to conclude that here the Giuntina of 1587 followed the 1559 text closely, making a few changes in spelling, sometimes being more faithful to the manuscripts but not hesitating to introduce 'corrections' such as 'la 'mpresa' for 'lampromessa' (correctly 'la promessa') and 'ne' fatti delle guerre' for 'ne casi delle guerre'.[70]

As was noted earlier in this chapter, Atanagi edited in Venice the continuation of Giovanni's chronicle by his brother Matteo. This was printed in 1562 by the Guerra brothers on behalf of the Giunta, who had the benefit of advice from Borghini. To the four books edited by Domenichi in 1554, Atanagi added four more and chapters 1–86 from the ninth. The dedication by Filippo and Iacopo Giunta declared that, in this new version, the 'treasure' ('tesoro') which this work constituted had now been cleansed of the stains borne by objects which have been shut up for a long time (an image similar to that of buried statues, used in the *Sonetti e canzoni* of 1527). There was a particularly bitter attack, too, on the Torrentino edition and on those who, for profit ('per vender' a minuto', literally 'to sell retail'), 'falsify writings, books and words no differently than certain plebeian artists do their merchandise' ('falsano non altrimenti gli scritti, i libri e le parole che si facciano certi artisti plebei le loro mercanzie'). This has a particularly Borghinian ring to it, since the dedication of his *Decameron* later railed similarly against the audacity and ignorance of those who had corrupted Boccaccio's purity 'perhaps to make their books more saleable to the general public' ('forse per rendere i libri loro più vendibili al vulgo'). The Giunta brothers tell us that 'a very old copy' ('un essempio antichissimo') was used; in an edition of 1577 they specified that they had been helped by a manuscript lent by Lodovico Castelvetro.

This 1577 edition contained the remainder of Matteo Villani's chronicle, from 9.87 onwards, together with the continuation by his son Filippo (11.61–102). The editor was Giuliano de' Ricci, who used a manuscript belonging to his family which had apparently been copied two hundred years before.[71] A letter to the readers argued that a study of the language of the work would both enrich present-day Florence and help in the understanding of the works revised in what was clearly seen as the renaissance of Florentine textual studies of the previous few years. The letter stressed, too, that nothing had been changed in the only source used. Ricci had preferred not to mend through conjecture ('racconciare per coniettura'), knowing the dangers of interference. 'You will find very few doubtful passages, and, if you find any,' readers were warned, 'rather than deciding that they are errors, strive to get the true sense out of those words or expressions' ('Troverrete pochissimi luoghi dubbii, e, trovandone, più presto che risolvervi siano errori, affaticatevi a trarre di quelle parole o maniere di dire il vero senso'). However, not every detail of the manuscript was followed.[72] Four years later Ricci's manuscript was used to revise Books 1–9.86 of Matteo's chronicle in an edition which contained yet another broadside against people in 'outside cities' ('città esterne') who perhaps thought they were putting old books to rights but had ruined them through their lack of understanding of the language and its past.

Two other volumes of previously unpublished Florentine historical works were printed by the Giunta brothers in 1568. One contained two works, the *Diario* of Biagio Buonaccorsi, which covered the period of Medicean exile from 1498 to 1512 and, perhaps to compensate for this, Niccolò Valori's life of Lorenzo il Magnifico. The other volume was the first edition of the *Historia antica* by Ricordano Malispini. This began with the foundation of Florence and went up to 1281, with a supplement which continued the story up to 1286. It is now thought that Malispini is a fictitious name and that the history was compiled, mainly from Giovanni Villani's chronicle, in the second half of the Trecento.[73] As the 1568 edition shows, however, Malispini was then thought to be a writer of the late Duecento, indeed to be 'perhaps the first Florentine writer' ('forse il primo Fiorentino scrittore'), to be treated therefore by the children of Florence with the same reverence which they would show to other vestiges of their past.

The dedication (by the printers to Duke Cosimo) and a detailed and passionate letter to the readers described the sources available and how they had been treated by the editors, to whom the printers refer in the plural. They had several copies, including one from Cosimo, 'perhaps among the oldest writings in Florence' ('forse delle più antiche scritture che siano in Fiorenza'), and had a faithful copy made which drew from the sources all their 'ancientness' ('vetustà'), keeping even errors, if errors they were. One very old manuscript in particular was thought to be the source of the others. But there was no attempt to prove this relationship, and perhaps the editors simply wanted to ease their problems by nominating a *codex optimus*. They debated whether to try to emend historical errors, such as the confusion of Attila and Totila, and linguistic offences in word separation and against well-ordered writing ('[il] regolato scrivere'). In a strongly Borghinian manner, they explained their decision to change as little as possible, explicitly appealing to the precedents of Torelli's Justinian and Vettori's Cicero. In due course, they argued, things which seemed strange might be confirmed as genuine; old texts such as this certainly helped one to understand obscurities in major writers such as Dante. Old usages might even come back into fashion, just as fashion in dress was cyclical (and the letter to the readers used a few archaisms such as 'conpassione' and 'trarrae'). Above all, one must not try to mingle the ancient and the new. If one did not like older writers, there were plenty of good modern ones. If anyone wanted to find out what his ancestors did and how they spoke and wrote about it, let him take one of these old writers and not try 'to mingle ancient deeds with new expressions' ('i fatti antichi co' modi di parlare novelli rimescolare'). They did not want to impose new rules on the pronunciation and writing of such an ancient work, other than helping the reader's comprehension by dividing some words and adding modern punctuation and accents. Inconsistencies in spelling had been tolerated, even though these were perhaps due to the scribe. No attempt had been made to remove the 'crudeness' ('rozzezza') typical of its times. In spite of this rather condescending comment, there was respect, even affection, for Malispini's 'ancient simplicity of speech' ('quella antica semplicità del parlare') and his 'pure eloquence' ('[la] pura eloquenza'). A similar respect had been shown for the 'ancient simplicity and purity' ('la simplicità e purità anticha') of Jacopone's *Laude*, a text even further outside the literary mainstream, in the edition of 1490.

One of the sources used for Malispini was written in the fifteenth century and is now, as Antonio Benci showed, BNF 2.4.28.[74] A note at the end records that it was approved for publication by the Inquisition in 1566.[75] Many meticulous corrections were made to punctuation, word division, capitalization and spelling. Sometimes, where the change to be made was evident, there was just a dot under a letter.[76] Four extra chapters (46-9) were inserted from another source. It looks as if it was intended to pass this manuscript on to the printers, for on f. 93r there is an instruction to use upper case and to start a new page. In the event, however, a new copy-text was prepared from this and other manuscripts. One of these additional sources, also identified by Benci, may have been what is now BNF 2.1.108. This contains a transcription made from BNF 2.4.28 in 1554, incorporating some minor spelling changes already introduced in its source and adding some new ones. A second hand then wrote in many alternative wordings or additional passages, using another fifteenth-century manuscript, now BLF Ashburnhamiano 510, in which a few corrections were written by the same hand from yet another source. The 1568 text followed these alterations, incorporating for instance a new chapter 12. The result was a combination of the text in BNF 2.4.28, which also contributed most of the chapter headings, and that of the Ashburnham manuscript, which in some cases seems to have been used in preference to BNF 2.1.108.[77] As for spelling, verb forms and so on, the editors regularized many of the idiosyncrasies of their sources. But their printed text still had some words left undivided (for example 'ella' = 'e la', 'allui' = 'a lui') and many spellings which Salviati, let alone a Venetian editor, might have regularized, such as 'leggieranno', 'mogle', 'aquistare', 'sollazo', 'chavallo'.

In 1598/9 the Giunta press printed a mid-fifteenth-century Florentine history, that of Poggio Bracciolini, in the translation by Poggio's son Iacopo. This translation had been printed previously, but not since 1492. The editor was Francesco Serdonati. Like some Venetian editors, he was a versatile figure: as well as editing Bracciolini, he published some translations and added some new *giunte* to Giuseppe Betussi's 1545 editions of Boccaccio's *De claris mulieribus* and *De casibus virorum illustrium*.[78]

The press of Bartolomeo Sermartelli provided, in 1584, another new history for the Florentine market, that of Iacopo Nardi, which covered the years

1494–1531. His source was a printing edited in Lyons in 1582 by Francesco Giuntini. The Florentine edition added summaries for each book and an introduction which listed historians from ancient Greece to the present and suggested some aspects of warfare which could be studied in histories. As for the text, Sermartelli, in his dedication, made the familiar claims that the previous edition was full of errors, which he had had corrected, but said that his editor had not altered the author's errors, either out of modesty or because he thought this best. However, the nineteenth-century editor of Nardi's *Istorie*, Lelio Arbib, noted that Sermartelli's editor modernized the Lyons text and accused him of changing passages without justification.[79]

In 1569 Sermartelli printed a religious work from the first half of the Trecento, Cavalca's *Disciplina degli spirituali*, edited by the Camaldolese monk Silvano Razzi. The dedication by a fellow monk explained in some detail how he and Razzi, rummaging among some old books, had come across a copy of a Florentine edition of 1487 (now unknown, if this date is correct). The work seemed essential to all clerics, and especially to the relatively uneducated ('a gl'idioti'). But, in order to remedy distortions in both spelling and sense, it was felt necessary not just 'to revise it roughly, as those in charge of correcting books or revising proofs seem mostly to do, but to rewrite it completely, to mend it and almost to give it, according to its need, another form' ('rivederla così alla grossa, come pare che facciano per lo più quelli che sono sopra o a correggere libri, o a rivedere stampe, ma di tutta riscriverla, racconciarla, e quasi darle, secondo che havea bisogno, un'altra forma'). Razzi eventually supplied a transcription which left some things in their original simplicity ('nella loro primiera simplicità'), since he did not want to change its concepts but to mend imperfections which made it hard to understand and which were due either to the carelessness of the first printers or, he acknowledged, perhaps to the author. A comparison with the Florentine printing of *c*. 1485 (*GW* 6396) shows that Razzi revised details of spelling, phonology and morphology and also rewrote sentences quite freely.[80]

Both the Giunta and the Sermartelli presses did something to foster another aspect of the Florentine literary tradition: the poetry in *ottava rima* of Luca and Luigi Pulci. In 1572 the Giunta brothers brought out the chivalrous poem *Ciriffo Calvaneo* together with other works attributed to Luca, presenting the edition as part of their family's long-standing efforts to rescue Florentine authors from oblivion (yet again the unburying metaphor is used) or from corruption. In the Venetian manner, each of the seven canti had a verse summary. There was none of the respect for antiquity shown in Florentine editions of Trecento prose: this edition had many changes to spelling and metre in comparison with the Florentine incunable.[81] The Sermartelli edition of Luigi Pulci's *Morgante* (1574) presented this epic as an important linguistic document which helped to show the continuity of Florentine from an earlier stage to the present. This tradition, it was argued, had shown variety, or rather a gradual enrichment, but Florentine had never lost 'a certain beautiful and sweet property' ('una certa sua vaga e dolce propietà') which had made it supreme in Italy.[82] Sermartelli's claim that the *Morgante* was now 'all repentant and emended of its past errors' ('tutto pentito et emendato da' suoi passati errori') probably referred in the first place to the censorship to which it had been subjected. References to sacred subjects were removed throughout the poem.[83] One passage affected was the description of Uliviero expounding Christian beliefs to the pagan Meridiana for carnal motives, because he would not make love to her unless she became a convert (8.8–13). The censor deleted the two stanzas in which Uliviero talked of the Trinity, Christ's miracles and the Crucifixion and in which the consummation of their love was described as a confirmation followed by the breaking of a fast ('e dopo a questo vennono alla cresima, | tanto che infine e' ruppono la quaresima'). In place of these stanzas was a new one in which the conversion was effected without recourse to theology and the love-making was described earthily ('vennono alle prese', 'they came to grips') but without sacrilege.

Censorship on religious grounds struck another late fifteenth-century work, the *Facezie* of the Piovano Arlotto, when this was printed by the Giunta press in 1565. Anecdotes which seemed too free to the Inquisitor had been removed, readers were told. Others were abbreviated or altered in various ways. In the first story, for example, Antonino, archbishop of Florence and a Dominican, became an anonymous 'arcivescovo di Firenze'. Equally radical was the stylistic revision. Readers were given the traditional excuse that the stories had become estranged 'from good language' ('dalla buona lingua') through the lack of diligence of earlier printers. Not only were there changes dictated by linguistic 'rules' (such as the elimination of *lui* as a subject pronoun), but whole sentences were rewritten in the interests of elegance and concision.

These examples of censorship remind us that editors in Florence were subject to at least one of the constraints which affected editors in Venice. In these last decades of the Cinquecento they also continued to show a common concern for making the interpretation of the text easier for the reader by revising punctuation and spelling and, in the genres of history and the epic, by adding summaries or marginal notes. Texts which occupied a low position on the literary scale, such as the *Facezie* just mentioned, were equally subject to manipulation on both sides of the Apennines.

In many respects, though, the Florentine scene was very different from that of the Serenissima. Editors in Florence were working in a strongly nationalistic and introspective climate. It is significant that many of the works mentioned in this chapter were dedicated to some member of the Medici family, often to the Duke or Grand Duke himself. When Florentine editors referred to their readers, these were envisaged as Florentine. Most of the works printed in Florence were of primarily Florentine interest and were presented as adding to the glory of the city. Contemporary Florentine was used to give more lustre even to masterpieces of the past on the grounds that its native speakers, having inherited the best linguistic qualities of the Trecento, were now raising the language to even greater heights.

But there were other much more positive characteristics in Florentine editing. One was the importance attached to the careful use of the earliest possible witness to a text. This attitude was clearly inspired by the examples of Vettori and Torelli, who in turn owed much to Poliziano. Of course, reliance on their methods, appropriate though these may have been in some cases, contained dangers. One was that a codex thought to be *optimus* might not prove as reliable as it seemed: this was the case, for instance, with the Mannelli manuscript of the *Decameron*. Another was that editors naturally found it hard to know how to use a group of sources when none of them stood out as superior, a problem which Bastiano de' Rossi and his colleagues faced after their collation of Florentine manuscripts of the *Commedia*.

A second important advance was that texts of the Due and Trecento now had editors who had an excellent knowledge of the language of that period. Moreover, they believed that, at least in principle, linguistic periods should remain distinct. 'Let the old times remain old and the new ones new', the editors of Malispini appealed ('gli antichi [tempi] antichi e i novelli, novelli si rimangano'.) Borghini's main battle was fought against those such as Ruscelli who modernized texts either because they failed to understand the language of the past or because they did not respect its right to be different from the linguistic ideal which they wanted to help their mainly non-Tuscan readers to master. In Florence, Borghini's point was taken for granted by the 1580s. Salviati hardly referred to Ruscelli at all in his *Avvertimenti*, and, when he did, he could afford to dismiss him peremptorily.

As a consequence of these new editorial criteria, older texts (although only Florentine ones) could now appear in a relatively genuine linguistic guise. Only relatively genuine: apart from the censorship of the Inquisition, texts were superficially modernized and standardized to varying extents, with Salviati the leading supporter of a more rational spelling. Nobody favoured the extreme conservatism advocated by Corbinelli in his French exile. But Borghini's methods, unlike those which had dominated Italian editing since the 1530s, made it more probable that, while the appearance of a text might be smoothed, its authentic 'substance' would still be preserved.

CONCLUSION

This book began with the example of Cristoforo Berardi, 'indegno correctore' of Dante's *Commedia* in 1477. Long before the end of the sixteenth century, the humility and deference implied by this self-description had become unthinkable. Editors had acquired a sometimes aggressive confidence in their part in the fabric of print culture, indeed of vernacular literary culture in general, in the Italian Renaissance. It is fitting to end by asking how far this self-confidence was justified. What significance did the work of Renaissance editors have for their contemporaries, and what legacy did they leave behind them?

The Cinquecento saw the wide acceptance in print of a variety of the vernacular, based on Trecento Tuscan, which achieved a considerable degree of standardization. The influence of editors was by no means the only factor involved in this process: the imitation of Tuscan outside Tuscany was already well established in the Trecento and its further spread in the age of printing coincided with a rebirth of vernacular literature, underpinned in the Cinquecento by the codifications of vernacular grammarians. But it seems probable that editors made a practical contribution to shaping a norm which struck a balance between the purist views expressed by Bembo and a rather less rigorous imitation of the Tuscan Trecento. It was particularly important that many editors shared the widespread antipathy to archaic or otherwise obscure forms, an antipathy found early in the century among supporters of a 'lingua cortigiana' and other opponents of Bembo and which persisted even after the 1530s, for example in writings on rhetoric. Editors were also instrumental in helping to make this usage accessible to a wider range of users than it might otherwise have reached, so that the print standard could become the basis of a general written norm. Print culture had by no means reached the stage where, as Ian Watt put it in his study of the rise of the novel in eighteenth-century England, the 'centre of gravity' of the reading public would shift to such an extent that a writer's main aim became the satisfaction of the middle class rather than of patrons and a literary elite.[1] The way in which some Cinquecento editors used both flattery and polemic attacks in dedications and annotations shows that they were still much concerned with their status in the world of letters. Nevertheless, this century did see the beginnings of a new attitude towards the consumers of literature. Many editors did their best from the 1530s onwards to make it easier for readers to understand and imitate the key model texts – Petrarch's *Canzoniere*, Boccaccio's *Decameron*, Ariosto's *Orlando furioso* – by providing glossaries, annotations and other sorts of practical help and advice. Explanations of the vocabulary of Boccaccio and Ariosto were often related directly to the readers' spoken linguistic experience by the use of dialect synonyms. Providing assistance to the reader became a distinctive speciality of editors in Venice, who had no reason to assume that their public had a thorough knowledge of Tuscan. In Florence, by contrast, linguistic annotations were exceptional; Ulivi's *Orlando furioso* of 1544 remained a complete anomaly.

Editors did not only help to form linguistic and stylistic tastes: they must also have had a notable influence in the Cinquecento on the formation of critical tastes. The material which they added to their volumes often guided readers towards a certain interpretation of the contents of a text or towards a judgement on its status within its genre. The main example of this was the way in which editors were in the vanguard of the efforts to justify Ariosto's *Orlando furioso* as an epic which was worthy of comparison with its classical ancestors and which had a sound moral message. Another example is the way in which introductory material was used to pass judgement on the portrayal of women in Boccaccio's *Corbaccio*. This work came out in 1516 with a defensive prologue by Castorio Laurario and in 1594 with a dedication by Filippo Giunta which dismissed accusations of misogyny and praised its 'most useful teachings' ('ammaestramenti utilissimi') on grounds similar to those used by Laurario. Conversely, Domenichi's dedication in 1545 suggested disfavour for the work because it was indeed insulting to women. Dolce's edition of Castiglione's *Cortegiano* showed that even something so apparently innocuous as a subject index could guide the reader to one aspect of a work at the expense of another. Critical attitudes could also be influenced by revisions of texts, in so far as these helped to define what was appropriate within genres such as the epic and the novella. And editors could point to certain ideals of both style and content through their selecting and ordering of letters, lyric poems and other pieces for the anthologies which became so popular from the mid 1540s. By shaping what was committed to print and by influencing its interpretation, then, Renaissance vernacular editors controlled the images of past and present writing which reached the reading public through the new medium. Since their readers included new writers, editors were also helping to shape what was yet to be written.

Renaissance editors, in collaboration with their printers, also made an enduring contribution to the way in which texts were presented. Between the 1470s and the mid Cinquecento, editions of the major literary works evolved from the unaccompanied text in a folio format, to a folio format with text surrounded by commentary, and then to smaller formats with empty or lightly used margins, notes placed at the end of each unit (a lyric poem, a novella or a canto), and separate glossaries and other appendixes.

More complex is the question of the contribution made by editors during the Italian Renaissance to vernacular textual criticism. At first sight, the picture may seem gloomy, because there was such a strong tendency throughout the period for the process of establishing the text to be dominated by arbitrary revision. When an editor spoke of the way in which he had prepared a text for printing, he might claim that, although the current vulgate had been corrupted by scribes or by careless and money-grabbing printers, he had restored the text to its original, pristine state. Yet often this restoration was guided principally or solely by the editor's own ideal of language and style. This must have been the case, for instance, with Francesco Alfieri's editions of Petrarch in the early Cinquecento or with the work of Rosello, Cassiodoro Ticinese and Gaetano in the 1520s. Their linguistic norms were in reality dictating the form of their texts, in spite of their boasts of having restored to their authors their original brightness, health or, quasi-miraculously, life itself. Sometimes editors, masquerading as scholars, made claims about the use of written evidence that were vague, misleading, or could even smack of charlatanry. These persisted to the very end of the Cinquecento. In a *Fiammetta* printed by the Giunta press in 1594, for instance, it was asserted that this and the other prose works of Boccaccio brought out by the press in that year were 'purged of every stain and bright and shining as they ever were' ('purgate da ogni macchia e chiare e lucenti come fur mai') with the help of manuscripts; in reality, the *Fiammetta* still bore the marks of the revisions of Gaetano.

Confidence in the rightness of imposing a linguistic ideal grew to such an extent that, from the 1540s onwards, some leading editors – Domenichi, Ruscelli, Porcacchi – explicitly declared that they had conferred a perfection which certain authors, writing in an age when the vernacular was still unpolished, had been unable to achieve. Domenichi declared in his *Orlando innamorato* of 1545, not that he was restoring the poem's pristine purity, but that he was bringing it from its original 'rozzezza' to a new 'pulitezza'. Textual evidence still mattered enough to Ruscelli for him to invent manuscript sources for some of his revisions to Ariosto's verse, but he did not trouble to do this with Collenuccio. In Florence in 1569, Silvano Razzi was prepared to admit that the errors in a text by Cavalca could be due to the author. 'Corretto' was now openly interpreted as 'corrected (according to current ideals)' rather than 'correct, authentic'.

How can one explain these practices, which could openly allow linguistic revisions to take over from the process of restoring the original? There were a number of causes involved. Editors had to take into account the interrelated influences of the habits of their readers and the commercial expectations of printers. There was a natural eagerness to advance the prestige of vernacular literature as it emerged from the shadow of Latin. Ruscelli prefaced his 1552 revisions of Collenuccio with remarks about the authority which the vernacular had acquired and two years later, editing Ariosto's *Satire*, talked of his desire 'to help raise our language to perfection' ('aiutar ... a metter in colmo la lingua nostra'). Florentine editors, especially under the Medicean duchy, wanted to assert the importance of their own variety of the contemporary vernacular.

Another factor of particular importance was the pervasive influence of contemporary aesthetic assumptions. We have seen how editors stressed that they wished to present something beautiful to behold, imparting 'nitidezza' or 'pulitezza' or purging stains. They also drew specific analogies with the world of fine art which are very revealing of their implicit aims. They occasionally compared their work with that of a restorer of antique statues, as in the *Sonetti e canzoni* of 1527. Delfino in 1516 saw himself as a portrait painter: just as Zeuxis combined the best features of his female models, so he chose from different texts and thus restored the 'prima bellezza' of the *Decameron*. While these comparisons should not be pushed too far, they must be taken seriously because of the close links which the Cinquecento considered to exist between art and letters. Dolce, for instance, the best-known editor of the mid Cinquecento, saw all writers as painters and 'every composition of learned men' ('qualunque componimento de' dotti') as painting.[2]

To what shared aesthetic principles, then, were editors alluding when they termed themselves restorers or painters? The references to antique statues point to the belief that it was better to remove all signs of imperfection when displaying an object from the past. A damaged statue was held to be improved by the restoration of missing limbs and by the polishing of its surface so that it appeared entire and new. Vasari, for instance, approved of the restorations carried out earlier in the Cinquecento by the Florentine Lorenzetto Lotti, which he considered to have more grace than the imperfect originals.[3] In Venice it was remarked that Giovanni Grimani had some antique sculptured heads 'repaired and made beautiful' for public display in 1587 'diligently ... so that they seemed quite different' ('[fece] con molto studio ... racconciar, et abbelir le teste ... in maniera che non parevano più quelle').[4] Similarly, it was believed that earlier paintings and frescoes could be improved by means of adaptations and reworkings.[5]

The comparison with portraiture seems to allude to the principle that an artist should take what was best from nature, rejecting defects and irregularities. One way of doing this was to combine the best features from different sources, depicting a figure theoretically possible but which had never existed. Alberti had used the example of Zeuxis in his *De pictura* to support his exhortation: 'let us always take the things we are to paint from nature, and let us always choose from them everything most beautiful and worthy' ('semper quae picturi sumus, ea a natura sumamus, semperque ex his quaeque pulcherrima et dignissima deligamus'). He repeated Quintilian's criticism of Demetrius, in contrast, for being 'more devoted to representing the likeness of things than to beauty' ('similitudinis exprimendae ... curiosior quam pulchritudinis'). 'Praiseworthy parts,' Alberti believed, 'should all be chosen from the most beautiful bodies' ('pulcherrimis corporibus omnes laudatae partes eligendae sunt'), though he recognized that this was very difficult 'because the merits of beauty are not all to be found in one place but are scattered and dispersed' ('quod non uno loco omnes pulchritudinis laudes comperiantur sed rarae illae quidem ac dispersae sint').[6] The same notions held good in the Cinquecento. Dolce used the Zeuxis story in his dialogue on painting, *L'Aretino*, to warn against deriving everything from direct imitation.[7] Vasari believed that the greatest beauty was achieved by joining together the most beautiful parts according to good judgement.[8] Raphael showed that he too embraced the notion of idealization, if in a rather different way, when he said in a letter to Castiglione of 1516 that a dearth of beautiful women meant that he could not use the method of Zeuxis; instead, he had to use 'a certain idea that comes into my mind' ('certa idea che mi viene nella mente').[9]

If we bear these principles in mind, it becomes easier to see how Renaissance editors could adopt methods which nowadays may seem strangely unsympathetic to the original aspect of their texts. In the Cinquecento, it was considered legitimate and praiseworthy for editors to use their own skills in order to give an improved representation of the original, in the same way as restorers used their techniques to make good the damage inflicted by time, and just as painters chose only the best features offered by nature, preferring, if necessary, *pulchritudo* to *similitudo*. Where the transmission of a text had introduced real imperfections such as incomplete verses, these were to be silently restored so that the reader was not presented with something fragmentary. Many faults, of course, existed only in the eye of the beholder, but they were removed nevertheless, even if the author was Petrarch or Boccaccio. When an editor had a number of sources for a text, none of which were judged perfect, the best method seemed to be that of Zeuxis: to select and combine the best parts from each. Dolce followed Delfino's example and said he had done just this in his own *Decameron*. It was generally assumed that in any case editors should not reproduce, like a Demetrius, apparent blemishes in their texts (inconsistent or non-standard linguistic usage, verses whose metre and rhyme did not meet with current standards, and so on). Some editors, indeed, took the process of revision so far that the evidence of their sources tended to be subordinated to their anachronistic ideal of perfect beauty.

Striving after beauty did not, however, always have to entail sacrificing truth. One can identify in Renaissance editing a current which was influenced to some extent, as was natural, by the predominant tendency to polish the surface of a text but which nevertheless stopped short of introducing major arbitrary improvements. A select group of editions from the first half of our period, notably the *Laude* of Jacopone printed in 1490, Bembo's Petrarch of 1501 and his Dante of 1502, Bardo Segni's *Sonetti e canzoni* and the Florentine

Decameron of 1527, showed how the preparation of a text for printing could give impetus to the search for the best available textual evidence and to the comparative and critical evaluation of this evidence. Then, for some three decades, this tradition seemed to be interrupted. No scholarly editions appeared in Florence, while in Venice first Domenichi and then Ruscelli openly scorned the evidence of what an author had written. It is true that a well-known voice was raised in Venice against Ruscelli's kind of revision: that of Lodovico Dolce. He denounced loudly and clearly, both in his editions and in his *Osservationi nella volgar lingua*, the practice of altering what the author had originally written. But his protestations, unsupported by a genuine scholarly method, did not have sufficient weight to convince any of his rivals in Venice to adopt more conservative methods.

By the 1560s, then, vernacular textual criticism had reached a crisis which was deepened by the interference of the Inquisition. On the one hand, it was generally accepted that there would often be a need to modify the surface of a text. On the other hand, there was still a recognition that the printed text should try to recover the pristine original. How could these two conflicting demands be reconciled? A possible way forward, the best that could be expected in the circumstances, was indicated in Florence by Borghini and his school. Like the finest vernacular editors before him, Borghini drew on the methods of the best classical scholars and learnt from them to show respect for the evidence of good manuscripts. He stressed the danger of rushing to emend an obscure reading: far better to try to understand it first, to look for instances in other texts, and to use conjecture only as a last resort. Like Bembo and his other predecessors, though, Borghini did not feel complete reverence for his sources. One had to eliminate the habits and what he believed to be the errors of individual scribes in order to give a true representation of the essence of a text. Making his own analogy with portraiture, he insisted that to represent the *vero* was not necessarily the same as representing the superficial, perhaps blemished *aparenza*. However, while the editor still had to use his critical judgement, this judgement was to be based on a greater sense of period in language, on an understanding and appreciation of earlier phases of the vernacular on their own terms, not (as in Ruscelli's case) on one's impressions or on modern, anachronistic tastes.

These methods were passed on through Salviati to Bastiano de' Rossi and the other members of the Accademia della Crusca who prepared the last important edition of the Cinquecento, the *Divina commedia* of 1595. Rossi was at pains to point out that their emendations were based on authority and reason ('l'autorità e le ragioni'), and that opinion ('[la] openione') had played no part in the selection of those readings which they recorded. Textual critics of later generations would, of course, have to give more thought to the question of how to go about the task of choosing between variants. But further progress would have been slower if the most authoritative editors of the

Renaissance had not shown, all the more laudably in that they were going against the trend of most editing of their period, that the subjective *openione* of the editor had to be set aside in favour of the *autorità* of one's sources. By establishing that the editor should be the servant of the text, not its master, as well as by developing new methods of presenting vernacular texts to the reading public, Renaissance editors laid solid foundations on which their modern counterparts could build.

NOTES

I PRINTERS, AUTHORS AND THE RISE OF THE EDITOR

1 Fowler, *Catalogue*, pp. 84–7; Trovato, *Con ogni diligenza*, p. 22.
2 G. Zappella, *Il ritratto nel libro italiano del Cinquecento*, 2 vols. (Milan, Editrice Bibliografica, 1988), I, 74–5.
3 The best-known example is that of Ariosto and the *Orlando furioso* of 1532, on which see Fahy, *L'"Orlando furioso' del 1532*.
4 On Malermi (or Malerbi) and his translations, see V. Meneghin, *San Michele in Isola di Venezia*, 2 vols. (Venice, Stamperia di Venezia, 1962), I, 139–46.
5 *Aldo Manuzio editore*, no. XLI. See also Dionisotti, *Gli umanisti*, pp. 12–14, and below, p. 59.
6 See below, pp. 14–15.
7 Pallavicino gives the pre-publication history in a letter to Sansovino in his *Delle lettere libri tre* (Venice, Rampazetto, 1566); see A. Quondam, 'La grammatica in tipografia', in *Le Pouvoir et la plume*, pp. 177–92 (pp. 187–90).
8 For claims of correctness in the prefaces and colophons of some other books from fifteenth-century Milan, see Rogledi Manni, *La tipografia a Milano*, pp. 65–7.
9 Letter to the readers in *I fiori delle rime de' poeti illustri* (Venice, G. B. and M. Sessa).
10 A. Quondam, 'La parte del volgare', in Sandal, *I primordi*, pp. 139–205 (pp. 162–3).
11 On the importance of textual completeness and presenting a text 'as new', see Lotte Hellinga, 'Manuscripts', and her 'Editing texts in the first fifteen years of printing', in *New Directions in Textual Studies*, edited by D. Oliphant and R. Bradford (Austin, University of Texas, 1990), pp. 127–49.
12 Examples are Bonaccorso and Alessandro Minuziano in fifteenth-century Milan; Peter Löslein, who had worked as a 'corrector' for Erhard Ratdolt and Bernhard Maler in fifteenth-century Venice; and, in the following century, Antonio Brucioli and Francesco Sansovino in Venice and Anton Francesco Doni in Florence.
13 Bernardo di Giunta's preface to the 1547 edition of the tragedies of Sophocles (see L. Perini, 'La stampa in Italia nel '500: Firenze e la Toscana', *Esperienze letterarie*, 15 (1990), 17–48 (p. 27)), and the letter on the text of the *Novellino* in the 1572 edition: see below, chapter 11, n. 18.
14 Scholderer, *Fifty Essays*, p. 82.
15 The pairing of the verbs was borrowed from a letter of the younger Pliny (4.26.1). On the meanings of *emendare*, *corrigere* and *recognoscere* for humanists, see Rizzo, *Il lessico*, pp. 249–80. Several different words are used in colophons to describe the work of correction, including Latin *castigare*, *corrigere*, *emendare*, Italian *castigare*, *correggere*, *emendare*, *vedere*, *rivedere*.

16 Fulin, 'Documenti', docs. 139 and 190, pp. 154, 178–9.
17 Fulin, 'Documenti', pp. 95–6; see also M. Sanudo, *I diarii*, XXI (Venice, 1887), cols. 484–5.
18 Quoted from Lucian, *Dialoghi*, in Quondam, 'La letteratura in tipografia', p. 657.
19 His critics, he said, should know that the prefaces they shun 'are sold by printers to their advantage at a higher price' ('ab impressoribus cum eorum utilitate pluris venundari'): Bussi, *Prefazioni*, p. 47.
20 Trovato, *Con ogni diligenza*, pp. 150–1.
21 On this edition, see below, p. 71.
22 Some information on the background and employment of proof-readers and editors is given in Hirsch, *Printing, Selling and Reading*, pp. 46–8.
23 Ghinassi, 'Correzioni', p. 37 n. 10.
24 Allenspach and Frasso, 'Vicende', p. 248; Di Filippo Bareggi, *Il mestiere*, p. 273.
25 Pre' Marsilio, an obscure figure, may have taught in Fano in 1505–6: A. Vernarecci, *Fossombrone dai tempi antichissimi ai nostri*, 2 vols. (Fossombrone, Monacelli, 1903–14), II, 479–81.
26 Bongi, *Annali*, I, xviii–xix.
27 Grendler, *Schooling*, p. 54.
28 An example from the Neapolitan press c. 1489 is Giovan Marco Cinico: M. Fava and G. Bresciano, *La stampa a Napoli nel XV secolo*, 2 vols. (Leipzig, Haupt, 1911–12), I, 64–74; Trovato, *Con ogni diligenza*, p. 112. For the involvement of Florentine scribes as Latin editors, see below, p. 43.
29 The first instance in Venice listed by Fulin is the request made in 1492 by 'Joannes Dominicus Nigro' for two editions of Latin works: 'Documenti', doc. 5, pp. 102–3.
30 T. Valenti, 'Per la storia dell'arte della stampa in Italia: la più antica Società tipografica (Trevi-Umbria 1470)', *LB*, 26 (1924), 105–27.
31 Allenspach and Frasso, 'Vicende', pp. 246, 264; Dionisotti, 'Fortuna del Petrarca', p. 63. Other Milanese examples relating to Latin incunabula are given in A. Ganda, 'Antonio Zarotto da Parma tipografo in Milano (1471–1507)', *LB*, 77 (1975), 165–222 (pp. 197–8).
32 Sartori, 'Documenti', pp. 141–2; Mardersteig, 'La singolare cronaca', pp. 255–8.
33 Fulin, 'Documenti', doc. 17, pp. 108–9.
34 M. Billanovich, 'Notizie e proposte per Innocente Ziletti', *IMU*, 26 (1983), 375–8.
35 Trovato, *Con ogni diligenza*, pp. 143–4, 159; see too p. 165 on Delfino's *Decameron* of 1516.
36 Belloni, 'Un eretico', pp. 55 n. 24 and 65 n. 49.
37 Balsamo, *Produzione*, pp. 29–34.
38 Trovato, *Con ogni diligenza*, pp. 252–4.
39 *BMC*, V, 174–5; Lowry, *Nicholas Jenson*, pp. 145–52. On Nievo, see G. Mantese, *Memorie storiche della chiesa vicentina*, III, Part II (Vicenza, Neri Pozza, 1964), 830–2.
40 *BMC*, V, 235–6; Lowry, *Nicholas Jenson*, pp. 163–4. Albignani's son Giovanni inherited an interest in editing and worked for Giovanni Giolito in Trino between 1508 and 1523: Bongi, *Annali*, I, xv.
41 For examples of editors' lack of involvement in the correction of Greek and Latin texts, see Kenney, *The Classical Text*, p. 51.
42 *La piazza*, p. 849.
43 Ceruti, *Lettere inedite*, pp. 71–4. The author was a member of the Mocenigo family,

perhaps the poet Iacopo, whose previously mutually beneficial relationship with Atanagi was mentioned by Bernardo Cappello in his *Rime* of 1560 (p. 262). The work, described as a *Retorica*, may never have been published in full, perhaps because of lack of payment by Mocenigo (see below, p. 18), but the 'tables' ('tavole') to whose printing Atanagi said he devoted much effort are probably the *Rhetoricorum Aristotelis . . . tabulae* printed by Domenico Nicolini in 1563.

44 Examples are Pietro Summonte's preparation of Pontano's *De prudentia* in Naples in 1508 and Gian Francesco Valerio's preparation of Castiglione's *Cortegiano* in Venice in 1528, on which see respectively W. H. Bond, 'A printer's manuscript of 1508', *Studies in Bibliography*, 8 (1956), 147–56 (on other manuscripts of Pontano prepared by Summonte, see Pontano's *Carmina*, edited by B. Soldati, 2 vols. (Florence, Barbèra, 1902), I, xix–xxix) and Ghinassi, 'L'ultimo revisore', p. 250 n. 50.

45 See below, pp. 57–8. Sclarici taught from 1481 to 1507: see A. Sorbelli, *Storia della Università di Bologna*, I (Bologna, Zanichelli, 1940), p. 244. He published in 1491 a collection of verse, entitled *Silvano*, which showed clear Petrarchan influence.

46 *I Marmi*, I, 177. Doni is accused of basing his 'Ragionamento della stampa' on an early version of the *Dialogo della stampa* of his rival Domenichi, a work printed only later (with attacks on Doni as editor, translator and author) in Domenichi's *Dialoghi* (Venice, Giolito, 1562), pp. 367–99 (the parallel passage is on p. 373). See Poggiali, *Memorie*, I, 260–1, and G. Masi, 'Postilla sull'*affaire* Doni-Nesi: la questione del *Dialogo della stampa*', *Studi italiani*, 4 (1990), 41–54.

47 Lowry, *The World of Aldus Manutius*, pp. 24–41.

48 Nuovo, *Alessandro Paganino*, pp. 57, 217.

49 Fahy, *Saggi*, pp. 155–68. The notes, on ff. 3A4v and 3A7r, probably refer to Ruscelli's criticisms of pressmen at the end of one of the two works which he edited for the Sessa brothers in 1558, *I fiori delle rime de' poeti illustri*, f. 2P7r–v (reproduced by Fahy, pp. 167–8) and ff. 2P2v and 2P8r.

50 Letter to Francesco Guarnieri in Perotti's *Cornucopiae* (Venice, Aldo Manuzio, 1499), p. 630, dated to 1473 by Giovanni Mercati in *Per la cronologia della vita e degli scritti di Niccolò Perotti arcivescovo di Siponto* (Rome, Biblioteca Apostolica Vaticana, 1925), pp. 90–2, but to 1470 by John Monfasani in 'The first call for press censorship: Niccolò Perotti, Giovanni Andrea Bussi, Antonio Moreto, and the editing of Pliny's *Natural History*', *Renaissance Quarterly*, 41 (1988), 1–31; see too Bussi, *Prefazioni*, p. li n. 38.

51 *Le tre fontane* (Venice, Gregorio de Gregori, 1526), ff. 26v–27r. Writing towards the end of the sixteenth century about the profession of 'correctors or censors' ('correttori, o censori'), Garzoni complained of their pedantry and inconsistency and spoke of the liberty with which some corrected much more than errors of language (*La piazza*, pp. 273–9); but he included in this category all critics of linguistic usage, whether they were involved in editing (like Ruscelli) or not (like Muzio).

52 Frey, *Der literarische Nachlass*, p. 215; Giovio, *Lettere*, II, 118.

53 See Ruscelli's letter of 3 April 1561 in his *Lettere di principi*, Book I (Venice, Ziletti, 1564), ff. 219r–28r.

54 Di Filippo Bareggi, *Il mestiere*, pp. 242–73.

55 The first, third and fourth of these letters are in B. Tasso, *Lettere inedite*, edited by G. Campori (Bologna, Romagnoli, 1869): letter XIV to Dolce, 6 April 1554,

pp. 96–8; letter XIX to Giglio, 5 March 1556, pp. 113–18; letter XXV to Ruscelli, 4 March 1557, pp. 145–55. The letter of 20 October 1554 is in Tasso's *Lettere*, 3 vols. (Padua, Comino, 1733–51), II, 144–5; that of 14 August 1557 is in Vol. II, 283–4. See too E. Williamson, *Bernardo Tasso* (Rome, Edizioni di Storia e Letteratura, 1951), pp. 20–1. On the Giglio–Ruscelli partnership, see Trovato, *Con ogni diligenza*, pp. 253–4. In the event, no further *Rime* by Bernardo were printed until Giolito's edition of 1560. Bernardo also had a difficult relationship with another editor, Atanagi, who helped him revise his *Amadigi*: Ceruti, *Lettere inedite*, p. 65.

56 Javitch, *Proclaiming a Classic*, pp. 46–7.
57 Bussi, *Prefazioni*, p. 20 (preface to Aulus Gellius, 1469): 'Haec enim mens mihi est, hoc unum imprimis studium ut, quod fieri possit, omnis homines Latinos me quoque ipso esse optem doctiores'. The claim that he worked 'gratis' comes in the concluding verses, line 8 (p. 27).
58 For Squarzafico, see Allenspach and Frasso, 'Vicende', p. 268. For fra Giocondo, see Fulin, 'Documenti', doc. 186, pp. 176–7. For Ruscelli, see his *Le imprese illustri* (Venice, Rampazetto, 1566), f. 2*IV.
59 Bussi, *Prefazioni*, pp. 70–1; Kenney, *The Classical Text*, p. 15.
60 Fulin, 'Documenti', p. 403.
61 Sartori, 'Documenti', pp. 141–2.
62 Mardersteig, 'La singolare cronaca', pp. 256–9.
63 Giorgio Merula earned 120 ducats a year; a top law professor might expect from 500 to 1,000 ducats. See Rigoni, 'Stampatori' (compositors earning 2½ to 4 ducats a month); Lowry, *Nicholas Jenson*, p. 187 (Merula and compositors); Stefano Pillinini, *Bernardino Stagnino: un editore a Venezia tra Quattro e Cinquecento* (Rome, Jouvence, 1989), p. 30 (law professors).
64 Bernardo Machiavelli, *Libro di ricordi* (Florence, Le Monnier, 1954), pp. 14, 35.
65 Allenspach and Frasso, 'Vicende', p. 236, and E. Menegazzo, 'Controversia giudiziaria per un codice di Livio', *IMU*, 25 (1982), 313–24 (p. 320 n. 16). Squarzafico had been working in that year for Windelin of Speyer, who had printed Livy's *Decades* in 1470; perhaps he had received a copy of Livy from Windelin in exchange for editorial help.
66 Rigoni, 'Stampatori'.
67 Sartori, 'Documenti', pp. 125–6. The arrangement appears to have been that these 10 ducats should come from the sale, by the printer, of ten copies to ten students to whom Vernia was forbidden to sell his copies.
68 A. Sorbelli, *Storia della stampa in Bologna* (Bologna, Zanichelli, 1929), pp. 16–17, 19–22; Balsamo, *Produzione*, pp. 22–3, 19–21.
69 E. Motta, 'Di Filippo di Lavagna e di alcuni altri tipografi-editori milanesi del Quattrocento', *Archivio storico lombardo*, 25 (1898), 28–72 (pp. 37–9, 54–7); C. Santoro, 'Due contratti di lavoro per l'arte della stampa a Milano', in *Miscellanea bibliografica in memoria di don Tommaso Accurti*, edited by L. Donati (Rome, Edizioni di Storia e Letteratura, 1947), pp. 185–92; Rogledi Manni, *La tipografia a Milano*, pp. 39–40, 56.
70 P. de Nolhac, 'Les Correspondants de Alde Manuce: matériaux nouveaux d'histoire littéraire (1483–1514)', *Studi e documenti di storia e diritto*, 8 (1887), 247–99 and 9 (1888), 203–48 (reprinted Turin, Bottega d'Erasmo, 1967), letter 86.
71 Pettas, *The Giunti*, pp. 128, 297. On Florentine coinage, see C. Cipolla, *La moneta a*

Firenze nel Cinquecento (Bologna, Il Mulino, 1987). A similar combination of cash and books used in Paris in 1545 is recorded in Parent, *Les Métiers*, pp. 115 and 224.
72 Carter, 'Another promoter'.
73 For further examples, see Di Filippo Bareggi, *Il mestiere*, pp. 265-71.
74 Ceruti, *Lettere inedite*, pp. 69-74. On Atanagi, see the entry by C. Mutini in DBI, IV (1962), 503-6. He specialized in collections of contemporary verse: see E. Albini, 'La tradizione delle rime di Bernardo Cappello', in *Studi di filologia e letteratura italiana offerti a Carlo Dionisotti* (Milan and Naples, Ricciardi, 1973), pp. 219-39; G. Rabitti, 'La "Vita di Giacomo Zane" scritta dal Ruscelli: prolegomeni per una monografia', *Quaderni veneti*, 11 (1990), 7-45; Bongi, *Annali*, II, 246-7 (poems of Berardino Rota, 1567); B. Croce in *Poeti e scrittori del pieno e del tardo Rinascimento*, Vol. I (Bari, Laterza, 1958), 365-75; E. Favretti, 'Una raccolta di rime del Cinquecento', *GSLI*, 158 (1981), 543-72; A. Jacobson Schutte, 'Irene di Spilimbergo: the image of a creative woman in late Renaissance Italy', *Renaissance Quarterly*, 44 (1991), 42-61.
75 That the situation was analogous in France is suggested by the high sum offered to an editor by a Parisian printer in 1555: Parent, *Les Métiers*, p. 121.

2 EDITORS AND THEIR METHODS

1 Folena, 'Überlieferungsgeschichte', pp. 374-96.
2 As in the collection made by the poet Nicolò de' Rossi of Treviso and the Escorial MS, probably copied in Padua, on which see Folena, 'Überlieferungsgeschichte', pp. 399-402; see also De Robertis, 'Il canzoniere Escorialense'. For an adaptation from Florentine to Neapolitan in a mid-Trecento manuscript, see the *Libro de la destructione de Troya: Volgarizzamento napoletano trecentesco da Guido delle Colonne*, edited by N. De Blasi (Rome, Bonacci, 1986), pp. 9-16, 39-40. For a reworking of a Tuscan text into Sicilian, see *La istoria di Eneas vulgarizata per Angilu di Capua*, edited by G. Folena, Pubblicazioni del Centro di studi filologici e linguistici siciliani (Palermo, 1956), in particular pp. xl-lxiv.
3 On Veneto manuscripts of the Bolognese *Fiore di virtù*, see M. Corti, 'Emiliano e veneto nella tradizione manoscritta del *Fiore di virtù*', *SFI*, 18 (1960), 29-68 (pp. 45-51), and 'Note sui rapporti fra localizzazione dei manoscritti e *recensio*', in *Studi e problemi*, pp. 85-91 (pp. 88-9). On the versions of the *Dialagu de Sanctu Gregoriu* by 'frati Iohanni Campulu de Missina', see the edition by S. Santangelo, supplement to *Atti dell'Accademia di scienze, lettere e belle arti di Palermo*, n. 2 (Palermo, Scuola tipografica 'Boccone del Povero', 1933), especially pp. v-xvi.
4 M. Corti, 'L'impasto linguistico dell'*Arcadia* alla luce della tradizione manoscritta', *SFI*, 22 (1964), 587-619.
5 Folena, 'Filologia testuale', pp. 29-31.
6 Folena, 'La tradizione', p. 54.
7 P. Rajna, 'Una canzone di Maestro Antonio da Ferrara e l'ibridismo del linguaggio nella nostra antica letteratura', *GSLI*, 13 (1889), 1-36; see also the *Rime* of Antonio edited by Laura Bellucci (Bologna, Commissione per i testi di lingua, 1967), pp. ccxxvii-ccxxviii.
8 *Istoria di Eneas* (see above, n. 2), pp. lx-lxiii; Folena, 'Filologia testuale', pp. 27-8.
9 Folena, 'La tradizione', p. 47.

10 Folena, 'La tradizione', pp. 4, 58; Petrocchi, *Introduzione*, pp. 17–47.
11 Folena, 'Filologia testuale', pp. 23–34; see Cavalcanti, *Le rime*, pp. 25–39.
12 V. Branca, 'Copisti per passione, tradizione caratterizzante, tradizione di memoria', in *Studi e problemi*, pp. 69–83 (pp. 69–77, 81–3).
13 A complaint made by Bussi: *Prefazioni*, pp. xli, 9.
14 Lowry, *The World of Aldus Manutius*, pp. 236–7.
15 Lowry, *The World of Aldus Manutius*, pp. 244–5; Kenney, *The Classical Text*, pp. 17–18.
16 Lowry, *The World of Aldus Manutius*, pp. 238–40.
17 Kenney, *The Classical Text*, p. 14. On the way in which humanists altered spellings and added their own conjectures, often using only one source (but also on the exceptions to this rule), see also R. Sabbadini, *Il metodo degli umanisti* (Florence, Le Monnier, 1922), pp. 56–60.
18 Lowry, *Nicholas Jenson*, p. 91.
19 An example from Jenson's 1470 edition of Cicero's letters is given by Lowry, *Nicholas Jenson*, pp. 89–90. In contrast, the much-maligned Bussi managed to improve the text considerably in the same year.
20 Lowry, *The World of Aldus Manutius*, pp. 224–5.
21 Luisa Capoduro, 'L'edizione romana del "De orthographia" di Giovanni Tortelli (Hain 15563) e Adamo da Montaldo', in *Scrittura, biblioteche e stampa a Roma nel Quattrocento*. Atti del 2° seminario, 6–8 maggio 1982, edited by M. Miglio and others (Città del Vaticano, Scuola Vaticana di Paleografia, Diplomatica e Archivistica, 1983), pp. 37–56 (pp. 50–1).
22 Reynolds and Wilson, *Scribes and Scholars*, pp. 157–8; Lowry, *The World of Aldus Manutius*, pp. 245–7.
23 Lowry, *Nicholas Jenson*, pp. 88–9, 92–3; Kenney, *The Classical Text*, p. 17.
24 Barbaro's respect for 'old readings' can be seen, for instance, in his *Castigationes plinianae*, edited by G. Pozzi, 4 vols. (Padua, Antenore, 1973–9), III 5,3 and IV 117.
25 C. Dionisotti, 'Calderini, Poliziano e altri', *IMU*, 11 (1968), 151–85 (p. 185).
26 Timpanaro, *La genesi*, pp. 4–6; Kenney, *The Classical Text*, pp. 5–10; Grafton, *Joseph Scaliger*, pp. 27–32; V. Branca, 'Il metodo filologico del Poliziano in un capitolo della "Centuria secunda"', in *Tra latino e volgare: per Carlo Dionisotti* (Padua, Antenore, 1974), pp. 211–43. Barbaro did not identify his sources as Poliziano did.
27 *Aldo Manuzio editore*, no. VIII. On classical and Biblical manuscripts used as printer's copy, see Reynolds and Wilson, *Scribes and Scholars*, p. 140; A. C. de la Mare and L. Hellinga, 'The first book printed in Oxford: the *Expositio symboli* of Rufinus', *Transactions of the Cambridge Bibliographical Society*, 7 (1978), 184–244, n. 79; Lowry, *The World of Aldus Manutius*, pp. 234–49; J. H. Bentley, *Humanists and Holy Writ: New Testament Scholarship in the Renaissance* (Princeton University Press, 1983), pp. 127–8. On manuscripts and books, mainly of Latin texts, used in the earliest years of Italian printing, see Bussi, *Prefazioni*, pp. xxxvii–xl; C. Frova and M. Miglio, 'Dal ms. Sublacense XLII all'*editio princeps* del "De civitate Dei" di sant'Agostino (Hain 2046)', and P. Casciano, 'Il ms. Angelicano 1097, fase preparatoria per l'edizione del Plinio di Sweynheym e Pannartz (Hain 13088)', both in *Scrittura, biblioteche e stampa a Roma nel Quattrocento: aspetti e problemi*. Atti del seminario 1–2 giugno 1979, edited by C. Bianca and others, Vol. I (Città del

Vaticano, Scuola Vaticana di Paleografia, Diplomatica e Archivistica, 1980), pp. 245–73 and pp. 383–94; M. Miglio, 'Dalla pagina manoscritta alla forma a stampa', in *Libri manoscritti e a stampa da Pomposa all'umanesimo*, edited by L. Balsamo (Florence, Olschki, 1985), pp. 149–56; G. Bertoli, 'I segni del compositore in alcune copie di tipografia di edizioni fiorentine del XVI secolo: un po' di casistica', *LB*, 91 (1989), 307–24; P. Scapecchi, 'An example of printer's copy used in Rome, 1470', *The Library*, 6th series, 12 (1990), 50–2. On the identification of vernacular manuscripts used as printer's copy, see Trovato, 'Per un censimento'.

28 M. H. Laurent, 'Alde Manuzio l'ancien, éditeur de S. Catherine de Sienne (1500)', *Traditio*, 6 (1948), 357–63 (p. 363).

29 Cases where new manuscripts were copied include the 1474 contract between Filippo da Lavagna and Bonaccorso (see above, p. 17), which stated that Filippo would have the *Scriptores Historiae Augustae* transcribed before printing (even though the source was probably a fifteenth-century MS), and Bembo's editions of Petrarch and Dante in 1501 and 1502 (see chapter 4).

30 'Caesaris Commentarios, librariis exemplar fidele poscentibus, tumultuaria festinatione recognitos, quantum otii a forensium actionum perturbatione sum nactus ... praecipiti quadam lectione percurri': Federici, *Memorie trevigiane*, doc. IX, pp. 197–8. On Bussi's haste, see Kenney, *The Classical Text*, pp. 13–15.

31 His source was probably De Gregori's 1508 edition: Fowler, *Catalogue*, p. 89.

32 Nuovo, *Alessandro Paganino*, pp. 43–5, 202–3.

33 On errata lists and corrections made during printing, see Trovato, *Con ogni diligenza*, pp. 86–93. On the latter, see too Fahy, *Saggi*, pp. 155–68. (They can be found even in works edited to a relatively high standard, such as the Jacopone of 1490, Bembo's Dante of 1502, and the *Decameron* of 1527.) For examples of last-minute addition of material at the Aldine press, see Richardson, 'The two versions'.

34 On this lack of specialization, see S. Noakes, 'The development of the book market in late Quattrocento Italy: printers' failures and the role of the middleman', *Journal of Medieval and Renaissance Studies*, 11 (1981), 23–55 (pp. 42–3).

35 A book-privilege application of 1492 shows that the printer Paganino Paganini planned to use four editors for a Latin Bible with commentary: Fulin, 'Documenti', doc. 9, pp. 104–5 (see also Lowry, *The World of Aldus Manutius*, p. 241).

36 'P. Stephanus Dulcinius elegantissimum hoc Manilii opus pro ingenii imbecillitate trecentis locis emendavit: quae autem ambigua videbantur, intacta reliquit. Doctores reliquum addant: et nihil deinceps, candide lector, desiderabis': *BMC*, VI, 721.

37 *Orality and Literacy* (London, Methuen, 1982), pp. 132–5.

38 Franciscus Argilagnes writing to the reader in his edition of the *Articella*, a collection of medical works: *BMC*, V, 325, on the 1487 edition of Battista Torti.

39 *BMC*, V, 179.

40 Astemio wrote on 18 May 1523 to Fredericus Nausea, in the latter's *Epistolarum... libri X*, that, as for marks of separation or period division ('de distinctionibus, sive commatibus'), printers or, as they are called, compositors ('impressores, sive (ut dicitur) compositores') set in print few of the things that editors noted (Basle, J. Oporinus, 1550, p. 23; but the date of 1525 given here should be 1523: Pesenti, 'Le edizioni veneziane', p. 308 n. 51).

41 Lowry, *The World of Aldus Manutius*, pp. 236–8, 246–7; Trovato, *Con ogni diligenza*, pp. 113, 131.
42 Hellinga, 'Manuscripts', pp. 4–6; Trovato, *Con ogni diligenza*, pp. 39–41.
43 Fahy, *Saggi*, pp. 167–8.
44 Trovato, 'Primi appunti', p. 3.
45 In England it appears that compositors might interfere with 'accidentals' such as spelling and punctuation but not with grammar: see H. T. Price, 'Grammar and the compositor in the sixteenth and seventeenth centuries', *Journal of English and Germanic Philology*, 38 (1939), pp. 540–8. On English practice in composition and proof-correction see too P. Simpson, *Proof-Reading in the Sixteenth Seventeenth and Eighteenth Centuries* (London, Oxford University Press, 1935); P. Gaskell, *A New Introduction to Bibliography* (Oxford, Clarendon Press, 1972), pp. 343–53; and Shakespearean studies such as C. Hinman, *The Printing and Proof-Reading of the First Folio of Shakespeare*, 2 vols. (Oxford, Clarendon Press, 1963).
46 Trovato, *Con ogni diligenza*, pp. 230–2.
47 Timpanaro, *La genesi*, p. 15 and pp. 3–4; Reynolds and Wilson, *Scribes and Scholars*, pp. 208–9.

3 HUMANISTS, FRIARS AND OTHERS: EDITING IN VENICE AND FLORENCE, 1470–1500

1 Scholderer, *Fifty Essays*, p. 88.
2 There is a useful survey of their editing in Lowry, *Nicholas Jenson*, chapters 3–5.
3 For example, works of Duns Scotus were edited by the Franciscan friars Antonio Trombetta, who taught metaphysics (*BMC*, v, 212), and Graziano da Brescia, a lecturer in theology (*BMC*, v, 401–2), while a Dominican lecturer in metaphysics, Francesco Securo da Nardò, edited a work on Aristotle by Antonius Andreae (*GW* 1656). (On Trombetta: Brotto and Zonta, *La facoltà*, pp. 203–7, and A. Poppi, 'Lo scotista patavino Antonio Trombetta (1436–1517)', *Il Santo*, 2 (1962), 349–67 (pp. 353–4); on Graziano and Francesco da Nardò: Brotto and Zonta, *La facoltà*, pp. 181–3, 195–7; on the latter, Gargan, *Lo studio*, pp. 114–15.) Volume v of the *BMC* provides several other examples of Franciscan editors who are not described as teachers, including a 'frater Rufinus' (p. 159), Francesco Benzoni da Crema (pp. 253, 276), Mariotto da Pistoia (p. 253), Luca da Suvereto (near Piombino) (p. 347), Francesco da Montefeltro (p. 441), and Giovanni Antonio da Padova (p. 457; see Brotto and Zonta, *La facoltà*, p. 233). Dominican editors of works by St Thomas Aquinas included Pietro da San Canciano and Giovanni Francesco da Venezia, both of whom worked for Jenson (*BMC*, v, 177–8, 181; Gargan, *Lo studio*, pp. 109–10, 117–18; Lowry, *Nicholas Jenson*, pp. 155–6), and Teofilo da Cremona (*BMC*, v, 342, 514–15).
4 *BMC*, v, 418–19.
5 *BMC*, v, 241. On fra Marino, see Gargan, *Lo studio*, p. 118.
6 *BMC*, v, 485 and 562. On the sources for the *Epistole*, see M. H. Laurent, 'Alde Manuzio l'ancien, éditeur de S. Catherine de Sienne (1500)', *Traditio*, 6 (1948), 357–63 (pp. 361–3).
7 On his career, see G. Ballistreri in *DBI*, xiv, 651–2; on Bruno and more generally on the Franciscan press of the Beretin Convento, see M. Lowry, '"Nel Beretin

Convento": the Franciscans and the Venetian press (1474–1478)', *LB*, 85 (1983), 27–40, and *BMC*, v, xvi, 238–9. In spite of the Tuscan links of the press, the language of the texts which it printed, other than *Geta e Birria*, has a strong Veneto character.

8 B. Nardi, *Saggi sulla cultura veneta del Quattro e Cinquecento* (Padua, Antenore, 1971), p. 71 and Plate 16.

9 The index was dated 1 November 1490. The volume, printed by Girolamo Paganini, contains other supplementary material such as a guide to the interpretation of Hebrew names, but it is not clear whether Bruno was responsible for this too. On the growth of editorial 'aids to the reader' in early Latin Bibles, see M. H. Black in *Cambridge History of the Bible*, 3 vols. (Cambridge University Press, 1963–70), I, 420–2.

10 *BMC*, v, 512. On the Brandolini brothers, see G. De Caro in *DBI*, xiv, 33–4, 35–7.

11 Most of the hypermetric lines can be reduced by apocope (lines 1, 2, 6, 14). However, line 4, with twelve syllables and stresses on the fourth and eighth, is of a sort described by G. E. Sansone in 'Appunti sul tredecasillabo e sull'endecasillabo ipermetro', *GSLI*, 128 (1951), 176–83.

12 C. Dionisotti, *Geografia e storia della letteratura italiana* (Turin, Einaudi, 1967), pp. 57–9.

13 On Mombrizio, a Milanese humanist (1424–82?) who edited several Latin texts, see T. Foffano, 'Per la data dell'edizione del "Sanctuarium" di Bonino Mombrizio', *IMU*, 22 (1979), 509–11. Bologni, a lawyer and Latin poet from Treviso (1454–1517), worked as an editor for the printer Michele Manzolo in the same city in the period 1477–81: see Federici, *Memorie trevigiane*, pp. 117–33, 188–9, 192–8, and Trovato, *Con ogni diligenza*, pp. 109–10.

14 Della Torre, 'Per la storia', pp. 193–213; Allenspach and Frasso, 'Vicende'. Among Squarzafico's vernacular editions was a revised version of Malermi's translation of the Bible (Gabriele di Pietro, 1477). He added the *Letter of Aristeas*, which claimed to explain the origin of the Septuagint, in Bartolomeo della Fonte's translation, thus imitating what had been done for the Latin Bible by Roman printers in 1471. See A. Vaccari, 'La fortuna della lettera d'Aristea in Italia', in his *Scritti di erudizione e di filologia*, 2 vols. (Rome, Edizioni di Storia e Letteratura, 1952–8), I, 1–23; Allenspach and Frasso, 'Vicende', pp. 251–2, 272–5; Trovato, *Con ogni diligenza*, pp. 105–6.

15 For the *Filocolo*, Squarzafico must have been using a manuscript, since his first edition of the work appeared in Venice in 1472 only eight days after the only previous one had been printed in Florence: see the 'Nota al testo' by A. E. Quaglio in G. Boccaccio, *Tutte le opere*, edited by V. Branca, I (Milan, Mondadori, 1967), 706–10. For the *Fiammetta*, Squarzafico's source was an earlier edition printed in Venice between 1472 and 1481: Quaglio, 'Per il testo', pp. 36–7, 130–9.

16 Allenspach and Frasso, 'Vicende', pp. 290–1.

17 In the 'Prologo', for instance, there is 'molto' > 'multo', 'prieghi' > 'preghi', 'cuore' > 'core', 'rifiutate' > 'rifutate', 'iddio' > 'idio', 'lagrimevole' > 'lachrimevole'.

18 The Venetian edition of 1472 is also the source for the Milanese edition of 1476, from which the Milan 1478 edition is derived in turn.

19 Wilkins, 'The fifteenth-century editions'. However, the Neapolitan edition of 1477,

which derives the arrangement of its material from earlier, Roman members of Wilkins's B family, surprisingly shares the text of his A family (Venice 1470 and Milan 1473). In the B family, the text of the Venetian edition of 1473 is independent both of the A family and of previous editions in its own family (Rome 1471 and 1473), and it is the source of the edition of *c.* 1474 (Wilkins's number VII). Windelin of Speyer's 1470 edition was prepared by a 'Christophorus', perhaps Berardi: Trovato, *Con ogni diligenza*, pp. 121–2.

20 'Corrections' of the 1474 edition include *Canz.* 53.85 'advien' for 'adiven', 219.2 'resentir' for 'retentir'. The 1477 edition reads 'retenir' in the latter case (which suggests that its source was the 1472 edition) and changes 'mi rammente' to 'mi lamente' at 311.6. The 1482 edition probably used that of 1472 as a model, but has the 'resentir' (219.2) of the 1474 edition. It introduces other changes such as 'ti chiede' at *Canz.* 53.106 and 'Capitoglio' at *Tr. Cup.* 1.14, both of which readings are found in the last two editions of the B family.

21 Trovato, *Con ogni diligenza*, pp. 123–8.

22 The Bolognese edition combined the first two *capitoli* of the *Triumphus Fame* (Wilkins, 'The fifteenth-century editions', p. 231), but this error was remedied in the 1478 edition, which made other corrections, e.g. at *Tr. Fame* 2.51 (hypermetric in the Bolognese edition) and *Tr. Cup.* 4.27 ('soave' > 'leggiadro' as in earlier editions). For a very helpful survey of Petrarchan editions see Fowler, *Catalogue*.

23 Centone died in 1527, according to B. Scardeoni, *De antiquitate urbis Patavii ... libri tres* (Basle, Nikolaus Episcopius the Younger, 1560), p. 255; see too Trovato, *Con ogni diligenza*, p. 127.

24 *Tr. Temp.* 124 'quantunque' (1490 'che ovunque'), 142 '*Tutto* vince' (1490 'Tanto').

25 Trovato, *Con ogni diligenza*, p. 134.

26 Peranzone refers to his long schoolteaching career in his *Vaticinium de vera futuri diluvii declaratione* (Ancona, Guerralda, [1523]); when he wrote this work he was teaching in Recanati and nearing old age. It is discussed in L. Thorndike, *A History of Magic and Experimental Science*, Vol. V (New York, Columbia University Press, 1941), 218–20. His interests in astrology and medicine led to his *Opusculum pulcherrimum de memoriae naturalis reparatione* which the same printer brought out in 1518 (BRF MS 707, ff. 126–43, is copied from this edition).

27 Trovato, *Con ogni diligenza*, pp. 134–6. Errors in Peranzone's *Canzoniere* which could have been corrected simply by looking at earlier Venetian editions include 'di lui io' for 'diluvio' (128.28) and 'de la sua gente' for 'de lassù, gente' (128.78).

28 Dionisotti, 'Fortuna del Petrarca', pp. 105–6; Trovato, *Con ogni diligenza*, pp. 128–31. The other Milanese edition of 1494, printed by Antonio Zarotto in August and edited by a certain 'Basilico', is based on the 1492 Venetian text. It makes a few corrections which suggest use of an earlier member of the D family, and corrects a few spelling errors.

29 Tanzio's edition shares errors with the 1490 edition at e.g. *Canz.* 53.47 ('sa sicura'), 53.76 ('non fanno'), 105.34 ('brama'), 105.45 ('al pasco'), 128.66 ('barbarico'), 136.13 ('al verno'), 219.2 ('risentir'), 219.4 ('inelli'), 311.10 ('Ove duo'). On the other hand, there are different readings at e.g. 53.8 ('è spenta' not 'aspenta'), 53.78 ('mancar' not 'manca'), 53.106 ('chier' not 'chie'), 136.7 ('di vin' not 'di vitii'), 311.6 ('mi ramenti' not 'mi ramente'), *Tr. Cup.* 1.96 ('Livia sua *pregnante*' not 'pregando'), 3.40 ('*poi guarda come Amor*' not 'et vedi'), *Tr. Fame* 2.51 ('et poi

cadde' not 'et cadde poi'), 113 ('*che col bel* viso et la ferrata coma' not 'che suo bel'), *Tr. Etern.* 3 ('*in che* ti fidi?' not 'in cui').

30 Quarta, 'I commentatori', pp. 301–16; Dionisotti, 'Fortuna del Petrarca', pp. 70–8; A. Pozone, 'Un commentatore quattrocentesco del Petrarca: Bernardo Ilicino', *Atti dell'Accademia Pontaniana*, n. s., 23 (1974), 371–90.

31 E. Raimondi, 'Francesco Filelfo interprete del *Canzoniere*', *Studi petrarcheschi*, 3 (1950), 143–64; Carrara, *Studi petrarcheschi*, pp. 100–2; Dionisotti, 'Fortuna del Petrarca', pp. 78–88; R. Bessi, 'Sul commento di Francesco Filelfo ai "Rerum vulgarium fragmenta"', *Quaderni petrarcheschi*, 4 (1987), 229–70.

32 Quarta, 'I commentatori', pp. 288–301; Dionisotti, 'Fortuna del Petrarca', pp. 88–9; Belloni, 'Les Commentaires', p. 150.

33 Lowry, *Nicholas Jenson*, p. 128.

34 Quarta, 'I commentatori', pp. 280–7; Della Torre, 'Per la storia', pp. 198–9, 202–13; Dionisotti, 'Fortuna del Petrarca', pp. 89–90; Allenspach and Frasso, 'Vicende', p. 256.

35 Wilkins, 'The fifteenth-century editions', p. 233. On the introduction of foliation in early books and its relation to indexing, see M. Smith, 'Printed foliation: forerunner to printed page-numbers?', *Gutenberg Jahrbuch* (1988), pp. 54–70. Marginal letters had already been used for cross-reference between text and commentary in the Milanese Dante of 1478, edited by Martino Paolo Nidobeato of Novara.

36 On the 'Antonio da Tempo' life, see Solerti, *Le vite*, pp. 329–38; Quarta, 'I commentatori', pp. 288–93, 320–2; Carrara, *Studi petrarcheschi*, pp. 95–7.

37 Readings found in manuscripts but not in previous editions include 'sie ombra o huomo' (*Inf.* 1.66), 'attraversato' (*Inf.* 31.9), 'da l'infima lacuna' (*Par.* 33.22), 'che 'l parlar mostra' (*Par.* 33.56).

38 Dionisotti, 'Dante nel Quattrocento', pp. 368–9.

39 Dionisotti, 'Dante nel Quattrocento', pp. 369–71.

40 On Landino's edition, see below, pp. 43–4. The Venetian edition of 1484 follows the 1481 text very closely, but introduces a slightly more northern colouring: Trovato, *Con ogni diligenza*, p. 131. The Brescia edition (1487) shows more independence, contaminating Landino's text with readings from editions of the 1470s (e.g. at *Inf.* 1.22, 33.26, 113) and some readings not found in earlier editions.

41 Trovato, *Con ogni diligenza*, pp. 132–3. Of the new readings, at least one finds support in authoritative manuscripts: 'che contraposto' at *Inf.* 34.113. Among others which may be conjectures are 'He quanto a dir' (*Inf.* 1.4), 'insollazava' (*Inf.* 34.50), and 'sovra suoi rivi freddi' (*Purg.* 33.111, which conflates the two main readings of previous editions, 'sovra suoi freddi rivi' and Landino's 'sovra suoi rivi verdi'). Fra Pietro also prefers readings from editions earlier than Landino's at *Inf.* 31.4 ('soleva la lancia', not 'solea far la lancia') and *Purg.* 32.102 ('di quella torma', not 'di quella Roma', which however remained in the commentary).

42 De Robertis, *Sonetti e canzoni*, I, 34–5. The appendix contained the first two *canzoni* from the *Vita nuova*, the fifteen *canzoni* of the collection made by Boccaccio, and the *discordo trilingue* 'Aï faux ris'. The text of the *Commedia*, curiously, was independent of that of the March edition, following quite closely that of 1484. It may be that, as Trovato suggests, this was really fra Pietro's first edition and that the date of March 1491 on the other should really be March 1492: *Con ogni diligenza*, pp. 132–3.

43 It follows the March 1491 text, for example, at *Inf.* 1.4 ('He quanto a dir') and 1.28 ('Poi ch'èi posato').
44 Estimates for the total number of Venetian incunabula vary from the 3,754 of J. M. Lenhart (Hirsch, *Printing, Selling and Reading*, pp. 58–9) to the 'some 4,500' of Scholderer (*Fifty Essays*, p. 205). On the basis of the entries for Italian libraries in Volumes I–V of the *IGI*, the proportion of Venetian incunabula is put at 37 per cent of about 8,100 printed in Italy: Quondam, 'La letteratura in tipografia', p. 584 n. 13, p. 588 and n. 1. The population of Italy was about 10 million at the start of the Cinquecento, with about 1.35 million living in Venetian territory and about 105,000 living in Venice in 1509: Beloch, *Bevölkerungsgeschichte*, III, 163, 17.
45 Examples of Tuscanizing changes: in phonology, 'devide' > 'divide', 'soto' > 'sotto', 'ochii' > 'occhi', 'cossa' > 'cosa', 'cum' > 'con', 'zò' > 'ciò'; in morphology, '*li* animali' > 'gli', 'a mi' > 'a me'. However, Teo appears not to have liked *altrui* (seen perhaps as too literary), for he changed 'de le altrui virtute' to 'de cotal virtute'. The hypometric line 'e per che e lo homo la morta' became 'et perché l'homo subito lamorta'. Non-Florentine stressed vowels include those in 'soa', 'doncha', 'benegno', 'nui', 'incomencia'. Teo's revised text was used for later Venetian editions of *L'Acerba* (1481 and 1487) and remained substantially unaltered when the work appeared in Venice and Milan in the early Cinquecento (with a commentary on the first part) edited by the same Niccolò Massetti who was responsible for Pulci's *Morgante* in 1502. On Teo's similar revision of *Atila flagellum Dei* (1477), see Trovato, *Con ogni diligenza*, pp. 110–11.
46 See G. Petrocchi in his edition of Masuccio's *Novellino* (Florence, Sansoni, 1957), pp. 604–13; F. Ageno, 'Per il testo di Masuccio Salernitano', *Romance Philology*, 14 (1960–1), 28–42; Gentile, *Repatriare Masuccio*; Trovato, *Con ogni diligenza*, pp. 112–13.
47 Tuscanizing changes in vowels include 'muglie' > 'moglie', 'ultre' > 'oltre', 'vulgo' > 'volgo', 'patruni' > 'patroni', 'ruzzi' > 'rozi', 'quisti' > 'questi', 'quilli' > 'quelli', 'littera' > 'lettera', 'giovene' > 'giovane'. However, the editor is uncertain about some cases of stressed *o* and *e*, and we find 'longo', 'gionto' (alongside 'iunto'), 'ponto', 'gionger', and 'venta' (= 'vinta'). Consonants are sometimes Tuscanized, as in 'scilenzio' > 'silenzio', 'iodece' > 'giudice', 'vicii' > 'vizii', 'robba' > 'roba', 'recrissimento' > 'rencrescimento', 'epsa' > 'essa'. Occasionally verb forms are altered in accordance with Trecento Tuscan usage: thus 'incominciorno' > '-arono', 'averrebbi' > 'avrebbe'. Exceptionally, a more poetic form is introduced in 'dubitare' > 'dottare'.
48 This could happen when the original form was not Tuscan, as in the changes (which take the language even further from Tuscan) 'Iorgi' > 'Zorzi', 'giudìo' > 'zudìo', or 'carrico' > 'cargo'. But the editor often altered forms which were consistent with Tuscan usage: rarely with vowels, as in 'suoi' > 'soi', more often with consonants, as in 'chiesa' > 'giesa' or 'giesia' or 'ghiesa', 'basciar' > 'basar', 'ciò' > 'zò', 'caccia' (noun) > 'caza', 'faccia' (verb) > 'faza', and 'proda' > 'prova', on which see Gentile, *Repatriare Masuccio*, pp. 171–7.
49 An example of regularization, where a verb and a participle become two parallel participles, is '*uscissene* fuore e chiamato' > 'uscitasene' (2.29).
50 On 'vivati': Gentile, *Repatriare Masuccio*, pp. 177–80.
51 Thus 'genonico uso' (for 'giunonico') > 'matrimonial corso' (14.25); 'demorcorno'

(for 'remorcorno') > 'menorno' (29.29); 'rausasse' (for 'causasse') > 'inanimasse' (43.17); 'inmicialto' (for 'inmiliciato') > 'relevato' (50.9).
52 Forms unchanged include 'medega' (for 'medica'), 'la rason', 'comenza', 'adoncha', 'a mi', 'nui siemo'. The later edition does not alter the morphology but introduces a few orthographical and phonological changes, such as 'deletta' for 'delecta', a final vowel in 'finire', some double for single consonants, 'alcuna' for 'alguna', 'ciascuno' for 'zascuno'.
53 D. E. Rhodes, *Studies in Early Italian Printing* (London, Pindar Press, 1982), pp. 6–13; A. Jacobson Schutte, *Printed Italian Vernacular Religious Books, 1465–1500: A Finding List* (Geneva, Droz, 1983), pp. 190–1; Sandal in I *primordi*, p. 245 n. 25. Changes in the *Monte* in the direction of Florentine include 'incomenza' > 'incomincia', 'mundo' > 'mondo', 'segniore' > 'signore', 'huomeni' > 'huomini', 'sui' > 'suoi', 'custui' > 'costui', 'ultra' > 'oltra'. 'Oretchi' (= 'oreci') becomes 'orechi'. Latinizing *ct* is introduced instead of *tt* in 'fatto', 'delettava' and so on. The change to the article in 'il quale' > 'el quale' and to the adjective in 'magiori chose' > 'magiore cose' could perhaps be seen as influenced by Quattrocento Florentine. But it seems unlikely that 'Ioannes' was concerned with editing in a serious way, since there are other changes that take the text further from Florentine, such as 'di' > 'de', 'venne' > 'vene', 'trattava' > 'tratava', 'ditto' > 'dito', and unchanged Veneto forms such as 'chariega', 'faza', 'zudei', 'zardino', 'dezuna' (= 'digiuna'). Near the end there is a lexical change in 'tu patchi' (= 'pasci') > 'tu sbevazi'.
54 *IGI* 8572 (Vol. v, Plate III) and 10336 (Vol. v, plate LXIV: see VI, p. 279 for the attribution).
55 Lepschy, '"Quel libro"'; Trovato, *Con ogni diligenza*, pp. 113–14.
56 For the Venetian figure, see n. 44 above. The figures for Florence and Milan are derived respectively from Rhodes, *Gli annali*, and Rogledi Manni, *La tipografia a Milano*. For Rome see Scholderer, *Fifty Essays*, p. 205, Hirsch, *Printing, Selling and Reading*, p. 59, and *La stampa a Firenze 1471–1550: omaggio a Roberto Ridolfi*, edited by D. E. Rhodes (Florence, Olschki, 1984), p. 21.
57 The Florentine figure is based on an estimate of 800 editions for a population of half a million. The population of Florence (city and suburbs) was about 62,000 in 1470 and 80,000 in 1520; that of the Florentine state was 584,576 in 1551: Beloch, *Bevölkerungsgeschichte*, II, 148, 202. (But we know that 100 copies of a book printed by the convent of Sant'Iacopo di Ripoli were ordered for sale in Milan: Nesi, *Il diario*, p. 50.) The Venetian figure is based on an estimate of 4,000 editions for a population of 1.4 million: see n. 44 above.
58 Rhodes, *Gli annali*, no. 723. See also Biagiarelli, 'Editori di incunabuli', pp. 211–14, and (on Cennini as scribe) De la Mare, 'New research', pp. 417, 445, 526–9.
59 Nesi, *Il diario*, p. 18. If the correction led to an edition, no copy of it is known.
60 *BMC*, VI, 670. On Vespucci, see B. L. Ullman and P. A. Stadter, *The Public Library of Renaissance Florence* (Padua, Antenore, 1972), pp. 38–43, 319–20; A. C. de la Mare, *The Handwriting of Italian Humanists*, I, fasc. I (Oxford, Association internationale de bibliophilie, 1973), 106–38; De la Mare, 'New research', pp. 417, 445, 447, 498.
61 Nesi, *Il diario*, p. 17. For more information on Della Fonte, see Trinkaus, 'A humanist's image', Biagiarelli, 'Editori di incunabuli', pp. 214–17, and De la Mare, 'New research', pp. 417, 445–6, 487–8.

62 C. Dionisotti, 'Cristoforo Landino', in *ED*, III, 566–8; Biagiarelli, 'Editori di incunabuli', pp. 215–17; M. Lentzen, *Studien zur Dante-Exegese Cristoforo Landinos* (Cologne, Böhlan, 1971), pp. 27–40; R. Cardini, *La critica del Landino* (Florence, Sansoni, 1973), pp. 113–232; Procaccioli, *Filologia*; Fabrizio-Costa and La Brasca, 'De l'âge des auteurs'.
63 *BMC*, IV, 66 and VI, 670; Quarta, 'I commentatori', pp. 277–8; Dionisotti, 'Fortuna del Petrarca', pp. 69–70, 90–1.
64 As in Della Fonte's dialogue *Pelago*: Trinkaus, 'A humanist's image', pp. 134–47.
65 Examples include 'talento' > 'bisognio', 'rintuzzato' > 'entrato', 'guari' > 'molto', 'altressì' > 'similmente': Trovato, *Con ogni diligenza*, p. 324 n. 58.
66 Richardson, 'The first edition', pp. 27–8. A contemporary Florentine work, Pulci's epic *Morgante*, may have undergone similar treatment: Trovato, *Con ogni diligenza*, pp. 308–11. On the Tuscanization in 1492, at another press, of a northern printing of the Mandeville *Trattato*, see Lepschy, '"Quel libro"', pp. 214–18.
67 Changes which reduce lines to eleven syllables include 'et che di lui *tractare* m'ha facto degno' > 'tractar' (2.6), 'ne' poggi fiesolani *dove* honorata' > 'ove' (9.3), 'lascinlo leggere alli *huomini* sottili' > 'huomin' (471.1). Florence still liked Latinizing (or pseudo-Latinizing) spellings, and forms of 'tutto' were changed to 'tucto'. However, *h* was eliminated in 'honor' > 'onor'. The same letter was removed as unnecessary in 'anticha' and 'pocha'. Double -*zz*- was spelled -*z*-, but double consonants were introduced in cases such as 'idea' ('goddess'), 'magiormente', 'del huomo' > 'dell'huomo'. Other changes included the elimination of the diphthongs in 'priegovi che prieghate', forms of 'obedire' > 'ubidire', 'viciosi' > 'vitiosi', '*rare* volte' > 'rade', and '*li* altri' > 'gli'. An arbitrary change *ad sensum* was made in 'dalla gente | la qual tu di' che non m'ha mai servito' > 'dalla gente | qual tu di' che non m'hanno mai servito' (473.3–4).
68 F. Ageno, 'Riflessioni sul testo del "Convivio"', *Studi danteschi*, 43 (1967), 84–114 (p. 85), and 'Per l'edizione critica del Convivio', in *Atti del Convegno internazionale di studi danteschi* [1971] (Ravenna, 1979), pp. 43–78 (pp. 60–9).
69 For a fuller account, see Richardson, 'The first edition'.
70 Carlo Dionisotti, 'Resoconto di una ricerca interrotta', *Annali della Scuola normale superiore di Pisa. Lettere, storia e filosofia*, 2nd series, 37 (1968), 259–69 (pp. 264–5).
71 See above, p. 23.

4 BEMBO AND HIS INFLUENCE, 1501–1530

1 On Augurello's influence: R. Weiss, 'Giovanni Aurelio Augurello, Girolamo Avogadro, and Isabella d'Este', *IS*, 17 (1962), 1–11 (pp. 1–2); Dionisotti in Bembo, *Prose e rime*, pp. 16–17; A. Balduino, 'Le esperienze della poesia volgare', in *Storia della cultura veneta*, Vol. III, *Dal primo quattrocento al Concilio di Trento*, Part I (Vicenza, Neri Pozza, 1980), pp. 265–367 (pp. 362–7). On the Petrarch edition, see especially Mestica, 'Il "Canzoniere"'; Vattasso, 'Introduzione'; C. Dionisotti, 'Pietro Bembo e la nuova letteratura', in Branca, *Rinascimento europeo*, pp. 47–59; Lowry, *The World of Aldus Manutius*, pp. 225–7; Belloni, 'Il commento petrarchesco'; Pillinini, 'Traguardi'; G. Frasso, 'Appunti sul "Petrarca" aldino del 1501', in *Studi in onore di Giuseppe Billanovich*, edited by R. Avesani and others (Rome, Edizioni di Storia e Letteratura, 1984), pp. 315–35. On the Dante edition: Barbi,

Della fortuna, pp. 107–10; Dionisotti, 'Dante nel Quattrocento', pp. 376–7. On both editions: Trovato, *Con ogni diligenza*, pp. 143–9.

2 Trovato, *Con ogni diligenza*, pp. 144–5.

3 Bembo's marginal notes refer to two manuscripts, an 'obiciano' and a 'thusco'. On the former see Weiss, *Un inedito*, pp. 67–8. For a description of Vat. lat. 3195 and 3197, see M. Vattasso, *I codici petrarcheschi della Biblioteca Vaticana* (Rome, Tipografia poliglotta vaticana, 1908), pp. 9–11, 15–17; for 3195, Vattasso, 'Introduzione', pp. vii–xii.

4 Vattasso, 'Introduzione', pp. xix–xxi, xxxi–xxxii, and P. Sambin, 'Libri del Petrarca presso suoi discendenti', *IMU*, 1 (1958), 359–69 (pp. 366–8).

5 Their acceptance led to the changes 'per cui ho' > 'per cu' i ho' 51.13 (really only a compromise with Petrarch's 'per cui i' ò'), 'lassan' > 'lascian' 94.4, 'quant'aria' > 'quanta aria' 129.60, 'et tutto' > 'e tutto' 131.12, 'baciale' > 'basciale' 208.12, 'mio' > 'mi' 216.12 and 217.6, 'ch'inanzi' > 'che 'nnanzi' 251.2, 'veduta' (?) > 'veduto' 281.12, 'febre' > 'febbre' 328.6, 'frode' > 'frodi' 355.4.

6 Examples of ignored readings, with Bembo's in parentheses, are 'pentersi' ('pentirsi' 1.13); 'que'' before *che* ('quel' 4.1, 44.1); 'altri homeri' ('altr'homeri' 5.8); 'non ò di ch'i'' ('non ho di cui' 311.7); 'vertute' ('virtute' 337.6, though 'virtuti' and 'virtù' are changed to 'vertuti' and 'vertù' elsewhere).

7 Similarly, in the 1538 edition of the *Prose* (3.58), Bembo emended 'Ma ben ti prego che 'n la terza spera' (287.9) and 'il dì sesto d'aprile, in l'ora prima' (336.13) in order to eliminate 'in la', which he considered 'incorrettamente scritto': *Prose e rime*, p. 274 (see too p. 693 on some further Petrarchan readings which Bembo thought could be improved). On f. 113v of Vat. lat. 3197, beneath the erased transcriptions of sonnets 265 and 266, Bembo noted two possible alternatives to 266.5 and 8: 'Quel' antico pensier ch'al cor mi spira' for Petrarch's 'Poi quel dolce desio, ch'amor m'inspira', and 'dovunque io son, tutthora mi martira' for 'dovunqu'io son, dì et notte si sospira'.

8 See especially Mestica, 'Il "Canzoniere"', pp. 317–24, and Pillinini, 'Traguardi'.

9 See T. J. Cachey Jr, '*Il pane del grano e la saggina*: Pietro Bembo's 1505 *Asolani* revisited', *The Italianist*, 12 (1992), 5–23.

10 In the categories of phonology and morphology, some of the many readings which Bembo prefers to 'P' variants are (with the latter in parentheses): 'fur' ('fuor' 11.9, 46.12); 'inchiostro' ('incostro' 23.99); 'riconobbe' ('ricognovve' 23.133); 'homai' ('hormai' 58.3, 270.75); 'havesser' ('havessir' 60.11); 'trahe' ('tra' 66.30, 71.93); 'frale' ('fraile' 63.5, 191.4); 'aguagli' ('avagli' 71.21); 'potrebbe' ('porrebbe' 71.84 etc.); 'farebbe' ('farrebbe' 71.85); 'rimansi' ('rimanse' 72.45); 'io fossi' ('io fusse' 73.15); 'sarebbe' ('sarrebbe' 80.5); 'fragil' ('fraile' 80.28); 'et spedita' ('expedita' 91.7); 'salendo' ('sallendo' 91.11); 'colui' ('collui' 92.3 etc.); 'guarrò' ('guerrò', 97.4); 'l'imaggini' ('l'imagine' 107.9); 'sappi' ('sapi' 112.1); '*me* 'l creda' ('mi' 129.40); 'più *spedito*' ('expedito' 129.54); 'per vui' ('per voi' 134.14, in 'Sicilian' rhyme); 'saranno' ('sarranno' 137.9); 'le torri' ('le torre' 137.10); 'et' + vowel ('ed' 138.3 etc.); 'si pò' ('se pò' 153.9); 'io fossi' ('i' fussi 166.1); 'mio' ('meo' 178.5 etc.); 'fosse' 3rd sing. ('fusse' 191.9); 'nessun' ('nesun' 200.5 etc.); 'dessi' ('desi' = 'si deve' 204.10); 'diman' ('deman' 237.39); '*buon*'alma' ('bona' 240.6); 'in *qua* che strani lidi' ('quai' 260.6); 'balia' ('bailia' 264.33); '*mi* pregio' ('me' 296.2); 'indi' ('inde' 325.20); 'lasciai' ('lassai' 325.37); 'havend'io' ('habbiend'io' 365.3). Syntactic

differences, some affecting meaning, are: 'dal lor' ('da lor' 73.69); 'così fu' io *da* begli occhi ... aggiunto' ('de' 110.13–14); 'come suol *chi* ... si pente' ('che' 131.7–8); '*da* le chiome stesse | lega 'l cor' ('de' 198.3–4); '*et nel* riprego' ('e 'l ne' 240.1); 'aprì 'l suo' ('apre il suo' 260.10); 'che 'n fin dal ciel' ('che 'nfin al ciel' 334.9).

11 Pillinini, 'Traguardi', pp. 74–5. At 270.12, for instance, Bembo keeps 'vali et', with synaloephe, ignoring Petrarch's 'val et'. Examples of truncation (with the omitted vowel in brackets) are: 'secur(o) senza sospetto: onde i miei guai' (3.7); 'l'ultimo stral(e) la dispietata corda' (36.10); 'col gran suono i vicin(i) d'intorno assorda' (48.10); 'beato venire men(o): che 'n lor presenza' (71.29).

12 Original readings include: in the *Tr. Cup.*, 'dicendo questo' for 'dicendo hor questo' (1.42); in the *Tr. Fame*, 'spesso' for 'stesso' (1.2), 'cotal' for 'così' (1.13), '*quant'ha* eloquentia' for 'quanto', 'quanta' or 'quanti' (3.20).

13 Weiss, *Un inedito*, pp. 11–12, 74–8.

14 On Bembo's punctuation and contemporary reaction to it, see Belloni, 'Il commento petrarchesco', pp. 464–77. He worked in close cooperation with Aldo, who had introduced acute accents (to indicate certain stressed vowels) and the semicolon in Bembo's *De Aetna*, printed in 1496 in a roman type cut by the same man who designed the new italic, Francesco Griffo. In 1501 Aldo used acute, grave and circumflex accents in his Latin grammar to help pupils' comprehension and pronunciation. However, he did not use accents in vernacular texts of this period which were not edited by Bembo.

15 'I added, as each place required, main and subordinate punctuation marks which, when well placed, act as a commentary' ('distinctiones subdistinctionesque, ut quisque locus exigebat, apposui, quae, cum bene collocatae sunt, commentariorum vice funguntur'): *Aldo Manuzio editore*, no. LXVII. This press also resisted the use of marginal notes: at the end of its 1522 edition of Budé's *De asse*, the value of keeping margins free for the reader's notes was stressed and the recent habit (said to be of Parisian origin) of printing in the margins was condemned.

16 Folena, 'La tradizione', pp. 57, 65; Petrocchi, *Introduzione*, pp. 89–91; G. Petrocchi, 'Dal Vaticano Lat. 3199 ai codici del Boccaccio: chiosa aggiuntiva', in *Giovanni Boccaccio editore e interprete di Dante*, edited by the Società Dantesca Italiana (Florence, Olschki, 1979), pp. 15–24.

17 Examples (with the rejected reading of Vat in parentheses): (*Inferno*) 'intese' ('intesi' 2.26), 'eterno' ('eterna' 3.8), 'fama' ('infamia' 3.36), 'fur fedeli a Dio' ('furo a Dio fedeli' 3.39), 'dov'io fossi' ('là 'v' i' fossi' 4.6), 'non havea' ('non ave' 4.26), 'incoronato' ('coronato' 4.54), 'molte volte' ('spesse' 25.27), 'abbarbicata' ('abarbacata' 25.58), 'com' più' ('con più' 26.12), 'così 'l sovran li denti' ('così l'un sovra l'altro i denti' 32.128); (*Purgatorio*) 'muoveti' ('moveati' 17.17), 'porti' ('porta' 18.12); (*Paradiso*) 'si vien' ('sen vien' 2.12), 'ne guati' ('agguati' 29.42), 'conchiuso tutto' ('tutto inchiuso' 30.17), 'non è 'l seguire al mi cantar' [Landino 'è el', 'mio'] ('nol seguirà il mi cantar' 30.30), 'che 'l mio' ('ke mio' 30.31), 'prefatii' ('profatii' 30.78), 'nel verde' ('ne l'erbe') and 'opimo' ('adimo', both 30.111), 'ch'ei sarà' ('k'el farà' 30.146), 'misericordia' ('e m.' 33.19), 'non siano' [Landino 'sieno'] ('noi siano' 33.30), 'memoria' ('materia' 33.57), 'et tutta nel mirar face'si accesa' [Landino 'faceasi'] ('& sempre di mirar facesi accesa' 33.99), 'che d'infante' ('ke d'un fante' 33.107), 'si spiri' ('s'aspiri' 33.120). For further examples, see Procaccioli, *Filologia*, pp. 22–30.

18 A comparison of Bembo's transcription of *Inferno* 1 with Vat shows changes including: $y > i$, $\varsigma > zz$; 'ke' > 'che'; 'facto', 'auctore' > 'fatto' and 'auttore'; h is added in forms of *avere* and in 'hora', 'allhor', 'anchor', 'humile'; both 'omo' and 'homo' > 'huomo'; 'rispuose' > 'rispose', 'intrai' > 'entrai', 'ala riva' > 'alla riva', 'coverta' > 'coperta', 'vollia' > 'voglia', 'tutti i' > 'tutt'i'; definite article 'li' > 'gli' before a vowel or before 'dèi'; 'rispuosi' (verb) > 'risposi' but, going against the norm established in Bembo's Petrarch, 'fosse' > the older 'fusse', evidently felt to be more appropriate for Dante. 'Anco' becomes 'anchor'. At *Inf.* 2.4, Landino's 'm'apparecchiavo' (first-person singular) is changed to the older form in '-ava'.

19 On the extent to which Bembo is independent of Vat, see also M. Roddewig, 'Bembo und Boccaccio unter dem Diktat von Vat. 3199: Qualität und Textabhängigkeit der Aldina-Ausgabe der Commedia', Deutsches Dante-Jahrbuch, 47 (1972), 125–62. Cases where Bembo's text departs from the reading of Vat (given in parentheses) in favour of a reading other than Landino's include: (*Inferno*) 'molte, et grandi' ('molte grandi' 4.29), 'con suo padre' ('con lo padre' 4.59), 'una palude *fa*' ('va' 7.106), 'nomar a l'altro' ('nomare un altro' 25.42); (*Purgatorio*) 'et de primi appetibili' ('et è prima appetibile' 18.57), 'lor mele' ('lo mele' 18.59); (*Paradiso*) 'al mi veder' ('il' 30.13), 'da questo *punto*' ('passo' 30.22), 'sole *il* viso' ('in' 30.25), 'in quel' ('et quel' 30.133).

20 For example, he changed the singular verb 'soverchia' for 'soverchian' at *Par.* 31.120 because of the plural 'le parti' which his father's manuscript had in the previous line; and he inserted 'gli' before 'conservi' at *Par.* 33.35 in an attempt to emend a line made hypometric by the error 'vuoi' for 'vuoli'.

21 In *Inferno* one finds (I give Bembo's text first, then the variant of Vat in parentheses): 'sbuffa' ('scuffa. Ant.' 18.104), 'ma' ('mai' 21.20), 'ribattendo' ('ribadendo' 25.8), 'camino' ('cammino, et ita fere semper' 25.28), 'ista' ('istra' 27.21), 'giamai' ('giammai, et ita semper' 27.64), 'punger' ('purger' 30.24). In *Purgatorio*: 'pareggiando' ('passeggiando' 17.10), 'che ne' ('chenne' 17.13), 'l'imaginar mio' ('l'imagine mia' 17.43), 've ne' ('vi ne' 17.132), '*che la* ragion' ('quella' 18.85), 'fervore' ('favore' 18.106), '*pur* che 'l sol' ('più' 18.110), 'et ei' ('et el' 19.29), 'm'è di là' ('di là m'è' 19.145), 'chiosa' ('cosa' 20.99), 'Anacreonte' ('Antifonte' 22.106), 'fongo' ('sfongo' 25.56), 'stendali' ('ostendali' 29.79), 'spola' ('stola. ant', 31.96). Bembo notes the error of Vat in inserting *Purg.* 29.9 at 25.126.

22 Examples: (*Inferno*) 'risonava in quel aer' > 'risonavan per l'aer' 3.23, 'piangen' > 'piangean' 14.20, 'gremito' > 'ghermito' 21.36 and 22.138, 'loro' > 'lor' 30.88, 'l'inverno' > 'di verno' 32.26; (*Purgatorio*) 'su' lume' > 'tu' (Landino 'tuo') 18.11, 'fumi' > 'fummi' (= 'mi fu', twice) 21.98, 'guance' > 'bocche' 23.108, 'fummo' (verb) > 'fumo' 26.88, 'giel' > 'ciel' 28.122; (*Paradiso*) 'chesto' > 'chiesto' 21.125, 'che spiro' > 'che giro' 23.103, 'tu *ne* vederai' > 'lo' 29.42, 'ubriferi' > 'ombriferi' 30.78. At *Purg.* 21.19 Bembo started by transcribing the reading of Vat, 'et poi andava forte'. He then hesitated between 'poi' and Landino's 'perché', and between 'andava' and Landino's 'andate'. The final reading was 'et parte andava forte'. At least one copy of the printed text (John Rylands, Christie 33 e 1) has Landino's 'per che andate forte'; others have 'et per che andava forte', which combines the readings of Vat and Landino. One reading was evidently a stop-press correction, made possibly by a proof-reader, but possibly by Bembo himself. 'Et parte' in the MS must have been written later and might be in another hand.

23 Examples: (*Inferno*) 'per altro' > 'per l'altro' 11.61, 'disse' > 'di, se' 16.67, 'vendetta' > 'giustitia' 24.119; (*Purgatorio*) 'solo et' > 'et solo' 20.73, 'fè già' > 'già fece' 25.63, 'vincer' > 'verde' 31.83, 'come *lo* specchio' > 'in lo' 31.121, 'hor con *altri*, hor con altri' > 'uni' (as in the Venetian edition of 1472) 31.123.

24 Examples from *Inferno*: 'prima' > 'innanzi' 4.33, 'con l'altre' > 'tra' 7.95, 'bige' > 'grige' 7.108, 'prima che' > 'anzi che' 8.54, 'capi' > 'lati' 13.41, 'dispetto' > 'dispregio' 14.70, 'divenimo' > '-immo' 14.76, 'si chiama' > 's'appella' 14.95, 'miran' > 'veggon' 16.119, 'quivi' > 'quindi' 18.103, 'sozza scapigliata' > 'sozza et scapigliata' 18.130, 'fuggir' > 'veder' 21.28, 'prima' > 'imprima' 24.21, 'fumavan ... fu*mo*' > '-mm-' 25.93, 'ch'ambedue' > 'c'hamendue' 25.101, 'da quel' > 'di quel' 27.109, 'teggia' : 'streggia' > 'tegghia' : 'stregghia' 29.74–6, 'che' > 'ei' 30.89, 'hoggimai' > 'horamai' 34.26.

From *Purgatorio*: 'quando quel' > 'allhor, che quel' 18.80, 'l'herede' > 'le rede' 18.135, 'ridotto' > 'renduto' 20.54, '*tu* vedrai' > 'ben' 21.24, 'qui' > 'quivi' 22.109, 'piaghe' > 'piage' 25.30, 'sentito poi' > 'sentiti puoi' 25.105, 'apparve fora' > 'apparse allhora' 26.27, 'con voce' > 'in' 27.9, 'poi che noi' > 'come noi' (but Bembo forgot to erase the *che*) 27.13, 'stanno' > 'fanno' 27.76, '*vi* chiami' > 'ne' 29.39, '*che* di tratti' > 'et' 29.75, '*di* porpora' > 'in' 29.131, 'di là' > 'di qua' 29.141, 'tre volte disse' > 'gridò tre volte' 30.12, 'mortali' > 'fallaci' 31.56, 'onde' > 'ove' 31.102, 'suoi' > 'lor' 32.42, 'ferì' > 'ferìo' 32.115, 'che 'l sol' > 'che sol' 32.159, 'porta' > 'reca' 33.78.

From *Paradiso*: 'tanto quanto' > 'tanto in quanto' 2.23, 'vede' > 'crede' 2.45, 'ella sorrise; et poi disse, s'egli erra' > 'ella sorrise alquanto; et poi, s'egli erra' 2.52, 'credo che 'l fanno' > 'credo che fanno' 2.60, '*con* bella' > 'di' 3.2, '*di* cappello' > 'del' 19.34, 'meco' > 'et meco' 20.55, 'donarmi' > 'largirmi' 23.86, 's'incoronava' > 'si coronava' 23.101, '*con la* melode' > 'ne la' 24.114, '*ad un* modo' > 'd'un' 28.56, 'triema' > 'trema' 30.25, 'disetti' > 'discetti' 30.46, 'obstante' > 'davante' 31.24, 'vedere' > 'volere' 33.103.

25 Dionisotti in Bembo, *Prose e rime* pp. 9–10.

26 Astemio had edited Latin texts since the 1470s. See the *DBI* entry by C. Mutini, IV, 460–1; G. Castellani, 'Lorenzo Abstemio e la tipografia del Soncino a Fano', *LB*, 31 (1929), 413–23, 441–60, and 32 (1930), 113–30, 145–60. Castellani (1929, p. 448) plausibly suggests that Astemio edited this Petrarch.

27 G. Fontanini, *Biblioteca dell'eloquenza italiana di monsignore Giusto Fontanini ... con le annotazioni del signor Apostolo Zeno*, 2 vols. (Venice, G. Pasquali, 1753), II, 21; Dionisotti, *Machiavellerie*, pp. 338–40; Trovato, 'Serie di caratteri'.

28 Fowler, *Catalogue*, pp. 86–92.

29 Fowler, *Catalogue*, p. 88; Mortimer, *Harvard College Library*, no. 372.

30 Changes include the removal or addition of final vowels in order to ensure metrical regularity, as in the internal rhyme 'cantare : sospirare' > 'cantar : sospirar' at 105.1 and 4, and 'aer tutto' > 'aere tutto' at 288.1. Changes in word division include 'ferir me' > 'ferirme' (3.13), 'se non se' > 'senonse' (22.2). Among other alterations are 'contra le *qual*' > 'qua' (*Tr. Cup.* 1.25), 'feriti di' > 'da' (*Tr. Cup.* 1.30) and '*qualunqu'e* si sia' > 'quandunque' (*Tr. Etern.* 84). Aldo adds some apostrophes to divide words, as in 'larme' > 'l'arme' (2.11), 'l'havessio' > 'l'havess'io' (53.14), '*de* colli' > 'd'e' (8.1, a spelling which Bembo had used in his Dante of 1502), '*che* cervi' > 'ch'e' (*Tr. Cup.* 4.4). Concern for metre is evident in

the use of the apostrophe to indicate some cases of apocope after a consonant (as 'consolar' i casi', *Tr. Cast.* 6) and after a vowel ('suoi argomenti' > 'suo' argomenti', *Tr. Cast.* 22), and in his introduction of a diaeresis, e.g. on 'rossigniüol' (311.1, but 'rosigniuol' at 10.10 is unchanged) or 'lacciüoli' (360.51), in order to indicate that this *u* between two other vowels is a semiconsonant. This usage originated with Bembo (Belloni, 'Il commento petrarchesco', pp. 477–8) and was later approved by Lodovico Martelli in Florence (Richardson, *Trattati*, pp. 59–60).

31 De Robertis, *Editi e rari*, pp. 27–49; Richardson, 'The two versions'.
32 Fowler, *Catalogue*, p. 90; De Robertis, *Editi e rari*, pp. 31–2.
33 Fowler, *Catalogue*, p. 90. On the revision of Paganini's edition, see above, p. 24.
34 On this series see Sturel, 'Recherches'; L. Balsamo, 'Intorno ad una rara edizione di Terenzio (Venezia 1506) e allo stampatore Alessandro Paganino', in *Contributi ... Donati*, pp. 11–25; and Nuovo, *Alessandro Paganino*, pp. 36–62, 152–64.
35 Sturel, 'Recherches', pp. 68–71.
36 'La non vulgar opera del vigilante impressore' consists, he told the reader, 'non solo in quadrare le inordinate linee da l'indotte mani d'alchuno rozo e semplice compositore pessimamente disposte, m'anchora in correggere l'inumerabili errori per la inscitia sì de tempi como de librarii ne li dotti poemati et historici cresciuti'.
37 Trovato, *Con ogni diligenza*, p. 167. On Fortunio, see below, pp. 66–7.
38 The *Trionfi* order was: *Tr. Cup.* 1 'Nel tempo', 2 'Era sì pieno', 3 'Stanco già', 4 'Poscia che'; *Tr. Cast.* 'Quando ad un giogo'; *Tr. Mortis* 1 'Quanti già', 2 'Questa leggiadra', 3 'La notte'; *Tr. Fame* 1 'Nel cor', 2 'Da poi', etc.
39 Fowler, *Catalogue*, p. 91; De Robertis, *Editi e rari*, p. 32.
40 L. Pertile, 'Le edizioni dantesche del Bembo e la data delle "annotazioni" di Trifone Gabriele', *GSLI*, 160 (1983), 393–402; Nuovo, *Alessandro Paganino*, pp. 49–51, 157–8, 186–7.
41 Trovato, *Con ogni diligenza*, p. 149.
42 Folena, *La crisi*, pp. 10–14; Trovato, *Con ogni diligenza*, pp. 154–5.
43 Trovato, *Con ogni diligenza*, pp. 156–57. Classicizing spellings eliminated include 'ad' > 'a' (before a consonant), -*y*- > -*i*-, 'vagabundo' > 'vagabondo', 'rumore' > 'romore', 'voluntario' > 'volontario', 'Iacobo' > 'Iacomo'. Bembo's spelling *d'e* is used for *de* (e.g. 'd'e faggi'). Tuscan double consonants are introduced, as in 'ucelli' > 'uccelli', 'idii' > 'Iddii', 'aghiaccio' > 'agghiaccio', 'coruccio' > 'corruccio', 'vene' > 'venne', and 'dele' becomes 'de le'; but there is also 'addiviene' > 'adiviene', 'appiè' > 'apiè'. A syntactic change is 'egli è *migliore* il poco terreno ben coltivare' > 'meglio'. In morphology, a rare change in a verb form is 'fusse' > 'fosse'. We also find 'violente *secure*' ('axes') > 'securi'. Here, though, one should note that the editor does not introduce 'scuri' but keeps the form in which one sees a dialect tendency to avoid syncope (Folena, *La crisi*, p. 36). Dialect 'sinestro' becomes 'sinistro'; but unaffected southern forms include 'roscigniuol', 'puten' (= 'potino', from *potare*), 'fiscelle', 'mantarro', 'duono' (noun).
44 Dionisotti, *Machiavellerie*, p. 342 n. 37. On Delfino, see Cicogna, *Delle inscrizioni*, III, 146–50, and S. Foà in *DBI*, XL (1991), 554–5. On the editing of the *Decameron* in the Cinquecento in general, see M. Ferrari, 'Dal Boccaccio illustrato al Boccaccio censurato', in *Boccaccio in Europe*, edited by G. Tournoy (Leuven University Press, 1977), pp. 111–33; Roaf, 'The presentation'; A. Stussi, 'Scelte linguistiche e

connotati regionali nella novella italiana', in *La novella italiana. Atti del Convegno di Caprarola... 1988*, 2 vols. (Rome, Salerno, 1989), I, 191–214 (pp. 194–8); Richardson, 'Editing the *Decameron*'.

45 Trovato, *Con ogni diligenza*, pp. 165–6.

46 Gaetano's comment comes in his preface to the *Teseida*, 1528. In 1518, before Delfino's book-privilege expired, Agostino de Zanni brought out an edition which, with its folio format and illustrations, closely resembled that printed in 1510 by Bartolomeo de Zanni, but whose text was that of the most recent alternative to Delfino's, the Florentine edition of 1516.

47 On Garanta, see Harris, 'L'avventura', pp. 97–8. On Astemio, see Pesenti, 'Le edizioni veneziane', pp. 307–15, and Richardson, 'An editor', p. 274 (but on the first letter mentioned there see Pesenti, p. 308). On Lorio: D. E. Rhodes, 'Lorenzo Lorio, publisher at Venice, 1514–1527', *LB*, 89 (1987), 279–83.

48 The printer was Guglielmo da Fontaneto. The selection was printed later in the year in Milan: Sandal, *Editori e tipografi*, I, no. 96. Poets other than those named in the title were included: see Barbi, *Studi*, pp. 77–86.

49 On Mezzabarba's MS (BMV, It. IX 191 (= 6754)) and on its links with the 1518 edition, see Barbi, *Studi*, pp. 8–31 and 80–6 respectively. The judgement on the 1518 'editore' is on p. 96. On Mezzabarba, see especially C. Frati, 'Antonio Isidoro Mezzabarba e il cod. Marciano Ital. IX, 203', *Nuovo archivio veneto*, n. s., 23 (1912), 189–99, and Weiss, *Un inedito*, pp. 75–6.

50 Works on Gualteruzzi are listed in G. Frasso, *Studi su i 'Rerum vulgarium fragmenta' e i 'Triumphi'. I: Francesco Petrarca e Ludovico Beccadelli* (Padua, Antenore, 1983), p. 5. On the 1525 edition, see B. Richardson, 'Criteri editoriali nella prima stampa del *Novellino*', *Lingua nostra*, 53 (1992), 4–7.

51 Dionisotti, 'Girolamo Claricio', pp. 329–30; Grafton, *Joseph Scaliger*, pp. 47–8.

52 *Le lettere*, edited by Guido La Rocca, I (Milan, Mondadori, 1978), 383–4.

53 Ghinassi's identification of Valerio (in 'L'ultimo revisore') is confirmed by a letter by Cristoforo Tiraboschi to be published by Fabio Bertolo in *Schifanoia*. The letter shows that the book was given to Valerio to be revised at the printers' request, though it is not absolutely clear who chose Valerio. On the background to the printing of the work, see also V. Cian, *Un illustre nunzio pontificio del Rinascimento: Baldassar Castiglione* (Città del Vaticano, Biblioteca Apostolica Vaticana, 1951), pp. 131–6. Bembo saw at least five 'quinterni' (here just 'gatherings': Rizzo, *Il lessico*, pp. 42–7) of the *Cortegiano* during printing: see his *Opere in volgare*, p. 730, and Cian, p. 135 n. 1. This represented nearly one third of an edition which contained sixteen folio gatherings, collating *4 a–o^8 p^6.

54 Ghinassi, 'L'ultimo revisore', pp. 241–9. Among the changes are *in lo* > *nello*; *li* and *gli* used as feminine indirect pronouns > *le*; and imperfects of the type (*io*) *avevo* > *aveva*. Forms which were, or which might appear to be, non-Tuscan were also eliminated: thus the use of double consonants is regularized; syncopated forms such as *biasmo, merta, desidrano* > *biasimo, merita, desiderano*; *ge lo* > *glielo*; conditionals such as (*noi*) *potressimo* > *potremmo*; and, in the third-person plural of the past historic, there is -*òrno* and -*àrno* > -*àrono*, -*irno* > -*irono*, and *forno* > *furono*. Only *potria* > *poria* goes against the advice of Bembo's *Prose* (3.43).

55 *Populo* and *populi* are nearly fifty times more common than *popolo* and its plural (discounting the quotation of these words in 1.39), and *regula* is nearly five times

more common than *regola* and *regole*. But the proportions are reversed in the cases of *pericolo* and *spettacolo*. As for other Latinisms: forms with the root *laud-* outweigh those in *lod-* in a proportion of over 32:1; the forms *vulgo, vulgare* and *vulgato* (rather than *volg-*) are constant; and the printed version bears out Castiglione's preference (stated in 1.35 and 39) for *causa, onorevole, patrone, satisfatto* rather than *cagione, orrevole, padrone, sodisfatto*. *Padre* and *madre* are constant, which suggests that Valerio chose to preserve *patrone*.

56 The diphthong is constant in *vien(e), tien(e)* and their compounds. Otherwise, undiphthongized forms are in a clear majority: there is only one *buono*, for instance, against well over 300 cases of *bon, bono*, etc. *Ancor, aver, dar, esser, far, fuor, gran, poter* and *talor* all outnumber the unapocopated forms, and there is only one instance of *pure* among over 300 of *pur*.

57 Ghinassi, 'L'ultimo revisore', p. 254 n. 56.

5 VENETIAN EDITORS AND 'THE GRAMMATICAL NORM', 1501–1530

1 Quaglio in Boccaccio, *Commedia delle ninfe*, pp. xix–xx, clxxxvi–cxcii. On Sanguinacci, see R. Sabbadini, 'Briciole umanistiche', *GSLI*, 43 (1904), 244–58 (pp. 246–7). The date '1501' which appears in at least one copy (British Library C.132.h.42) is clearly an error; the correct date of late 1503 is confirmed by Zilius's statement that Boccaccio died 129 years previously in 1375.

2 The Tuscanization of Boiardo's lyric poetry in the Venice 1501 edition was only partial and superficial: M. M. Boiardo, *Opere volgari*, edited by P. V. Mengaldo (Bari, Laterza, 1962), pp. 333–4, 355–7; Trovato, *Con ogni diligenza*, pp. 149–50.

3 Examples include: (northern forms) 'fameglio', 'la rasone', 'le lege', 'aqua', 'marchexe', 'volendose', 'soi'; (Latinisms) 'eximie virtù', 'dilecto', 'subiecto', 'dextra'.

4 This has Florentine forms such as 'huomo', 'luogho', 'piglierò', 'volgar', 'pistola' (as well as 'epistola') and 'vogliamo' alongside non-Florentine forms such as 'ragi' (= 'raggi'), 'sullime' (= 'sublime'), 'disdegnarai legere'.

5 P. Stoppelli, 'Preliminari per una nuova edizione delle "Porrettane"', in *Letteratura e critica: studi in onore di Natalino Sapegno*, 5 vols. (Rome, Bulzoni, 1976), III, 145–58 (pp. 152–7); Trovato, *Con ogni diligenza*, pp. 153–4. Bartolomeo de Zanni printed the *Novellino* of Masuccio Salernitano in 1503 and probably again in 1510. But, though these editions were produced in parallel with the *Porretane* of 1504 and another edition of 1510, they show only superficial changes from Quattrocento Venetian editions, of the type 'como' > 'come', 'cossì' > 'così'.

6 See below, p. 76 and n. 33.

7 For the period 1505–15, see I. Rocchi, 'Per una nuova cronologia e valutazione del "Libro de Natura de Amore" di Mario Equicola', *GSLI*, 153 (1976), 566–85 (pp. 576–7), and Dionisotti, *Gli umanisti*, pp. 116–30.

8 Nuovo, *Alessandro Paganino*, pp. 55–61, 159, 213–18; B. Richardson, 'Le edizioni del *Corbaccio* curate da Castorio Laurario', *LB*, 94 (1992), 165–9.

9 C. Dionisotti, 'Ancora del Fortunio', *GSLI*, 111 (1938), 213–54, and 'Il Fortunio e la filologia umanistica', in Branca, *Rinascimento europeo*, pp. 11–23; Belloni, 'Alle origini'.

10 The only factor which could override the principle of abiding by rules was the evidence of rhyme: Belloni, 'Alle origini', p. 194.

11 See especially E. Raimondi, 'Il Claricio: metodo di un filologo umanista', *Convivium*, 1–3 (1948), 108–34, 258–311, 436–59; Dionisotti, 'Girolamo Claricio'; Trovato, *Con ogni diligenza*, pp. 167–9; and (for the *Ameto*) Quaglio in Boccaccio, *Commedia delle ninfe*, pp. cclxxxiii–cccv. On the controversy about the authenticity of the redaction of the *Amorosa visione* published by Claricio, see Branca in Boccaccio, *Tutte le opere*, III (Milan, Mondadori, 1974), 541–9.
12 The grave accent was used in cases such as à, ciò, là ('there'), the verbs è, fè, vè ('vedi'), on the name *Orphèo*, and on oxytones such as colà, però, piè, placò, verrò. The apostrophe was used liberally to indicate elision. There were no accents or apostrophes in the *Ameto*.
13 On Zoppino's career, see Harris, 'L'avventura', pp. 88–94.
14 See Bruni and Zancani, *Antonio Cornazzano*, pp. 170–2, and also pp. 188–90 on the *Opera nova* edited by 'B. L.' for Zoppino in 1517. On Bartolomeo Laurario, see G. Vedova, *Biografia degli scrittori padovani*, 2 vols. (Padua, Minerva, 1832–6), I, 494–5.
15 For Sarti's editorial career, see especially J. Hill Cotton, 'Alessandro Sarti e il Poliziano', *LB*, 64 (1962), 225–46, and P. Veneziani, 'Platone Benedetti e la prima edizione degli "Opera" del Poliziano', *Gutenberg Jahrbuch*, 1988, pp. 95–107.
16 Two works printed independently by Rusconi in 1518 did not show any more consistent standards of textual criticism and linguistic revision. The unidentified editor of Boccaccio's *Ninfale fiesolano* tried to improve the text in the eclectic manner of Delfino's *Decameron* but with less assiduity: he turned occasionally to an undated Venetian incunable (*GW* 4492) to emend his main source, the Venetian edition of 1477. While he corrected some errors, he also replaced existing errors with different ones, and the revision was abandoned after the first hundred stanzas or so: Balduino, 'Per il testo', 1967, pp. 149–51. The text of the *Opera nova ... in laude de Clitia* of the Florentine poet Francesco Cei (1471–1505) was substantially the same as that of the previous Florentine edition (Filippo di Giunta, 1514) and left intact the plural *canzone* (condemned by Bembo in the letter to the readers in his 1501 Petrarch), a number of Latinizing spellings (such as 'epsi', 'instrumento', 'texendo'), and some popular Florentine forms ('alle mane mia', 'drieto'). Only some minor and inconsistent tinkering with the language was carried out: a few Latinizing spellings were removed (as in 'sonecto' > 'sonetto', 'saxo' > 'sasso') but others were introduced ('costringe' > 'constringe' and, coinciding with northern usage, 'riferire' > 'referire', 'dubbii' > 'dubii'), and the popular form 'drento' became 'dentro'.
17 For example, 'doi' (= 'due'), 'adonque', 'exallare' (= 'esalare'), 'comertio', 'habiando', 'exercir', 'prehendendo', 'adviene', 'pegio', plural 'cicatrice'.
18 Sandal, *Editori e tipografi*, I, no. 103 and III, no. 598. In Zoppino's edition there has been some reduction in Latinizing spellings (as in 'facto' > 'fatto', 'prompti' > 'pronti', 'recognobbi' > 'reconobbi'; 'picti' > 'pincti' results in a compromise between Latin 'picti' and Italian 'pinti'). A Florentine vowel is introduced in articulated prepositions such as 'dil' > 'del'. On the other hand, the change 'volse' > 'vuolse' (= 'volle') shows uncertainty about Tuscan diphthongization. Most of the changes towards Tuscan concern consonants: double and single consonants, as one would expect (as in 'aqua' > 'acqua', 'aparechiata' > 'apparecchiata', as well as 'doppo' > 'dopo'), the sibilant in 'ambasiatrice' > 'ambasciatrice', the affricate in 'acontio' > 'aconcio' (but 'io comenzava' remains).

19 On this revision and on Benivieni's views, see Dionisotti, *Machiavellerie*, pp. 346–52, and Trovato, *Con ogni diligenza*, pp. 178–83. Examples of Cassiodoro's corrections of Latinisms are 'tucti' > 'tutti', 'scriptura' > 'scrittura', 'transumpto' > 'trasunto', 'Ioanni' > 'Giovanni'; morphological corrections include 'doverebbano' > '-ebbono', 'alle cose *mia*' > 'mie'.

20 De Gregori's 1526 edition did not go much further. It began by replacing *-zi-* with *-ti-* in words such as 'letizia', 'licentia' but abandoned this revision from about gathering E, nearly half way through the text.

21 These include 'el buso' ('the hole'), 'como', 'segurtà', 'missia' (= 'mischia', verb), 'butar' ('to cast'), 'lassa sugare', 'cortello', 'staga' (= 'stia'). 'Apizare' ('to light' [a fire]) is a Venetian version of the southern type *appiccià*, used in preference to Venetian *impizar*.

22 On Guazzo, see E. Pasquilini, *Un guerriero-letterato del Cinquecento: M. Guazzo*, Part 1 (Uderzo, Bianchi, 1903); on the Tebaldeo and Sannazaro editions, Trovato, *Con ogni diligenza*, pp. 80–1, 197–9. Guazzo's *Filocolo* was also printed by Bindoni and Pasini, with a book-privilege, in March 1530: Quaglio, *Scienza e mito*, pp. 226–9.

23 The Proemio, for example, had 'la rasone', 'faciamo sugetti', 'dele lege', 'recordo', 'eximie', 'de impazarme' (= 'di impacciarmi'), 'fabula del populo'.

24 Changes include 'rasone' > 'ragione', 'asumpto' > 'asunto', 'passorono' > 'passarono' (the older form). The replacement of 'che persentendo lui' with 'ch'egli persentendo' appears to result from the strictures of Fortunio and Bembo on the use of *lui* as subject. Among unaltered forms are 'ale quale', 'constrinseno', 'consequi', 'soi', and the same 'de impazarme' mentioned in the previous note.

25 G. Arbizzoni, 'Appunti sulle stampe delle "operette amorose" di Olimpo da Sassoferrato', *Res publica litterarum*, 10 (1987), 9–20.

26 In one version of a stanza, we have the Tuscan articulated prepositions 'nella' and 'alla' but the northern 'fazza' ('face', 'appearance') in rhyme with 'piazza' and 'razza'. In the other version, we find northern 'nela' and 'ala', and Tuscan 'faccia' in rhyme with the spurious 'piaccia' and 'raccia'. See Harris, 'L'avventura', pp. 95–100 and *Bibliografia*, pp. 92–5; Trovato, *Con ogni diligenza*, pp. 174–5.

27 Brown, *The Venetian Printing Press*, pp. 73–7, 207, 209–10.

28 M. Beer, *Romanzi di cavalleria: Il 'Furioso' e il romanzo italiano del primo Cinquecento* (Rome, Bulzoni, 1987), pp. 149–51; Trovato, *Con ogni diligenza*, pp. 173–4. Beer (pp. 168–85) studies the rewriting of another chivalrous poem, *La Dragha de Orlando innamorato* by Francesco Tromba da Gualdo: this was first printed in Perugia in 1525 in a language with strong Umbrian colouring, but then the first book of the poem was printed by Bindoni and Pasini in Venice a year later with some corrections.

29 Richardson, 'An editor'.

30 He replaced Latinisms (as in 'auxilio' > 'aiuto', 'ariete' > 'montone') and popular or dialectal words (as in 'zochi' > 'zoccoli', 'scammatico' > 'schiamazzo'), and he introduced terms with a rare or archaic flavour, such as 'prence' for 'principe' or 'amendue' for 'tutti dui'.

31 See especially Ghinassi, 'Correzioni'.

32 Important studies by Quaglio are 'Prime correzioni al "Filocolo": dal testo di Tizzone verso quello del Boccaccio', *SB*, 1 (1963), pp. 27–252, and *Scienza e mito*, pp. 217–29. On the sources of Gaetano's *Teseida*, see Ghinassi, 'Correzioni', pp. 90–3.

33 See B. Migliorini, *Storia della lingua italiana*, fifth edition (Florence, Sansoni, 1978), pp. 343–8, and Trovato's edition of Machiavelli, *Discorso*.
34 Ghinassi, 'Correzioni', pp. 45–6. Gaetano has diphthongized forms such as 'priego' (noun) and 'puoco' in contrast with undiphthongized ones such as 'prego' (verb) and 'poca'.
35 Fortunio recommended *havessono* and used *havessero* in his text; Bembo preferred the *-ero* ending to *-ono*, and considered *-eno* and *-ino* to be outside Tuscan usage, even though Petrarch had used them (*Prose*, 3.44).
36 This led in the *Teseida* to corrections such as 'Tornò accompagnato dal suo padre' > 'Andonne accompagnato da suo padre' (2.96.2); see Ghinassi, 'Correzioni', pp. 69–83.
37 Such changes include 'somno' > 'sonno', 'transcorra' > 'trascorra'; 'discuopre' > 'discopre', 'piatose' and 'piatà' > 'pietose' and 'pietà', 'troverrete' > 'troverete', 'uficio' > 'officio'. He separated prepositions from the following element as in 'acciò' > 'a ciò', 'nelle quali' > 'ne le quali'. On this edition, see Quaglio, 'Per il testo', pp. 141–3.
38 For example, in the Prologue the subjunctives in 'quando di sé discernano o sentano' became indicatives in *-ono*; 'con quel cuore che sogliono' was rewritten as 'col quale'; intransitive *menomare* was made reflexive; and 'li dilicati visi *con lagrime* bagnerete' became 'di lagrime'.
39 Thus, for instance, 's'avanzi' became 's'aumenti', 'disiri' was replaced with 'desii'. In the phrase 'tutte insieme adunate' the last word was omitted as redundant, as was *in* from 'in quanto io posso'. On the other hand, 'mi piace... *di farvi*... pietose' was expanded to 'tentare di farvi'. Two more radical pieces of rewriting in the 'Prologo' are seen in the following examples, where the Giunta text precedes Gaetano's. In the first case, Gaetano must have thought his version more elegant: 'priegovi che d'haverle [= le lagrime] non rifiutiate' > 'priegovi adunque che quelle non ritengate'. In the second he probably judged Boccaccio's period too complex: 'pensando che, sì come li miei, così poco sono stabili li vostri casi, li quali se alli miei simili ritornassero ... [le vostre lagrime] chare vi sarebbero rendendolevi' > 'pensando che, se a' miei casi, che così poco stabili sono, i vostri mai simili divenisseno ... caro vi sarebbe, che io ve le rendessi'. Gaetano's version of Boccaccio's letter to Pino de' Rossi was, like the *Fiammetta*, a revision of a Giunta edition, that of 1516. The changes introduced were similar, involving syntax (as in 'avisando ... voi *havete* chiusi' > 'haver'), vocabulary (such as 'inanzi' > 'prima', 's'affatica' > 'si fatica'), and a slightly greater use of apocope than before.
40 Cicogna, *Delle inscrizioni*, IV, 95–100 and VI, 819–20; Mestica, 'Il "Canzoniere"', pp. 309–12; Baldacci, *Il petrarchismo*, pp. 59–62; Belloni, 'Un eretico'.
41 Solerti, *Le vite*, pp. 361–77; Baldacci, *Il petrarchismo*, p. 52.

6 STANDARDIZATION AND SCHOLARSHIP: EDITING IN FLORENCE, 1501–1530

1 D. E. Rhodes, 'The printer of Ariosto's early plays', *IS*, 18 (1963), 13–18; Varasi and Casella in Ariosto, *Commedie*, pp. 792–5, 804–7.
2 Biagiarelli, 'Editori di incunabuli', pp. 217–20. Bernardo's publishing career extended at least to a volume of poems by Cristoforo Fiorentino ('l'Altissimo') in 1525.

3 See G. Folena's 'Nota al testo', 'Apparato' and 'Appunti sulla lingua' in his edition of the *Motti e facezie del Piovano Arlotto* (Milan and Naples, Ricciardi, 1953). References below are to this edition.
4 However, Fortunio's work may have helped to inspire preparations for a revised edition, apparently never printed, of Benivieni's *Canzoni e sonetti dello amore*: manuscript revisions to a copy of the 1500 edition, made perhaps by the poet's grandnephew Lorenzo Benivieni (born 1495 or 1496 and therefore of a generation more receptive to the new ideas on language emerging from the Veneto), include the alteration of *lui* and *lei* as subject pronouns to (respectively) *egli* and *esso*, *ella* and *essa*, of the articles *el* and *e* to *il* and *i*, of plurals such as *le luce* to *le luci* (even when these were in rhyme, so that some rewriting was involved), of the subjunctive form *fussero* to *fossero*, and the reduction of hypermetric verses. However, Latinizing spellings (involving *x*, *ct*, *mn* and so on) were not affected. See R. Ridolfi, 'Girolamo Benivieni e una sconosciuta revisione del suo Canzoniere', *LB*, 66 (1964), 213–34, Dionisotti, *Machiavellerie*, pp. 349–52, and Trovato in Machiavelli, *Discorso*, pp. xix–xxi.
5 Examples are numbers 53 (a brief *motto* on the penis) and 75, in which the priest makes love with a passionate nun; 103, on how good priests are at begging; 118, 143 and 147, which concern the priest's love of the tavern; 162, which ends with the priest addressing to a cook the blasphemous punchline 'Se ti fusse arrecato Cristo, fàllo arrosto'; 54, which mentions excretion, and 125, on the harm done by a horse-ride to the priest's 'culo'.
6 One such is number 81, which praises Federigo da Montefeltro as Duke of Urbino, and was perhaps excluded from an edition dedicated to Pietro Salviati, grandson of Lorenzo il Magnifico and nephew of Pope Leo X, because of the Pope's campaign in 1516 to oust one of Federigo's successors, Francesco Maria della Rovere, from the duchy and to install in his place the young Lorenzo, also grandson of il Magnifico. If so, the edition may be datable to this year.
7 Although one finds the occasional change (such as 'oldendo' > 'odendo', 'zente' > 'gente'), the poem remained full of forms such as 'soi', 'giaque', 'caciare', 'zorno', 'alazar' (= 'allacciare'). On this work see C. Dionisotti, 'Appunti su cantari e romanzi', *IMU*, 32 (1989), 227–61 (pp. 229–50).
8 Changes are made to vowels (as in 'vole' > 'vuole', 'lassaren' > 'lasseren', 'desio' > 'disio'). Double consonants are introduced (as in 'giaque' > 'giacque'), 'zorni' becomes 'giorni', and so on with parallel forms in *z-*. Venetian 'oldendo' becomes 'odendo'. 'Diece migliara de pagani' is altered to 'diecimila buon pagani'. Examples of morphological changes are 'io *te* ringratio' > '*ti*', 'serò' > 'sarò', and (bringing the text closer to contemporary Florentine) 'fusse' > 'fussi' (third-person singular), definite articles 'il' > 'el' and 'i' > 'e'. Both morphological and syntactic changes (away from Venetian usage) are seen in 'se i arbor fusse penne' > 'se gli arbor fusson penne'. As for Latinisms, some are eliminated (as in 'summo' > 'sommo', 'advocato' > 'avocato') but others are introduced (as in 'aiuto' > 'adiuto', 'tutto' > 'tucto', 'disse' > 'dixe').
9 On Filippo and Lucantonio, see P. Camerini, *Annali dei Giunti*, I (Florence, Sansoni, 1962), pp. 21–43; on the cultural importance of the Giunta press, see Dionisotti, *Machiavellerie*, pp. 177–92; on the legal disputes between Aldo and the Giunta press, see Lowry, *The World of Aldus Manutius*, pp. 156–8.

10 On Filippo's editors of Greek and Latin texts, see Dionisotti, *Machiavellerie*, p. 190; on those of Filippo and Bernardo, see Pettas, *The Giunti*, pp. 40–85.
11 Dionisotti, 'Dante nel Quattrocento', p. 378, and 'Girolamo Claricio', p. 324.
12 As in the Giuntine Petrarch of 1504, acute accents indicated oxytone forms such as 'fará', 'canteró', 'lasció'. Accents were also used in apocopated forms ('tu sè', 'poté' = 'potei', and 'drizzámi' = 'mi drizzai') and on letters before which Bembo would have used an apostrophe, as in 'chè' = 'ch'è', 'mèra' = 'm'era'.
13 For Benivieni's linguistic views, see above, chapter 5 n. 19.
14 Dionisotti, *Machiavellerie*, pp. 186–92, 342–4.
15 Branca, 'Introduzione', pp. lxiii–lxxxii, lxxxiv–lxxxvi (on Mannelli's manuscript); Trovato, *Con ogni diligenza*, pp. 177–8 and n. 42.
16 Quaglio concludes that this *Fiammetta* was based on the same Venetian edition, dating from between 1472 and 1481, as that of Squarzafico, but that the editor also used a manuscript as well as conjecture to correct errors or to provide alternative readings: Quaglio, 'Per il testo', pp. 139–41. The changes listed in the errata confirm the editor's intention of bringing the spelling more closely into line with Florentine pronunciation. The Latinisms 'iudicio' and 'consilio' become 'giudicio' and 'consiglio'; non-Florentine 'gionta', 'longho', 'pigliarete' become 'giunta', 'lungo', 'piglierete'; 'novo' becomes 'nuovo'. A verb ending later approved by Bembo is introduced in 'stetteno' > 'stettero'. The *Ninfale fiesolano* of 1518, whose main source was one of the three incunabula printed in Florence by Bartolomeo de' Libri (see Balduino, 'Per il testo' (1967), pp. 147–9), was printed in roman type, without punctuation (other than a full stop at the very end), and was presumably not seen as a product of the same philological standing as the series of texts in italic.
17 See Quaglio's edition, pp. ccvi–ccxvi, ccxlvii–ccxlviii.
18 Dionisotti, *Machiavellerie*, pp. 352–62.
19 For instance, a double *-gg-* was introduced in 'sogetti' and 'lege' (verb), Latinizing 'prompto' and 'docta' became 'pronto' and 'dotta', 'el mondo' became 'il mondo', feminine plurals in *-e* such as 'madre' were changed to *-i* and the unstressed pronoun 'se' to 'si'.
20 Examples are the change of 'schiera' to 'stiera' and certain verb endings, *-orono* or *-oro* for the third-person plural of the past definite of first-conjugation verbs (as in 'appelloron', 'andoro') in contrast with Fortunio's *-arono*, and 'diventassi' (replacing 'venissi') in the third-person singular of the imperfect subjunctive, where Fortunio had given *-asse*.
21 A hendecasyllable with its stresses on the third, seventh and tenth syllables, 'Trecento anni poi qua stette al più scarso' becomes 'Stiè trecento anni poi quivi al più scarso'. One of the examples of the avoidance of diaeresis on nouns in *-ione* is 'la visïone del pampino sparso' > 'la visione del pampin largo e sparso'.
22 De Robertis, *Editi e rari*, pp. 32–3; Dionisotti, *Machiavellerie*, pp. 341–2.
23 The *Triumphus Mortis* had two *capitoli* numbered I, the fragment 'Quanti già' and 'Questa leggiadra', followed by 'La notte che seguì', numbered II. Likewise the *Triumphus Fame* opened with 'Nel cor pien' and 'Da poi che morte', as in the earlier non-Aldine editions, but these were given as alternatives, both numbered I, with 'Pien d'infinita' and 'Io non sapea' as *capitoli* II and III respectively.
24 For example, the Giuntina has 'cantare' at *Canzoniere* 105.1 (= 1501), not 'cantar' (= 1514 and 1521); 'stesso' at *Tr. Fame* 1.2 (= 1514 and 1521), not 'spesso' (=

1501); 'advien' at *Canzoniere* 53.85 (= 1521), not 'adiven' (= 1501 and 1514). Examples of readings which differ from all three Aldines are 'n'assedemmo' (*Tr. Cup.* 1.51, not 'n'ascendemmo'), 'Tutte' (*Tr. Temp.* 142, not 'Tanto'), and 'a me mi volsi e dissi' (*Tr. Etern.* 3, not 'mi volsi e dissi, guarda').

25 Fowler, *Catalogue*, p. 92; De Robertis, *Editi e rari*, pp. 32–3.

26 De Robertis, *Sonetti e canzoni*, p. 24; Dionisotti, *Machiavellerie*, pp. 335–7.

27 Trovato, *Con ogni diligenza*, pp. 183–4. The corrected copy of the 1522 edition is now BNF 22.A.4.2.

28 There is a facsimile reproduction and a very valuable introduction in De Robertis, *Sonetti e canzoni*. In quoting from the edition I have kept to its system of accentuation.

29 De Robertis, *Sonetti e canzoni*, pp. 18–19; Dionisotti, *Machiavellerie*, pp. 334–5.

30 De Robertis, *Sonetti e canzoni*, pp. 23–7 and 83–5.

31 Grave and acute accents are found on many monosyllables ending in a vowel, except for definite articles, unstressed pronouns and the preposition *di*. There is a distinction between some pairs of words which would otherwise be homographs: *è* 'and', *è* 'is'; the interjection *ò*, the conjunction *ò*; the conjunction *sè*, the pronoun *sè*. The grave accent is used (a) on some prepositions or articulated prepositions, which are written separately from the article; for example *à*, *à i*, *à gli*, *à le*, but *a 'l*; *dè l'*, but *di*; *dà*, *dà la*, but *da 'l*; *nè l'* but *ne 'l*; *frà*; however, there is no accent on *su*: the rule seems to be that only *a* and *e* are accented but that the accent is not used before the apostrophe of *'l*; (b) on the conjunctions *ò* and *nè* as well as *è*; (c) on other words such as *mà*, *chè* (conjunction and pronoun, but interrogative *che*), *chì*, *ohimè*, partitive *nè*, *nò* followed by a verb (but *nó* is also found). The acute is used in other cases, including (a) verbs, for example *é*, *hó*, *há*, *fó*, *fá*, *dá*, *só*, *vó* (from *andare*, but *vo'* from *volere*), *vá*, *fú*, *puó*; (b) stressed pronouns, for example *mé*, *tú*, *té*, *sé*; and (c) several other words such as *giá*, *lá*, *tré*, *quí*, *sí*, *ció*, *piú*, *sú*, *ú* 'where', *pró*. Some indications of vowel values (*ohimè* apparently with a closed *e*, *tre* and the pronouns *me* and *se* apparently with an open *e*) differ from modern values and, at least in the cases of *me* and *se*, also from sixteenth-century Tuscan ones, according to the evidence of Trissino (Richardson, *Trattati*, p. 7).

32 De Robertis, *Sonetti e canzoni*, pp. 48–55, and 'Il canzoniere escorialense', pp. 62–137.

33 De Robertis, *Sonetti e canzoni*, pp. 42–5.

34 Thus in the poems from Dante's *Vita nuova* in Book I he has 's'é (com'io credo) in ver di mé adirata' for 'sì com'io credo, è 'nver di me adirata' (12.11, l. 12); 'ch'affogherieno il cor' for 'che sfogasser lo cor' (32.5, l. 8); and 'mosso é dà gli occhi' for 'mosse dagli occhi' (38.10, l. 13): De Robertis, *Sonetti e canzoni*, pp. 31–2. Among adjustments in Cavalcanti's poems are 'tutto mi struggo' > 'tanto mi struggo' and 'e 'l cor' > 'il cor' in the sonnet 'A me stesso di me': De Robertis, *Sonetti e canzoni*, p. 55. Segni also reconstructs conjecturally Cavalcanti, IX.52 (a line missing in his source) as 'per via troppo aspra, è dura'. On the preparatory work for the Cavalcanti section, see Favati in Cavalcanti, *Le rime*, pp. 87–104, 154–7.

35 M. Picone, 'Filologia cinquecentesca: i Giunti editori di Guittone', *Yearbook of Italian Studies*, 2 (1972), 78–101; De Robertis, *Sonetti e canzoni*, pp. 64–6.

36 Part of BLF Mediceo Palatino 119, dating from the early fifteenth century: S. Debenedetti, *Nuovi studi sulla Giuntina di rime antiche* (Città di Castello, Lapi, 1912), pp. 49–69; De Robertis, *Sonetti e canzoni*, pp. 83–5.

37 One finds 'cierto' > 'certo', 'chaldo' > 'caldo', 'folgle' > 'foglie', 'ongni' > 'ogni', 'onbra' > 'ombra'; medial double consonants were often introduced, as in 'abandona' > 'abbandona'; words were separated, as in 'elluna' > 'e l'una', 'delle' > 'dè le'; and etymological initial *h* was added in words such as 'herba'.

38 Examples are 'sança' > 'senza', 'virtute' > 'vertute', 'maravigli' > 'meravigli', 'sprendor' > 'splendore', and the restoration of the full form of some falling diphthongs, as in 'i'" > 'io', 'sare"' > 'sarei', 'chu' i" > 'chui io'.

39 These include the feminine plural adjective in 'nobile donne' > 'nobili', the definite article 'e' > 'i', the indirect feminine singular pronoun in 'starli' > 'starle', 'verun'erba' > 'nessuna herba', past definite 'viddi' > 'vidi', and the present subjunctive (third-person singular) 'facci' > 'faccia'. See respectively *Prose* 3.5, 9, 19, 24, 31, 45.

40 To avoid such synaloephe, in II.19 he had simply to introduce an elision ('che io' > 'ch'io'), but in II.4 he had to omit a word, changing 'chui i' son' to 'cui son', and in I.19 he had to rearrange the word order, so that 'Io avea il chor duro chom' una pietra' became 'Io havea duro il cor come una pietra'.

41 On annotated copies of the Giuntina: De Robertis, *Sonetti e canzoni*, pp. 22–3. On the 'Raccolta Bartoliniana': Barbi, *Studi*, pp. 119–214.

7 TOWARDS A WIDER READERSHIP: EDITING IN VENICE, 1531–1545

1 For the dates of these presses, see Ascarelli and Menato, *La tipografia*, pp. 321–84.

2 The details on Liburnio's readers are on f. A2r-v; the comments on Bembo are on f. D7r-v. The fundamental study on Liburnio is C. Dionisotti, 'Niccolò Liburnio e la letteratura cortigiana', *Lettere italiane*, 14 (1962), 33–58.

3 Grendler, *Schooling*, pp. 174–88; R. Black, 'The curriculum of Italian elementary and grammar schools, 1350–1500', in *The Shapes of Knowledge from the Renaissance to the Enlightenment*, edited by D. R. Kelly and R. H. Popkin (Dordrecht, Kluwer, 1991), pp. 137–63.

4 This Florentine was, however, still interested in the text of the *Canzoniere*. For example, he defended 'desir' (22.24) against 'destin' and 'grama' (105.34) against 'brama'. (Bembo's 1501 text had 'destin' and 'grama'.) He preferred 'vede' to 'vide' at 259.14, on the grounds of sense and manuscript evidence.

5 Dionisotti in Bembo, *Prose e rime*, pp. 48–9.

6 Aretino, *Sei giornate*, p. 212.

7 B. Migliorini, 'Sulla lingua dell'Ariosto', in his *Saggi linguistici* (Florence, Le Monnier, 1957), pp. 178–86.

8 Belloni, 'Un eretico', pp. 65–70.

9 Baldacci, *Il petrarchismo*, pp. 65–8; Belloni and Trovato, 'Sul commento'.

10 At least one of these variants found its way into his own text: at 105.16 he rejected 'I' die' in guarda a san Pietro' in favour of 'Dio ne guardi e san Pietro', on the grounds that the latter was found in the old manuscripts which he had seen and that a similar expression was still alive in his own day.

11 On Gesualdo's edition, see Belloni, 'Di un "parto"'.

12 For example, he identified as Tuscan 'dèi' rather than 'devi' or the older 'debbi', 'sarebbe' rather than Petrarch's 'fora' (which the *Prose*, 1.10, linked with Provençal

but Gesualdo considered 'of the common Italian tongue'), and 'saglita' rather than Petrarch's 'salita'.

13 Some examples from *Canzoniere* 70–100: *appannare* is glossed as 'to veil and conceal' ('velare e celare') and compared with the Neapolitan sense of 'to leave ajar'; the Italian usage of *donna* to mean 'woman' is compared unfavourably with its Petrarchan, and also Neapolitan, sense of 'noble lady'; the Neapolitan use of conditional *poria* in the first- and third-persons singular is contrasted with its Tuscan use in the third person only; the imperfect *credia*, which Bembo linked with Provençal (*Prose*, 1.10), is seen rather as a form found in parts of Italy and especially Calabria and Sicily; *haggio* from Latin *habeo* is compared with Neapolitan *raggia* from *rabies* and seen as derived from southern *haiio*, while the people of Sessa have *hao*, whence came Tuscan *ho*; and Latin *findo* is translated with Neapolitan *spaccare*.

14 A Neapolitan link is suggested in Trovato, 'Serie di caratteri', p. 101.

15 Accents were used on oxytone words and to distinguish between monosyllables such as é 'is', è 'and', é 'he', e (article). But in other cases Greek usage is respected and grave and acute accents are not linked with differences of meaning: thus for example both *dí* and *dì* mean 'day', *dí* is the imperative of *dire*, *di* is the preposition.

16 The editor could have known of Gesualdo's commentary in Neapolitan circles or perhaps through Bembo, who had been asked to scrutinize it before printing: Belloni, 'Di un "parto"', pp. 363–4. The derivation of *agognare* from *agoniare* corrected Vellutello's fanciful derivation from *augurare*: Belloni, 'Un eretico', p. 72 n. 73.

17 This defence of Petrarch is of interest for contemporary debates on the vernacular and its verse. Against those who despise this tongue, it is asserted that Petrarch is superior as a love poet to any Greek or Latin author; that Petrarch is not to be condemned as lascivious; and that the study of Tuscan can help with that of Latin, Greek and Hebrew because of its conformity with them. One should ignore those who are Tuscan 'by birth or by declaration' ('ò di natione ò di professione') and who mock others by saying that rules and 'osservationi' on Trecento authors are of no use but that one must be born in Tuscany. Similarly, one should ignore those who reject Tuscan in favour of another regional variety or a koinè: Tuscan is incontrovertibly the most cultivated language of Italy and, unlike other varieties, already has a body of literature.

18 On his life (*c.* 1508–68) and works see E. A. Cicogna, 'Memoria intorno la vita e gli scritti di messer Lodovico Dolce letterato veneziano del secolo XVI', *Memorie dell'I. R. Istituto veneto di scienze, lettere ed arti*, 11 (1862), 93–200, and G. Romei in *DBI*, XL (1991), 399–405. For descriptions of this and other editions of the *Furioso*, see Agnelli and Ravegnani, *Annali*, I; on the editorial contributions, see too G. Fatini, *Bibliografia della critica ariostea (1510–1956)* (Florence, Le Monnier, 1958).

19 For comparative costs, see Grendler, *The Roman Inquisition*, pp. 12–13.

20 *AIS* map 1420.

21 On Dolce's edition, see Javitch, *Proclaiming a Classic*, pp. 31–6, 50–3. For examples of Dolce's linguistic revisions in 1535 and 1542, see Trovato, *Con ogni diligenza*, pp. 222–4.

22 Agnelli and Ravegnani, *Annali*, I, 114.

23 Domenico's annotations were ready by 12 March 1540, the date of his own dedication. On him, see Belloni and Trovato, 'Sul commento'; on his annotations,

see Javitch, *Proclaiming a Classic*, pp. 49–50. Giolito's dedication to the Dauphin is dated 31 May 1542. The 1542 quarto edition of Bindoni and Pasini must be later, since Pasini's dedication refers to someone else having heard of Fausto's work and hence having already followed in his footsteps. The octavo version is dated October 1542.

24 An extract from the letter is given in Trovato, *Con ogni diligenza*, p. 222. In a sample passage from Canto 6, one error of Giolito's earlier editions, 'mormori' for 'mormorii' (6.24), was corrected in 1544, but the error 'verde e giallo' for 'verdegiallo' (6.13) remained. Merato was the dedicatee of the *Corbaccio* edited by Domenichi for Giolito in 1545.

25 Examples of local terms, sometimes given in Tuscanized form, are *stuare* (*AIS* map 921) for Boccaccio's *attutare* ('to calm, extinguish'), *versoro* (*AIS* map 1434) for *aratro* ('plough'), *scorlare* (*AIS* map 1256) for *crollare* ('to shake'), *persegher* (*AIS* map 1283) for *pesco* ('peach tree'), *follo* (*AIS* map 935) for *soffione* ('bellows'), *sparagnare* (*AIS* map 282, also found in southern Italy) for *risparmiare* ('to save'), *sopiare* (*AIS* maps 168, 936, also found in Emilia) for *soffiare* ('to blow'), *ameda* (*AIS* map 20) for *zia* ('aunt'), *cioto* (*AIS* map 191) for *zoppo* ('lame'). Minerbi's text followed the Florentine 1527 edition for the Proemio and introduction to Day 1, but then Delfino's edition.

26 Studies of Brucioli's life and works include G. Spini, *Tra Rinascimento e Riforma: Antonio Brucioli* (Florence, La Nuova Italia, 1940) and 'Bibliografia delle opere di Antonio Brucioli', *LB*, 42 (1940), 129–80; Dionisotti, *Machiavellerie*, pp. 193–226; A. Landi, 'A proposito di Antonio Brucioli', *Archivio storico italiano*, 146 (1988), 331–9.

27 However, *pavero* is now restricted almost entirely to the Veneto: *AIS* map 908.

28 *AIS* maps 1507, 847.

29 Mortimer, *Harvard College Library*, no. 540; Trovato, *Con ogni diligenza*, pp. 76–7.

30 The costs were shared by Curzio Navò, as the title pages of some copies show.

31 For more details, see Trovato, *Con ogni diligenza*, pp. 216–18.

32 In 1542 Dolce edited the *Fiammetta* for Giolito with no greater originality, using Gaetano's text and adding only an index of noteworthy passages.

33 Dionisotti in Bembo, *Prose e rime*, pp. 50–1.

34 On Domenichi (1515–64), see especially Poggiali, *Memorie*, I, 221–93; R. Bruni, 'Polemiche cinquecentesche: Franco, Aretino, Domenichi', *IS*, 32 (1977), 52–67; A. D'Alessandro, 'Prime ricerche su Lodovico Domenichi', in *Le corti farnesiane di Parma e Piacenza, 1545–1622*, edited by M. A. Romani and A. Quondam, II (Rome, Bulzoni, 1978), 171–200; A. Piscini in *DBI*, XL (1991), 595–600.

35 Belloni, 'Les Commentaires', pp. 152–3.

36 Recent accounts are C. Bologna, 'Tradizione testuale e fortuna dei classici', in *Letteratura italiana*, edited by A. Asor Rosa, VI (Turin, Einaudi, 1986), 445–928 (pp. 510–11), and V. Grohovaz, 'Prime note sul commento al Petrarca attribuito a Giulio Camillo Delminio', *Studi petrarcheschi*, n. s., 4 (1987), 339–47. Camillo may have been the source of another set of notes once attributed to Bembo: see G. Belloni, 'Appunti su una recente fototipica della terza aldina petrarchesca', *Rivista di letteratura italiana*, 4 (1986), 175–90, and T. Badalin, 'Appunti sulle chiose attribuite al Bembo (Padova, Biblioteca del Museo Civico, C. P. 1156)', *Studi petrarcheschi*, n. s., 4 (1987), 325–37.

37 On Daniello and Petrarch, see Weiss, *Un inedito*, pp. 23–5; E. Raimondi, 'Bernardino Daniello e le varianti petrarchesche', *Studi petrarcheschi*, 5 (1952), 95–130; Baldacci, *Il petrarchismo*, pp. 63–5; G. Belloni, 'Sul Daniello commentatore del "Canzoniere"', *Lettere italiane*, 32 (1980), 172–202 and in 'Les Commentaires', p. 154.
38 Daniello transposed the first two lines of *Canz.* 210. He gave three reasons for this in a letter to Benedetto Varchi in the 1541 edition: the reading of a 'very old' manuscript belonging to Girolamo Molino, the rhyme scheme (which would otherwise not match the second quatrain), and the flow of Petrarch's thought. It was in this letter that he discussed the order of the *Trionfi*: see G. Frasso, 'Francesco Petrarca, Trifon Gabriele, Antonio Brocardo', *Studi petrarcheschi*, n. s., 4 (1987), 159–89.
39 Barbi, *Della fortuna*, pp. 118–19; C. Dionisotti, 'Alessandro Vellutello', in *ED*, v, 905–6; Procaccioli, *Filologia*, pp. 20–1; Fabrizio-Costa and La Brasca, 'De l'âge des auteurs'.
40 Readings from outside Bembo's and Landino's texts but found in other editions include *Inf.* 32.109 'che più favelle', 34.113 'che opposito' (also in Florence 1506) and *Par.* 33.89 'quasi conflati'. They are shared by the editions of Foligno 1472, Naples 1477 and Venice 1477, but Vellutello could have been using a combination of other sources. His 'Poi che posato un poco' at *Inf.* 1.28 may have come from a manuscript.
41 Some emendations were collected for him in 1535 in Rome by Onorato Fascitello: Ceruti, *Lettere inedite*, pp. 85–6.
42 As was the case with the revision of Aretino's *Ragionamento della Nanna e della Antonia* carried out in 1535–6, perhaps by Francesco Coccio (see Aretino, *Sei giornate*, pp. 418–31), and with that of Mario Equicola's *Libro de natura de Amore* for the Nicolini edition of 1536: Trovato, *Con ogni diligenza*, pp. 200–1.
43 E. Weaver, '"Riformare" l'"Orlando innamorato"', in Bruscagli, *I libri*, pp. 117–44 (pp. 137–40); Harris, *Bibliografia*, pp. 159–71.
44 Examples, with the text of a recent edition (Nicolini 1539) first, are: 'Bucifal havia nome quel roncione: | così scritto era in quella dipintura; | sopra vi era Alessandro in su l'arcione, | e già passato ha il mar sanza paura. | Qui son battaglie e gran destruttïone' (2.1.23.1–5) > 'Bucifallo il caval era chiamato: | ... | ... Alessandro ben armato, | ... | Qui son battaglie, e rovine di stato'; 'né più potea campare ad altra guisa: | arso era tutt'insino a la camisa' (3.1.20.7–8) > '... | sendo a suo scampo ogni strada precisa', where a Latinism ('precisa' = 'cut off') is introduced; 'gioglia | ch'indi partirsi mai non gli vien voglia' (3.1.55.7–8) > 'gioia | ch'ivi star sempre non sarebbe noia', where the syntax has also been revised.
45 This last change could again lead to rewriting when in rhyme, as in 'se con quiete qui attenti m'ascoltati' (2.1.3.4) > 'se d'audïenza mi sarete grati'.
46 Examples from 2.1 are 'gioso' > 'qua giù', 'travarcato' > 'esser passato', 'fiumana' > 'fiume' (2.1.25.2, in rhyme, part of a wider change also affecting lines 4–6: 'et ha d'intorno la gente villana. | Ma quel roina il muro in ogni lato | sopra nemici, e quella terra spiana' > 'et gente ha intorno di villan costume. | Ma quel rovina ... | né parte, che la terra non consume'), 'in guerra a gran mortoro' > 'senza alcun ristoro'; from 3.1, 'gorgogliava' > 'mormorava', 'aciaffa' > 'afferra', 'roncion' > 'destrier', 'divampare' > 'infiammare'. 'Fiata' was eliminated in 'basandoli la

bocca alcuna fiata' (3.1.23.2) > 'et poi che gli hebbe la bocca basciata'. The change 'prodo campïone' (3.1.23.5) > 'gagliardo campione' could have been made to avoid a diaeresis, but Domenichi did not normally try to do this.

47 For example, diaeresis is avoided in 'Quand'io *varai* mia barchetta prima' > 'disciolsi' (1.4.1) and in 'fannosi feste e *cose triümphale*' (1.9.5) > 'mostra ogniun che vale', and what were felt to be awkward cases of synaloephe were eliminated with 'e pe boschetti le nimph*e innamora*' (1.3.4) > 'e pei boschi ogni Nympha s'innamora' and 'c'havessi diligentia havut*o e* ingegno' (1.5.4) > 'ma pien d'alto sapere e d'alto ingegno'. Examples of other changes from the first few stanzas of Canto 1, giving first the readings of a recent Venetian edition, that of 1537: 'appresso a Dio' > 'appresso Iddio', 'e 'l verbo lui' > 'e il verbo in lui' (preferring to restrict *lui* to oblique cases), 'de tuo servi pietosa' > 'di tuoi servi piatosa', 'come gli hebbe un Ormanno el suo turpino' > 'non già Normando come fu Turpino', 'così *intervien chi* vive con sospetto' > 'avviene a chi'.

48 Among other things, plurals such as 'gente' and 'dolce' (feminine) now ended in -*i*; the article *el* became *il*; the masculine subject pronoun *e'* was eliminated; the endings -*iàn* and -*irèn* (first-person plural) became -*iàm*, -*irèm*; in the past historic, the 'appiccoe' type became 'appiccò' and the 'corsono' type became 'corsero'; in the present subjunctive endings of the second and third conjugations, *a* replaced *i* ('possi' > 'possa', 'venghin' > 'vengan'); third-person singular imperfect subjunctives in -*ssi* were changed to -*sse*.

49 P. Fiorelli, 'Pierfrancesco Giambullari e la riforma dell'alfabeto', *SFI*, 14 (1956), 177–210.

50 Giovanni could perhaps have been a son, now very old, of Luigi's younger brother Luca, who died in 1470 and whose children became Luigi's wards. Luigi is not known to have had children of his own.

51 Examples, with the 1537 and Domenichi corrections in brackets, are 'Gan *traditore* lo condusse alla morte' 1.8.3 ('traditor'), 'A Roma tutti *andare* vogliamo, Orlando, | Ma per molti *sentieri* n'andiàn cercando' 2.7.7–8 ('andar', 'sentier'), 'aiuta *anchora* con tue virtù divine' 18.1.7 ('anchor'). An exception is 'alzava con due *man* la spada forte' 3.2.7, which agrees with Domenichi against the original 'mani'.

52 There are minor variations in spelling but also wider discrepancies; for instance, in the first stanzas of Canto 19 Domenichi's text and (with some minor variants) the Venetian editions of 1507 and 1537 are all closer to Francesco di Dino's text in reading 'da hora *a sempre*' (Comin 'e sempre'), 'chiunque e' *sia*' (Comin 'si sia'), '*vidon al fine* scapigliata e scalza' (Comin 'vidonla afflitta'), '*si gli* raccomanda' (Comin 'se gli').

53 As in *stelle* for *scheggie* ('chips of wood'), or *frignocola* for the snapping of fingers.

54 Three years earlier it had seemed that the Giunta press in Venice was going to obtain a good manuscript from Florence with a view to improving on the earlier, incorrect Venetian printings of this work, for it was appreciated by connoisseurs of early prose as well as by those who sought practical advice on agriculture: Trovato, *Con ogni diligenza*, pp. 203–4. On contemporary readers, see M. Ambrosoli, 'Lettori e chiosatori delle edizioni a stampa di Pier de' Crescenzi tra 1474 e 1561', *Rivista storica italiana*, 96 (1984), 360–413, and S. Polica, 'Lettori cinquecenteschi di Pietro de' Crescenzi', *La Cultura*, 24 (1986), 231–55.

55 Similarly, the Venice 1533 edition of Benivieni's translation of Savonarola's *Della*

semplicità della vita cristiana followed Florentine editions. But the text was revised more thoroughly for the Venetian edition of 1547 (printed 'al segno della Speranza'), so that Latinizing spellings were removed, *el* became *il*, and so on.

56 Examples of changes, in comparison with earlier Venetian printings of 1501 (edited by Squarzafico) and 1516, are (punctuation and spelling) 'delhumana' > 'dell'humana', 'chose' > 'cose', 'piace' > 'piacie', 'extrinsece' > 'extrinseche'; (phonology) 'drento' > 'dentro'; (articles) 'el' > 'il'; (nouns) 'le laude' > 'le laudi'; (pronouns) 'lui' as subject > 'egli'; (verbs) 'affirmorono' > 'affirmarono', 'finxono' > 'finsero', 'dimostrerremo' > 'dimostreremo', 'apparischino' > 'appariscano'; (Latinizing vocabulary) 'preterea' > 'oltre di ciò'. The editor also claimed to have improved on Landino's version by using a better Latin text and to have added a new index, which was not however printed.

57 *Api* ('bees') is glossed with the typically Florentine *pecchie* (*AIS* map 1152), but north-eastern or Veneto glosses include *buovoli* for *chiocciole* ('snails', *AIS* map 459), *incalmare* for *innestare* ('to graft', *AIS* map 1255), *marangoni* for *legnaiuoli* ('carpenters', *AIS* map 219), *fregole* for *minuzzoli* ('crumbs', *AIS* map 991), *bugancie* for *pedignoni* ('chilblains', *AIS* map 383).

58 Frasso, 'Sebastiano Fausto', pp. 367–9. A book-privilege was obtained for this translation and another by Dolce on the strength of their being unpublished. Fausto added an index, marginal *notabilia* and an appendix of historical material from Pius II; but the translation was only a revision of the Milanese edition of 1490, with some quite free rewriting on the first page and thereafter revisions of the punctuation and the language (such as, at the start of Book 2, 'piccolo' > 'picciolo', 'cavagli' > 'cavalli', 'erono' > 'erano', 'vennono' > 'vennero', 'et approximavasi el fine' > 'et il fine s'approssimava', 'interim' > 'tra tanto', 'reciproci odii' > 'uguali odij').

59 Gerber, *Niccolò Machiavelli*, II, 48–57; Trovato, *Con ogni diligenza*, pp. 193, 211.

60 On the role of Cicero's letters: Grendler, *Schooling*, pp. 217–33. It was suggested as early as 1535 that Bembo might provide a vernacular equivalent: Moro, 'A proposito di antologie', pp. 99–100.

61 On these anthologies, see A. Jacobson Schutte, 'The "Lettere volgari" and the crisis of Evangelism in Italy', *Renaissance Quarterly*, 27 (1975), 639–88; *Le 'carte messaggiere'. Retorica e modelli di comunicazione epistolare: per un indice dei libri di lettere del Cinquecento*, edited by A. Quondam (Rome, Bulzoni, 1981); Moro, 'Introduzione' to the *Novo libro di lettere*; J. Basso, *Le Genre épistolaire en langue italienne (1538–1622)*, 2 vols. (Rome, Bulzoni, 1990). Anthologies could also satisfy other types of curiosity, as in the volume of *Viaggi fatti da Vinetia* collected by Antonio Manuzio in 1543.

62 A similar need for vernacular models lay behind the *Orationi diverse di diversi rari ingegni*, a collection of seven recent speeches compiled by Clario and printed by Griffio in 1546.

63 On Nardi's assistance: P. Trovato, 'Intorno al testo e alla cronologia delle "Lettere" di Jacopo Bonfadio', *Studi e problemi di critica testuale*, 20 (1980), 29–60 (pp. 32–5). A work on the Turks by Ramberti (1503–46) was printed by the Aldine press in 1539, and Paolo Manuzio dedicated Cicero's *De officiis* to him in 1541.

64 Trovato, *Con ogni diligenza*, pp. 213–15.

65 N. Cannata Salamone, 'Per un catalogo di libri di rime 1470–1530', in Santagata and Quondam, *Il libro di poesia*, pp. 83–9.

66 A. Caro, *Lettere familiari*, edited by A. Greco, 3 vols. (Florence, Le Monnier, 1957–61), I, 49–50, 57.
67 See most recently R. Fedi, 'Bembo in antologia', in his *La memoria della poesia: canzonieri, lirici e libri di rime nel Rinascimento* (Rome, Salerno, 1990), pp. 253–63, and Clubb and Clubb, 'Building a lyric canon'.

8 THE EDITOR TRIUMPHANT: EDITING IN VENICE, 1546–1560

1 Quondam, 'La letteratura in tipografia', pp. 578–87, 644; Grendler, *The Roman Inquisition*, pp. 132, 229.
2 On Doni and Domenichi: Grendler, *Critics*, p. 56, and see above, chapter 1, n. 46. On Dolce and Ruscelli: Trovato, *Con ogni diligenza*, pp. 241–52; Clubb and Clubb, 'Building a lyric canon', pp. 337–8.
3 *Con ogni diligenza*, pp. 241–97.
4 A. Carlini, 'L'attività filologica di Francesco Robortello', *Atti dell'Accademia di scienze, lettere e arte di Udine*, 7th series, 7 (1967), 53–84; Kenney, *The Classical Text*, pp. 29–36.
5 On Francesco Sansovino, see Cicogna, *Delle inscrizioni*, IV, 32–91, and three articles by Christina Roaf: 'Francesco Sansovino e le sue *Lettere sopra le diece giornate del Decamerone*', *Quaderni di retorica e poetica*, 1 (1985), 91–8; 'The presentation'; 'Cultura e conoscenze di un giovane del Cinquecento: Francesco Sansovino e le "Lettere sopra le diece giornate del Decamerone"', in *Omaggio a Gianfranco Folena*, 3 vols. (Padua, Editoriale Programma, 1993), II, 1107–16. Sansovino accused a Florentine enemy of his, employed by the printer Baldassare Costantini to correct these *Lettere* while Sansovino was away from Venice, of deliberately mauling the text to such an extent that he no longer recognized it: Quondam, 'La letteratura in tipografia', pp. 668–9.
6 Quaglio in Boccaccio, *Commedia delle ninfe*, pp. xxiii–xxiv, ccxvi–ccxx. This became the standard edition for the rest of the century, with the addition of some marginal *notabilia* in Giolito's edition of 1558. Sansovino copied Gaetano's text of the *Filocolo* (rebaptized *Filocopo*) for Giovita Rapirio in 1551.
7 A similarly disparaging attitude towards one of Boccaccio's minor works is evident in Domenichi's dedication of the *Corbaccio* printed by Giolito in 1545, where he quotes the printer as saying that, if only one copy of such a misogynistic work existed, it might not be wrong to burn it.
8 As in 'in cruccio' for 'in iscretio' (8.2.46), 'giuocata' for 'giucata' (9.4.22).
9 In an edition of 1549, printed by Giovanni Griffio, Sansovino identified the sources of these variants as recent Florentine and Venetian editions: Roaf, 'The presentation', p. 113. I accept Christina Roaf's argument here that Dolce was probably not involved in the editing of the 1546 text.
10 For instance, he gave Lombard *peschiera* for *vivaio* ('fish-pond') and *chiapare* for *incogliere* ('to catch'); *aggiugnere*, he added, is what 'we' say (i.e. Venetian *azonzer*). Similarly, he identified with a Venetian readership when he translated *manicaretto* ('tasty dish') with *guazzetto* and said that this was also called *intingolo* 'by us'. For *lattime* ('milk crust') he gave Venetian *brozze* (*AIS* map 689). *Colto* was defined as the fields worked around a house and compared with Venetian *bruolo* (*AIS* map 1415). Under *bamba* he gave the Venetian word for 'doll' as *puavola* (modern *piavola*: *AIS* map 750).

11 Examples of Sansovino's readings can be found at 1 intr. 2 and 97, 1.3.12, 1.4.11, 1 concl. 11, 2.7.20 and 81, 6.7.18, 6 concl. 47, 7.9.4, 10.6.4, 10.8.5.
12 At the end of his life, Ruscelli declared a preference for 'Ieronimo' rather than the more 'corrupt' Tuscan form 'Girolamo' (*Le imprese illustri*, 1566).
13 The reading 'habitari' for 'habituri' is one of the aspects of Ruscelli's *Decameron* criticized by Lodovico Castelvetro (*Opere varie critiche ... non più stampate* (Berne [Milan], 1727), pp. 106–8). Grazzini attacked Ruscelli's presumption and ignorance in the sonnet 'Com'hai tu tanto ardir, brutta bestiaccia': *Rime*, 2 vols. (Florence, Moücke, 1791–2), I, 107–8.
14 Richardson, 'Editing the *Decameron*', p. 27.
15 Trovato, *Con ogni diligenza*, pp. 247–8.
16 On Clario, see Giacomo Moro in the *Novo libro di lettere*, pp. lxxiii–lxxvi. Among the other works edited by him was Speroni's tragedy *Canace* (for Valgrisi, 1546). The edition made very few corrections to the autograph manuscript and did not even introduce apostrophes and accents: C. Roaf, 'A new autograph of Sperone Speroni's *Canace* and its relationship to the textual tradition of the play', in *Essays in Honour of John Humphreys Whitfield*, edited by H. C. Davis and others (London, St George's Press, 1975), pp. 137–54.
17 I quote from the 1562 octavo printing, which represents Dolce's final version of the *Osservationi*.
18 He used commas instead of some semicolons, more apostrophes (as in *be'* rather than *be*), more grave accents (as on the preposition *à* and the noun *dì*), and spellings without etymological *h* or Greek digraphs (as in *ancora*, *Zefiro*, *Triomfo*).
19 Trovato, *Con ogni diligenza*, pp. 278–80, 287–8.
20 C. Dionisotti, 'Lodovico Dolce', *ED*, II (1970), 534–5.
21 For example, among the variants quoted are a reading from Boccaccio's *Esposizione* ('coperte', *Inf*. 1.17) and readings of the 1481 edition ('Poi posato ebbi un poco', *Inf*. 1.28, and 'famoso e saggio', *Inf*. 1.89). Dolce followed Vellutello with 'ch'è opposito' at *Inf*. 34.113 (where he was suspicious of the Aldine 'che dè opposto') and at *Purg*. 32.38 ('cerchiar' against the Aldine 'cerchiaro'). His 'li vidi 'nsin là' at *Inf*. 32.34 could have been derived from a manuscript.
22 Trovato, *Con ogni diligenza*, pp. 299–300.
23 Trovato, *Con ogni diligenza*, pp. 212–13, 316–20. Giolito's debt to Virginio for the extra stanzas and 'some comedies and other fragments' (which he intended to publish shortly) was mentioned in a letter in his 1549 edition, f. *2r–v.
24 Trovato, *Con ogni diligenza*, pp. 222–4.
25 Javitch, *Proclaiming a Classic*, pp. 36–9; E. Falaschi, 'Valvassori's 1553 illustrations of *Orlando furioso*: the development of multi-narrative technique in Venice and its links with cartography', *LB*, 77 (1975), 227–51. On Clemente Valvassori, a judge who later became a Carthusian monk, see Grendler, *The Roman Inquisition*, pp. 317–19. On the Florentine *Furioso* of 1544, see below, pp. 133–4.
26 Trovato, *Con ogni diligenza*, pp. 282–6; Javitch, *Proclaiming a Classic*, pp. 39–43, 54; P. Coccia, 'Le illustrazioni dell'*Orlando Furioso* (Valgrisi 1556) già attribuite a Dosso Dossi', *LB*, 93 (1991), 279–309.
27 Valgrisi's 1558 edition added a long *Dechiaratione* of ancient and modern historical references by Eugenico (ff. d1r–k1r).
28 Ruscelli, *Tre discorsi* (Venice, Pietrasanta, 1553), pp. 144 (on 'gl'incudi'), 171–8.

29 Giorgio Masi supports Ruscelli's case in his review of Fahy, *L'"Orlando furioso' del 1532* in *Studi italiani*, 5 (1991), 166–78 (pp. 174–8).
30 On all these editions see the 'Note ai testi' by Casella and Ronchi in Ariosto, *Commedie*, pp. 795–844. Giolito's editions of *La Cassaria* and *La Lena* can be compared with earlier manuscript or printed versions. Such a comparison is not possible for the other two plays, but an editorial intervention seems very probable.
31 On Doni (1513–74), see Grendler, *Critics*, pp. 49–65. On the revision of the *Satire*, see Bongi, *Annali*, I, 280–6, Tambara's edition of the *Satire*, pp. 23–9, and C. Segre, 'Storia testuale e linguistica delle "Satire"', in *Ludovico Ariosto: lingua, stile e tradizione*, edited by C. Segre (Milan, Feltrinelli, 1976) pp. 315–30. It seems that Sansovino added *argomenti* to the *Satire* for Giolito's second edition of 1553 and a reprint in 1556: Agnelli and Ravegnani, *Annali*, II, 12–13, 15.
32 Trovato, *Con ogni diligenza*, pp. 275–8.
33 Agnelli and Ravegnani, *Annali*, II, 61–2.
34 See the edition by A. Saviotti (Bari, Laterza, 1929), pp. 330–3, and Trovato, *Con ogni diligenza*, pp. 269–72.
35 Ruscelli also made additions of his own to Paolo Giovio's treatise on *imprese* when this was printed by Ziletti in 1556: Trovato, *Con ogni diligenza*, pp. 281–2.
36 As in 'loco' > 'luogo', 'sequenti' > 'seguenti', 'allhora' > 'allora', 'ruinò' > 'rovinò', 'esperientia' > 'isperienza', even 'magistrato' > 'maestrato'.
37 Gerber, *Niccolò Machiavelli*, II, 75–6.
38 Gerber suggested that the same editor and printer may have been responsible at about this time for reissues of four earlier Venetian editions of Machiavelli's prose works. In each case the first sheet was reprinted, with the Aldine device on the title page, and with some revisions: Gerber, *Niccolò Machiavelli*, II, 63–5.
39 Ossola, '"Il libro del Cortegiano"', pp. 43–50; Guidi, 'Reformulations', pp. 143–51; V. Cox, *The Renaissance Dialogue: Literary Dialogue in its Social and Political Contexts, Castiglione to Galileo* (Cambridge University Press, 1992), pp. 106–7.
40 G. Ferroni, 'Les genres comiques dans les commentaires', in *Les Commentaires*, pp. 63–70 (pp. 68–70).
41 J. Quétif and J. Echard, *Scriptores ordinis praedicatorum*, Vol. II (Paris, Ballard and Simart, 1721), 259–60, 825; Bongi, *Annali*, I, 462–3 and *ad indicem*.
42 Bruni and Zancani, *Antonio Cornazzano*, pp. 90–2, 201–2.

9 IN SEARCH OF A CULTURAL IDENTITY: EDITING IN FLORENCE, 1531–1560

1 On the fortunes and misfortunes of the Academy, see especially Plaisance, 'Une première affirmation' and 'Culture et politique'; on the links with the Paduan academy, see Samuels, 'Benedetto Varchi'.
2 D. Moreni, *Annali della tipografia fiorentina di Lorenzo Torrentino impressore ducale*, second edition (Florence, Daddi, 1819); Biagiarelli, 'Il privilegio'.
3 Pettas, *The Giunti*, pp. 87–8.
4 Biagiarelli, 'Il privilegio', pp. 347–51.
5 Gigli, *Studi*, pp. 321–58; Barbi, *Della fortuna*, pp. 113–14; G. Manacorda, 'Benedetto Varchi: l'uomo, il poeta, il critico', *Annali della R. Scuola Normale Superiore di Pisa*, Filosofia e filologia, 17 (1903) (including references to alternative Dante

readings in Varchi's works); G. Vandelli, 'Il più antico testo critico della Divina Commedia', *Studi danteschi*, 5 (1922), 41–98; Folena, 'La tradizione', pp. 49–51; Petrocchi, *Introduzione*, pp. 76–8. On Martini, see too Plaisance, 'Une première affirmation', p. 363 n. 3; Samuels, 'Benedetto Varchi', pp. 625–7.

6 Letters of this period to Giovanni Taddei, Varchi and Cellini are in Bembo's *Opere* (Milan, Classici italiani, 1810), VII, 400–8. Letters from the Florentine group to Bembo are in Book 3 of *Delle lettere da diversi re ... a Mons. Pietro Bembo scritte* (Venice, Sansovino, 1560). On questions of dating see F. Piovan, 'Per la datazione del sonetto del Bembo al Varchi', *IMU*, 27 (1984), 311–29; Moro, 'A proposito di antologie', pp. 99–105.

7 *Prose fiorentine*, I, letters 2–3, 5–11, 14.

8 Samuels, 'Benedetto Varchi', pp. 616–28.

9 See Bembo's letter of 8 March 1533 to Ramusio in his *Opere in volgare*, pp. 736–7.

10 Grafton, *Joseph Scaliger*, p. 56; see pp. 52–70 on Vettori's critical method in general.

11 *Prose fiorentine*, I, letter 5.

12 Niccolai, *Pier Vettori*, p. 158; Martinelli, 'Contributo', pp. 203–4. On Vettori's long feud with Manuzio, see Grafton, *Joseph Scaliger*, pp. 88–93.

13 Grafton, *Joseph Scaliger*, pp. 63–8.

14 W. Rüdiger, *Petrus Victorius aus Florenz* (Halle a. S., Niemeyer, 1896), p. 49.

15 A. Porro, 'Pier Vettori editore di testi greci: la "Poetica" di Aristotele', *IMU*, 26 (1983), 307–58.

16 U. Pirotti, *Benedetto Varchi e la cultura del suo tempo* (Florence, Olschki, 1971), pp. 109–16.

17 *I Marmi*, I, 94, 132.

18 Lelio Torelli calculated in 1563 that, if the Giunta firm were granted the tax exemptions which they had requested, they would import three times as many books as they exported: Biagiarelli, 'Il privilegio', pp. 349–50.

19 Among the changes are 'el' > 'il', subject pronouns 'lui' and 'le' > 'egli' and 'elle', masculine plural 'mia' > 'miei', third-person subjunctives in *-ssi* and *-ssino* > *-sse* and *-ssero*: Gerber, *Niccolò Machiavelli*, II, 27–30.

20 Gerber, *Niccolò Machiavelli*, II, 34; Trovato, *Con ogni diligenza*, p. 211. The 1532 Giunta edition of Machiavelli's *Istorie fiorentine*, derived from a manuscript revised for the printers, had kept the subjunctives in *-ssi* and *-ssino*, but these were corrected for the 1537 edition: Trovato, *Con ogni diligenza*, pp. 196–7; Gerber, *Niccolò Machiavelli*, II, 40–1.

21 G. Fatini, 'Per un'edizione critica delle opere di Agnolo Firenzuola', *SFI*, 14 (1956), 21–175. On Scala, see Plaisance, 'Culture et politique'.

22 Among these changes are 'anticha' > 'antica', 'antiqui' > 'antichi', 'dishonore' > 'disonore', 'exprimere' > 'esprimere', 'istare' > 'stare', 'virtuti' > 'virtù', 'lo suo' > 'il suo', 'usono' > 'usano', 'faccino' > 'facciano'.

23 See Rosanna Bettarini's comments in Vasari, *Le vite*, Testo, I, xi–xiii. See too Frey, *Die literarische Nachlass*, letters 122–36, and J. Bryce, *Cosimo Bartoli (1503–1572)* (Geneva, Droz, 1983), pp. 52–5.

24 G. Nencioni, 'Premesse all'analisi stilistica del Vasari', *Lingua nostra*, 15 (1954), 33–40 (p. 34). In 1559 Varchi told Cellini that he preferred the language of the artist's *Vita* in its purity ('in cotesto puro modo') rather than revised by another.

25 Ulivi's edition is dated 1544 and the dedication was written on 3 January 1544. This was probably not according to the Florentine style of dating in which the year began from the Incarnation (25 March), and which would mean that the volume was printed early in 1545. Firstly, it is much more likely that Ulivi's brief list of descriptions was expanded by Dolce later in 1544 than that Ulivi shrank Dolce's list. Secondly, Ulivi used a Giolito edition earlier than that which Giolito brought out after 1 March 1544 (the date of Dolce's letter to Giolito). This is shown by similarities with the 1543 quarto subject index rather than with the slightly revised 1544 index; one example is 'Legge di Scotia, la quale *dannava* a morte ciascuna donna, che *con un* suo amante fosse trovata' where the later edition had 'condannava', 'con alcuno'.
26 On the earlier editions and on Grazzini's, see A. Virgili, *Francesco Berni* (Florence, Le Monnier, 1881), pp. 514–23; Berni, *Poesie e prose*, edited by E Chiòrboli (Geneva and Florence, Olschki, 1934), pp. 377–88 (and the apparatus criticus); Plaisance, 'Culture et politique', pp. 216–18.
27 For instance, the plurals 'torre' and 'ciglie' > 'torri' and 'ciglia', subjunctive 'facci' > 'faccia', the indirect pronoun 'gli' (= 'to her') > 'le', 'presto' > 'tosto'.
28 Other frequent revisions were the addition of final -*e* where this had been apocopated, *uo* for *o* in words such as 'gioco' and 'vole', and 'moro' > 'muoio'.
29 The collection was revised and enlarged by Domenichi in 1562 and in 1564, and then again after his death by his friend Porcacchi. See G. Folena, 'Sulla tradizione dei "Detti piacevoli" attribuiti al Poliziano', *SFI*, 11 (1953), 431–48; T. Zanato's edition of the *Detti piacevoli* (Rome, Istituto della Enciclopedia Italiana, 1983), pp. 27–42; A. Fontes-Baratto, 'Pouvoir(s du) rire. Théorie et pratique des facéties aux xve et xvie siècles: des facéties humanistes aux trois recueils de Lodovico Domenichi', in *Récritures III: commentaires, parodies, variations dans la littérature italienne de la Renaissance* (Paris, Université de la Sorbonne Nouvelle, 1987), pp. 9–100.
30 Grendler, *Critics*, p. 66.
31 Such as 'aveano' > 'havevano' (f. 9r of the MS), 'lo chavallo' > 'il cavallo', 'cierchi' > 'cerchi', 'sança' > 'senza', 'li rispuose' > 'gli rispose' (f. 19r). For the contents, see C. Ricottini Marsili-Libelli, *Anton Francesco Doni, scrittore e stampatore* (Florence, Sansoni, 1960), pp. 40–3.
32 Dante, *Epistolae*, edited by P. Toynbee, second edition (Oxford, Clarendon Press, 1966), pp. xxxi–xxxvi, 211–13; Boccaccio, *Opere latine minori*, edited by A. F. Massèra (Bari, Laterza, 1928), pp. 307–8, 323–4.
33 Compared with the Peri edition, Grazzini's text introduces punctuation and accents, regularizes the metre, and makes changes to spelling, phonology, morphology and vocabulary (such as 'svemorati' > 'sventurati', 'la poesia *contende* col rasoio' > 'combatte').
34 *Canti carnascialeschi*, edited by C. Singleton (Bari, Laterza, 1936), pp. 467–8, 471–80.
35 Rhodes, *Gli annali*, nos. 13–16.
36 The original version is quoted here from the edition numbered 16 by Rhodes. *IGI*, 1, Plate III shows a copy of another of the earlier editions with pen corrections which could have been made with a view to a later printing. Some but not all of the changes coincide with those made in the 1560 edition.
37 Luiso, 'Le edizioni', pp. 296–309; Porta, 'Censimento', 1976, p. 102, and

'L'ultima parte', p. 25; Trovato, 'Primi appunti', pp. 2-3 and *Con ogni diligenza*, pp. 83-5.
38 See the edition by Enzo Esposito (Ravenna, Longo, 1974), pp. xxxix-xlii.
39 A few of Nannini's marginal notes referred to variants from a manuscript.
40 Luiso, 'Le edizioni', pp. 309-15.

10 PIETY AND ELEGANCE: EDITING IN VENICE, 1561-1600

1 Ruscelli had been bedridden with dropsy for about nine months: see L. Groto, *Lettere famigliari* (Venice, Valentino, 1606), letter of 29 April 1566, ff. 39v-40r, and the undated letter of Damiano Zenaro in Ruscelli's *Imprese illustri* (Venice, Rampazetto, 1566), pp. 565-66; this says Ruscelli died on Thursday 9th of the previous month, which in that year must have been May.
2 Grendler, *The Roman Inquisition*, pp. 129-34.
3 Bongi, *Annali*, I, xxxii-xli; Quondam, 'La letteratura in tipografia', pp. 641-7.
4 Grendler, *The Roman Inquisition*, pp. 225-33; Quondam, 'La letteratura in tipografia', p. 584.
5 For what follows, see especially Brown, *The Venetian Printing Press*, chapters 13-17; Sorrentino, *La letteratura italiana*, pp. 21-61; Grendler, *The Roman Inquisition*, chapters 2 and 3; C. Fahy, 'The *Index librorum prohibitorum* and the Venetian printing industry in the sixteenth century', *IS*, 35 (1980), 52-61; A. Rotondò, 'La censura ecclesiastica e la cultura', in *Storia d'Italia*, Vol. 5, Part II (Turin, Einaudi, 1973), 1397-1492.
6 Among their victims were Ruscelli and Pietrasanta, tried by the Venetian Holy Office for publishing an obscene work in 1555: Trovato, *Con ogni diligenza*, pp. 253-4.
7 J. M. de Bujanda, *Index de Rome, 1557, 1559, 1564* (Sherbrooke, Centre d'études de la Renaissance, 1990), p. 827.
8 Brown, *The Venetian Printing Press*, pp. 92-5; Grendler, *The Roman Inquisition*, pp. 145-61.
9 Ceruti, *Lettere inedite*, p. 60.
10 Agnelli and Ravegnani, *Annali*, I, 126.
11 Brown, *The Venetian Printing Press*, pp. 144-7.
12 Giovio, *Lettere*, I, 4-11, and for instance I, 118-23 (an anti-Spanish passage omitted on pp. 121-2) and 166 ('fin nel cul ai cani' > 'fin nelle spelunche de' timidi pesci').
13 For this censorship and for the linguistic revision of their source (the Florentine edition of 1561) carried out for the two Rampazetto editions of 1561-2 and 1566, see Grazzini's *Teatro*, edited by G. Grazzini (Bari, Laterza, 1953), pp. 590-2.
14 Moro, 'A proposito di antologie', pp. 71-9.
15 B. Castiglione, *Opere volgari e latine*, edited by G. A. and G. Volpi (Padua, Comino, 1773), pp. 3-6; V. Cian, 'Un episodio della storia della censura in Italia nel secolo XVI: l'edizione spurgata del *Cortegiano*', *Archivio storico lombardo*, 14 (1887), 661-727; Ossola, '"Il libro del Cortegiano"', pp. 47-50; Guidi, 'Reformulations', pp. 162-84.
16 See the *DBI* entry by Nicola Longo: XXV (1981), 353-5.
17 Both the contents and the typography of the unexpurgated 'Venice 1587' edition

of the *Cortegiano* attributed to Domenico Giglio make it probable that this is a false imprint.

18 Sorrentino, *La letteratura italiana*, pp. 198–203; J. Tschiesche, 'Il rifacimento del Decamerone di Luigi Groto', in Brunello and Lodi, *Luigi Groto*, I, 237–71.

19 For details of the contents see G. B. Passano, *I novellieri italiani in prosa*, 2nd edition, 2 vols. (Turin, Paravia, 1878), I, 541–54.

20 Fabrizio-Costa and La Brasca, 'De l'âge des auteurs'.

21 He pointed out, for instance, that 'caro' in the sense of 'lack' (*Purg.* 22.141) was used by Villani and in the *Novellino*, and that 'gorna' was the Lombard equivalent of Dante's 'doccia' ('channel', *Inf.* 14.117, 23.46).

22 'Huopo' is explained as meaning *bisogno* ('need'); it is an old word, says Sansovino, and much used in Bembo's *Prose*, 'but certainly one should use it rarely' ('ma certo si dee l'huomo servirsi di lei poche volte').

23 C. Dionisotti, 'Bernardino Daniello', *ED*, 2 (1970), 303–4. A modern transcription is *L'espositione di Bernardino Daniello da Lucca sopra la Comedia di Dante*, edited by R. Hollander and others (Hanover and London, University Press of New England, 1989), on which see L. Pertile, 'Il Daniello di Dartmouth', *IS*, 46 (1991), 102–9.

24 *AIS* maps 477, 1497.

25 Mambelli, *Gli annali*, pp. 44–5, 51–2.

26 Baudrier, *Bibliographie lyonnaise*, IX, 246–7. On Bevilacqua, see the *DBI* entry by A. Cioni, IX (1967), 798–801.

27 The notes are not attributed to Ridolfi, but, in the exchange of letters contained in Angelieri's edition of 1586 mentioned below, he does not deny Alfonso Cambi Importuni's suggestion that he was their author. Examples of topics covered are the use of *lo* as an article, the differences between the pronouns *il* and *lo*, *mi* and *me*, the irregular plural *le mani* (*Prose*, 3.5), and the archaic use of *haggio* (*Prose*, 3.50).

28 Baudrier, *Bibliographie lyonnaise*, IX, 297–8; Baldacci, *Il petrarchismo*, pp. 72–3, 175–7.

29 V. Monforte, 'L'attività di Luigi Groto in margine ai Cinque canti dell'Ariosto', in Brunello and Lodo, *Luigi Groto*, I, 273–88.

30 Javitch, *Proclaiming a Classic*, pp. 77–80; on Anguillara, see the *DBI* entry by C. Mutini, III (1961), 306–9.

31 Di Filippo Bareggi, *Il mestiere*, pp. 37, 84–6, 164–5. On his editing of Sannazaro's *Arcadia* (1566), see Trovato, *Con ogni diligenza*, pp. 300–1, 306–7.

32 On the prominence given to Statius, see Javitch, *Proclaiming a Classic*, pp. 54–60, where the expansion is attributed to Dolce.

33 Javitch, *Proclaiming a Classic*, pp. 60–70.

34 Agnelli and Ravegnani, *Annali*, II, 50–2.

35 Agnelli and Ravegnani, *Annali*, II, 63–4. Turchi also edited a number of religious works for Giolito and was at least partly responsible for the second book of *Lettere facete et piacevoli* (Venice, Andrea Muschio, 1575), whose compilation had been begun by Atanagi and in which Aldo Manuzio was also involved: see Giovio, *Lettere*, I, 11–14; Renouard, *Annales*, pp. 220–1.

36 Agnelli and Ravegnani, *Annali*, II, 67–8.

37 Tambara in Ariosto, *Satire*, pp. 32–3.

38 Harris, *Bibliografia*, pp. 194–5, 209–11, 215–16.

39 H. F. Woodhouse, *Language and Style in a Renaissance Epic: Berni's Corrections to*

Boiardo's 'Orlando innamorato' (London, MHRA, 1982); on the structural changes see too Harris, *Bibliografia*, pp. 137-8.
40 Rotondò, 'Nuovi documenti', pp. 155-6.
41 P. V. Mengaldo, *La lingua del Boiardo lirico* (Florence, Olschki, 1963), pp. 163, 355.
42 V. De Maldé, 'Il postillato Manuzio delle "Rime": contributo alla storia dell'editoria e della tradizione tassiana', in *Studi di letteratura italiana offerti a Dante Isella* (Naples, Bibliopolis, 1983), pp. 113-43 (pp. 124-5 and n. 50), where the point is also made that Aldo corrected some genuine errors on the basis of a careful comparison of Tasso's usage in other poems.
43 P. Grendler, 'Francesco Sansovino and Italian popular history 1560-1600', *Studies in the Renaissance*, 16 (1969), 139-80 (pp. 158-61).
44 See Bongi, *Annali*, I, xxxvii-xxxix, II, 295-7 (on the historical series); I, xl-xli, II, 273-6 (on the 'garland'); II, 276, 341 (on the 'tree'); II, 388-9 (on the 'staircase').
45 Bongi, *Annali*, II, 201-2; P. Guicciardini, 'Le prime edizioni e ristampe della *Storia d'Italia*: loro raggruppamento in famiglie tipografiche', *LB*, 49 (1947), 76-91.
46 Ridolfi, *Studi guicciardiniani*, pp. 204-5, 217-18.
47 Bongi, *Annali*, II, 256-61; C. Angeleri, 'Il Guicciardini nella "Vita" di fra Remigio Fiorentino', in *Francesco Guicciardini nel IV centenario della morte* (Florence, Centro nazionale di studi sul Rinascimento, 1940), pp. 213-29.
48 Frasso, 'Sebastiano Fausto', pp. 372-3.
49 Porcacchi added marginal *notabilia* and a brief summary of the last years of Lodovico Sforza and the subsequent fortunes of the state of Milan.
50 Examples of Porcacchi's superficial changes are 'rovine' > 'ruine', 'iurisditione' > 'giurisdittione', 'gli' [= 'alla città.'] > 'le', 'habbino' > 'habbiano', 'persi' > 'perduti', 'dappoi' > 'dipoi', 'sbassati' > 'abbassati'.
51 The *Vocabolario* was, however, not published until 1584, after Porcacchi's death: L. Coglievina, 'Il *Vocabolario Nuovo* di Tomaso Porcacchi', *Lingua nostra*, 26 (1965), 35-8.
52 *Gomitolo* ('ball of string') and *vizzo* ('withered') are glossed respectively as *gemo d'acce* and *carne fiappa* in Venetian usage.
53 Bongi, *Annali*, II, 264; see also II, 213-14 on Nannini's edition.

11 A 'TRUE AND LIVING IMAGE': EDITING IN FLORENCE, 1561-1600

1 A. Panella, 'L'introduzione a Firenze dell'Indice di Paolo IV', *Rivista storica degli archivi toscani*, 1 (1929), 11-25; Biagiarelli, 'Il privilegio', pp. 343-7. See too J. A. Tedeschi, 'Florentine documents for a history of the *Index of Prohibited Books*', in *Renaissance Studies in Honor of Hans Baron*, edited by A. Molho and J. A. Tedeschi (Florence, Sansoni, 1971), pp. 577-605, and M. Plaisance, 'Littérature et censure à Florence à la fin du xvie siècle: le retour du censuré', in *Le Pouvoir et la plume*, pp. 233-52.
2 See Niccolai, *Pier Vettori*, pp. 224, 227-8 for letters of 1543 and 1544; for 1546-65, see Martinelli, 'Contributo'. A very good summary of Borghini's life and work is provided by Folena in the *DBI*, XII (1970), 680-9.
3 Martinelli, 'Contributo', pp. 194-6.
4 See above, p. 133. Borghini and Silvano Razzi also revised the proofs of the 1568 edition.

5 For Guicciardini, see below, pp. 174-5. For Grazzini, see Plaisance's edition (Abbeville, Paillart, 1976), pp. 15-18, and his apparatus criticus.
6 *Prose fiorentine*, pp. 171-9.
7 However, Borghini thought it essential to be completely faithful to the original spelling when transcribing ancient inscriptions, and in a letter to Onofrio Panvinio he criticized Paolo Manuzio for failing to do so: Ceruti, *Lettere inedite*, p. 41.
8 See for example Pozzi, 'Il pensiero', 1971, p. 293 and 1972, pp. 209-10, 232-5.
9 Borghini, *Scritti inediti*, p. 373. On *mora* in *Purg*. 3.129, see these *Scritti*, p. 40. See too C. Arlia's edition of Borghini's attacks on Ruscelli, the *Ruscelleide, ovvero Dante difesa dalle accuse di G. Ruscelli* (Città di Castello, Lapi, 1898).
10 In his copy (see n. 14 below), Borghini commented on one of Nannini's marginal notes that 'this poor annotator cannot talk of Florentine matters' ('questo povero huomo chiosatore non sa ragionare delle cose di Firenze'), and he called Nannini a 'sheep' ('pecora') for his poor knowledge of Trecento Tuscan.
11 Borghini, *Scritti inediti*, pp. 232-3, 257-61 (and p. lxvi for the possible dating *c*. 1559). On usage and rules, see too Pozzi, 'Il pensiero', 1972, pp. 207-19.
12 Gigli, *Studi*, pp. 269-85. On the dated manuscript see Petrocchi, *Introduzione*, p. 67.
13 Pozzi, 'Il pensiero', 1971, pp. 219-20; Folena in the *DBI*, xii, 683.
14 BMF R.O.304. Some of the readings are inserted in another hand.
15 *Opuscoli inediti o rari di classici o approvati scrittori*, Vol. 1 (Florence, Società poligrafica italiana, 1844), pp. 15-41. On the date: Carrai and Madricardo, 'Il "Decameron" censurato', p. 239. The letter shows that the preparation of the 1573 *Decameron* was still in its early stages.
16 See especially G. Biagi, *Le novelle antiche dei codici panciatichiano-palatino 138 e laurenziano-gaddiano 193* (Florence, Sansoni, 1880), and L. Di Francia's edition (Turin, UTET, 1930), which gives both the 1525 text and the stories added in 1572.
17 The first story, for instance, opened by quoting a saying of Christ, but in Borghini's edition this became just a 'common opinion and a true one' ('comune sententia et verace'). In his story 54, the adjective 'luxoriosa' ('lustful') became 'fresca' ('fresh').
18 Two drafts with notes in his hand survive among his papers in Florence: *Scritti inediti*, pp. lxxi, lxxiii, 11-14. The preface is republished in Di Francia's edition, pp. 167-71.
19 See Di Francia's edition, p. 54.
20 For example, Borghini pointed out that 'diliverroe', meaning 'diliberrò', had the same use of *v* for *b* still found among peasants near Florence, who say *liverare* for *liberare*. Salviati confirmed in his *Avvertimenti* on the *Decameron* that Borghini wrote these notes (Vol. 1, 2.12).
21 In the first story, for example, the manuscript provided 'baldanza' for 'abondanza', 'fra gli altri' for 'infralli altri', 'si può huomo parlare' for 'si può parlare', 'risponsi' for 'risposi', 'assimigliare' for 'somigliare'. In story 68 it provided 'lo buono calore naturale viene meno' instead of 'il buon calore naturale e non venga meno'. At a preparatory stage Borghini copied variants for twenty stories from his manuscript into a copy of Gualteruzzi's undated edition, now BNF Landau Finaly stampe 262. A few variants are in a second hand attributed, in a note by an early owner, to Vettori.
22 Thus in his story 6 'die di Pasqua' became 'Pasqua', 'ne fo quello chio giudico'

became 'ne foe quello ch'io vi dico', and repetitions of 'lo 'mperadore' were eliminated. In his story 54, changes include 'per alquanto tempo rimasa' > 'rimasa', 'scordichare tutto, cioè levare il chuoro dadosso' > 'scorticare', 'ciaschuno molto se ne maravigliava' omitted; 'Nel tempo antico' > 'Fu già tempo in Roma che'; 'era pocho tempo dimorata' > 'poco tempo era dimorata'; 'una ... donna, ... la quale ... era molto giovane *donna*' > 'd'anni'.

23 Pozzi, 'Vincenzio Borghini e la lingua', pp. 294–304; the quotations are from p. 301.
24 Biagi, *Aneddoti*, p. 293. On Florentine attempts to produce an expurgated *Decameron* in the 1560s, see Brown, *Lionardo Salviati*, pp. 161–2.
25 This copy is now BNF 22.A.4.1. Manrique's *licenza* is in Tapella and Pozzi, 'L'edizione', pp. 55–6.
26 Tapella and Pozzi, 'L'edizione'. For Vettori: *Prose fiorentine*, p. 241.
27 Pozzi, 'Vincenzio Borghini e la lingua', pp. 295–301.
28 Tapella and Pozzi, 'L'edizione', p. 58.
29 *Prose fiorentine*, p. 241; for the date 1573, see Tapella and Pozzi, 'L'edizione', p. 517.
30 V. Branca, 'Per il testo del "Decameron": testimonianze della tradizione vulgata', *SFI*, 11 (1953), 163–243 (pp. 168–9, 183–9). On the relationship of Mn, the 'Deo Gratias' and Beccadelli's manuscript see too Salviati's *Avvertimenti*, Vol. I, 1.5. Branca (pp. 186–7) repeats the criticism of the Deputati's excessive faith in Mn made by Michele Barbi in *La nuova filologia*, pp. 35–85 (pp. 35–6).
31 Tapella and Pozzi, 'L'edizione', p. 516; *Prose fiorentine*, p. 240.
32 In Borghini's *Scritti inediti* (p. 87) this point is explicitly linked with Ruscelli's preference for *habitari*.
33 Borghini, *Scritti inediti*, p. 31.
34 Much has been written about this aspect of the edition; see especially Sorrentino, *La letteratura italiana*, pp. 151–85; Biagi, *Aneddoti*, pp. 282–326; G. Lesca, 'Vincenzo Borghini e il "Decameron"', in *Studii su Giovanni Boccaccio* (Castelfiorentino, Società storica della Valdelsa, 1913), pp. 246–63; Mordenti, 'Le due censure'; Tapella and Pozzi, 'L'edizione'; and Carrai and Madricardo, 'Il "Decameron" censurato'.
35 Tapella and Pozzi, 'L'edizione', pp. 64–74.
36 *Prose fiorentine*, p. 239.
37 Examples are the tirade against friars in 3.7.32–47 and much of the story of how the monk Rustico taught Alibech to 'put the devil into hell' (3.10.9–31, 32–5).
38 Tapella and Pozzi, 'L'edizione', pp. 367–75.
39 Tapella and Pozzi, 'L'edizione', pp. 382–3.
40 *Prose fiorentine*, pp. 238–9.
41 Rotondò, 'Nuovi documenti', pp. 152–3; Carter, 'Another promoter'. The reaction seems to have begun in Venice, where Sansovino and fra Remigio may have been trying to produce a pirated edition, and then in Ferrara: Tapella and Pozzi, 'L'edizione', p. 517.
42 Tapella and Pozzi, 'L'edizione', pp. 517–44.
43 Now respectively BNF 25.28 and 2.3.350: Trovato, 'Per un censimento', pp. 47–8.
44 The restorations included 'i quali' > 'li quali', 'un dì' > 'uno dì', 'mandò' > 'mandoe', 'fu' > 'fue'.
45 See for example his *Scritti inediti*, p. 203.
46 Trovato, *Con ogni diligenza*, p. 101 n. 102.

47 Tapella and Pozzi, 'L'edizione', p. 196.
48 Borghini's copy, now incomplete, is divided between BNF 6.112 and BLF 42.12. Another copy made from this transcription is BLF 42.11. The original is lost. On the projected edition, see Barbi, *La nuova filologia*, pp. 87–124, and F. Ageno, 'Per il testo del "Trecentonovelle"', *SFI*, 16 (1958), 194–274.
49 See the *DBI* entry by G. Benzoni, XXVIII (1983), 750–60, with earlier bibliography; M. Plaisance, 'Jacopo Corbinelli: de l'exclusion à l'exil. La rupture avec Florence', in *L'Exil et l'exclusion dans la culture italienne* (Aix-en-Provence, Université de Provence, 1991), pp. 67–76. On the letter of 1579: P. Soldati, 'Jacopo Corbinelli e Lionardo Salviati', *Archivum romanicum*, 19 (1935), 415–23 (p. 419).
50 In the dedication of his *Corbaccio* (Paris, Morel, 1569).
51 This case is developed in the preface to the *Bellamano* (see n. 54 below); an extract is given in Trovato, *Con ogni diligenza*, pp. 12–13.
52 The title was *L'Ethica d'Aristotile ridotta in compendio da ser Brunetto Latini. Et altre traduttioni, e scritti di quei tempi. Con alcuni dotti avvertimenti intorno alla lingua.*
53 On the Dante edition, see *Il trattato De vulgari eloquentia*, edited by P. Rajna (Florence, Le Monnier, 1896), pp. xi–xxxi, lxix–lxxxv. The *Ricordi* included some accents, but there were none in *La fisica*.
54 Barbi, *Studi*, pp. 288–91, 294–301; L. Quaquarelli, '"Quelle pochette annotationi dalla margine tirate del libro mio": la *princeps* bolognese della *Bella mano* di Giusto de' Conti nelle postille dell'editore cinquecentesco Jacopo Corbinelli', *Schede umanistiche*, 2 (1991) 51–79.
55 See the article by Soldati (n. 49 above); P. M. Brown, 'Jacopo Corbinelli and the Florentine "Crows"', *IS*, 23 (1971), 68–89; Brown, *Lionardo Salviati*, especially pp. 31–9, 186, 195.
56 P. M. Brown, 'I veri promotori della "rassettatura" del "Decameron"', *GSLI*, 134 (1957), 314–32; Carter, 'Another promoter'.
57 *Avvertimenti*, Vol. I, 1.2.
58 *Avvertimenti*, Vol. I, 2.18. Branca's critical edition agrees with Salviati at 3.5.10 and 6.9.10 but disagrees with him at 10.8.82.
59 This point is illustrated by Corbinelli's comments on why he had left the spelling 'superbio' in the *Ethica* of 1568 (p. 32). It is deep folly to want to teach how to pronounce verse through writing, he says; rather, everything should perhaps be pronounced as it is written ('è forte stultitia volere insegnare pronuntiare il verso con la scrittura; anzi forse tutto s'harebbe a pronuntiare sì come scrivere').
60 *Discorso storico sul testo del Decamerone*, in *Saggi e discorsi critici*, edited by C. Foligno (Florence, Le Monnier, 1953), pp. 301–75 (p. 324).
61 *Lionardo Salviati*, p. 167. On Salviati's censorship, see especially Sorrentino, *La letteratura italiana*, pp. 186–98; P. M. Brown, 'Aims and methods of the second "rassettatura" of the *Decameron*', *Studi secenteschi*, 8 (1967), 3–41; Mordenti, 'Le due censure'.
62 J. Usher, 'The fortune of "Fortuna" in Salviati's "rassettatura" of the *Decameron*', in *Renaissance and Other Studies: Essays Presented to Peter M. Brown*, edited by E. A. Millar (University of Glasgow, 1988), pp. 210–22.
63 See Barbi's edition of the *Vita nuova* (Florence, Bemporad, 1932), pp. xc–xcvi.
64 As in 'continuamente' > 'continovamente', 'elli' > 'egli', and, in the imperfect indicative, first-person singular *-a* > *-o* and third-person singular *-ea*, *-ia* > *-eva*, *-iva*.

65 Barbi, *Della fortuna*, pp. 122–7; L. Donati, *Chi furono gli Accademici della Crusca che prepararono la Divina Commedia del 1595?* (Florence, Sansoni, 1953).
66 Ridolfi, *Studi guicciardiniani*, pp. 193–6. C. Panigada lists twenty-four censored passages in his edition (5 vols., Bari, Laterza, 1929, Vol. v, 325–6): the main ones concerned the alleged incest of Lucrezia Borgia (3.13), the origin of papal temporal power (4.12), the implications of a psalm for the non-Christians found by the voyages of discovery (6.9), and the 'tyranny of priests' (10.4). See too P. Guicciardini, 'La censura nella *Storia guicciardiniana*', *LB*, 55 (1953), 134–56.
67 For Borghini, see above, pp. 157, 158. Salviati compared Nannini's text unfavourably in his *Avvertimenti* (Volume I, 2.12) with a manuscript owned by Speroni.
68 Porta, 'Censimento', 1979, pp. 110–11. BNF MS 2.1.289 may once have belonged to the Tornaquinci family: Porta, 'Censimento', 1976, p. 89.
69 This is now in the British Library, 177.g.16. The corrections are in two hands, one of them Agnolo's. Up to Book 3 the corrections are very detailed, as if it had been intended to use the copy as the basis of a new edition.
70 Porta, 'L'ultima parte', p. 25 n. 1.
71 Now BNF Palatino E.B.10.3: Porta, 'Censimento', 1976, pp. 117–18. Ricci planned an expurgated edition of his grandfather Machiavelli's works which was ready in 1573. On him see G. Sapori, 'Giuliano de' Ricci e una sua cronaca inedita', in *Studi in onore di Armando Sapori*, 2 vols. (Milan, Cisalpino, 1957), II, 1061–70.
72 There was the usual modernization of spelling and other corrections such as, in the Prologue to Book 10, 'nonnè da maravigliare' > 'non è da maravigliarsene', 'numerabile exercito' > 'innumerabile esercito'.
73 See for example C. T. Davis, 'The Malispini question', in *Dante's Italy and Other Essays* (Philadelphia, University of Pennsylvania Press, 1984), pp. 94–136; J. C. Barnes, 'Un problema in via di chiusura: la *Cronica* malispiniana', *Studi e problemi di critica testuale*, 27 (1983), 15–32.
74 See Benci's edition of the *Storia fiorentina*, 2 vols. (Livorno, Masi, 1830), I, lviii–lix.
75 A passage concerning the clergy was censored (in chapter 223 in the 1568 edition). The printed edition had a later licence, dated 1567.
76 For instance, there were dots under the italicized letter in 'al*ch*uno', 'a*b*iamo', 'inpercio', 'meg*l*o', 'brievemente'.
77 For example, the 1568 editors preferred the Ashburnham reading near the start of chapter 56. Their text has 'abitare in Fiorenza, Fiorenza molto cominciò a multiplicare di popolo, e di gente', as in this manuscript (apart from minor spelling details and a small gap in the manuscript between the first 'Fiorenza' and 'molto'); BNF 2.4.28 (after correction) and 2.1.108 have 'abitare in Firenze, sì come di sopra è detto, la detta città cominciò molto a multiplichare di gente e di popolo' (2.1.108 'Fiorenza', 'giente'). In BNF 2.1.108, chapters 17 and 18 were crossed out with the comment that they were unworthy of the author and irrelevant; but these and some other deleted passages were retained in the printed text.
78 L. S. Camerini, *I Giunti tipografi editori di Firenze 1571–1625* (Florence, Giunti Barbèra, 1979), nos. 152, 223, 229, 254, 265.
79 See Arbib's edition (2 vols., Florence, Società editrice delle storie del Nardi e del Varchi, 1842), I, iv–v. Arbib also pointed out that both the Lyons and Florence editions omit passages, including two from Book 6 on the superstition of certain clergymen and on magic (Vol. II, 40–1, 43).

80 An example from the first chapter is 'Dobbiamci adunque sollecitare che, poi che abbiamo rinuntiato al mondo e dallo stato della frigidità del peccato ci siamo partiti, di procedere a lo terzo stato del fervore dello spirito, acioché non rimagnamo nello stato tiepido di mezzo, perciò che il tiepido è degno d'essere da Dio vomitato' > 'Poi che dunque habbiamo rinunziato al mondo, e siamoci partiti dallo stato della frigidità del peccato, dobbiamo sollecitamente procedere e cercare di pervenire al terzo stato del fervore dello spirito; acciochè, rimanendo nello stato tiepido e di mezzo, non meritiamo per ciò di essere da Dio vomitati'.

81 Comparing the first canto of the pre-1490 edition (Rhodes, *Gli annali*, no. 556) with that of 1572, one finds, for example, 'adveduto' > 'avveduto', 'saxo' > 'sasso', 'mi donoe per segnio' > 'mi donò per segno'; hypermetric lines are reduced as in 'et prestandomi *il celo* qui del suo aiuto' > 'il ciel'.

82 For these ideas of sweetness and continuity, see P. M. Brown, 'The conception of the literary "volgare" in the linguistic writings of Lionardo Salviati', *IS*, 21 (1966), 57–90, and J. R. Woodhouse, 'Vincenzo Borghini and the continuity of the Tuscan linguistic tradition', *IS*, 22 (1967), 26–42.

83 Sorrentino, *La letteratura italiana*, pp. 326–41.

CONCLUSION

1 *The Rise of the Novel* (Harmondsworth, Penguin, 1963), pp. 36–61.
2 *Trattati d'arte*, I, 155; Lee, '*Ut pictura poesis*', p. 197.
3 *Le vite*, Testo, IV, 307.
4 M. Perry, 'Cardinal Domenico Grimani's legacy of ancient art to Venice', *Journal of the Warburg and Courtauld Institutes*, 41 (1978), 215–44 (pp. 227–8). On contemporary attitudes to the restoration of limbless statues, see O. Rossi Pinelli, 'Chirurgia della memoria: scultura antica e restauri storici', in *Memoria dell'antico nell'arte italiana*, edited by S. Settis, 3 vols. (Turin, Einaudi, 1984–6), III, 183–250.
5 See for instance A. Conti, *Storia del restauro e della conservazione delle opere d'arte* (Milan, Electa, [1982]), pp. 38–49, 210–13.
6 L. B. Alberti, *De pictura*, III 55–6, in *Opere volgari*, edited by C. Grayson, Vol. III (Bari, Laterza, 1973), 95–7; see too Blunt, *Artistic Theory*, pp. 15–17, and Panofsky, *Idea*, pp. 48–9.
7 *Trattati d'arte*, I, 172; Lee, '*Ut pictura poesis*', p. 204.
8 Blunt, *Artistic Theory*, pp. 90–1.
9 Panofsky, *Idea*, pp. 59–60.

SELECT BIBLIOGRAPHY

This list gives details of works cited in the notes which are of general relevance to the subject of this book or which are mentioned more than once.

Agnelli, Giuseppe and Ravegnani, Giuseppe, *Annali delle edizioni ariostee* (Bologna, Zanichelli, 1933)
Aldo Manuzio editore: dediche, prefazioni, note ai testi. Introduzione di Carlo Dionisotti. Testo latino con traduzione e note a cura di Giovanni Orlandi, 2 vols. (Milan, Il Polifilo, 1976)
Allenspach, Joseph and Frasso, Giuseppe, 'Vicende, cultura e scritti di Gerolamo Squarzafico, alessandrino', *IMU*, 23 (1980), 233–92
Aretino, Pietro, *Sei giornate*, edited by G. Aquilecchia (Bari, Laterza, 1969)
Ariosto, Lodovico, *Commedie*, edited by A. Casella, G. Ronchi and E. Varasi (Milan, Mondadori, 1974)
  Satire, edited by G. Tambara (Livorno, Giusti, 1903)
Ascarelli, Fernanda and Menato, Marco, *La tipografia del '500 in Italia* (Florence, Olschki, 1989)
Atti del Congresso internazionale di studi danteschi (20–27 aprile 1965) (Florence, Sansoni, 1965)
Baldacci, Luigi, *Il petrarchismo italiano nel Cinquecento* (Milan and Naples, Ricciardi, 1957)
Balduino, Armando, 'Per il testo del "Ninfale fiesolano"', *SB*, 3 (1965), 103–84 and 4 (1967), 35–201
Balsamo, Luigi, *Produzione e circolazione libraria in Emilia (XV–XVIII sec.): studi e ricerche* (Parma, Casanova, 1983)
Barbi, Michele, *Della fortuna di Dante nel secolo XVI* (Florence, Bocca, 1890)
  Studi sul Canzoniere di Dante (Florence, Sansoni, 1915)
  La nuova filologia e l'edizione dei nostri scrittori da Dante a Manzoni (Florence, Sansoni, 1938)
Baudrier, Henri and Julien, *Bibliographie lyonnaise*, 12 vols. (new impression, Paris, Nobele, 1964)
Belloni, Gino, 'Di un "parto d'elephante" per Petrarca: il commento del Gesualdo al "Canzoniere"', *Rinascimento*, 2nd series, 20 (1980), 359–81
  'Un eretico nella Venezia del Bembo: Alessandro Vellutello', *GSLI*, 157 (1980), 43–74
  'Il commento petrarchesco di Antonio da Canal e annesse questioncelle tipografiche e filologiche sull'aldina del 1501', in *Miscellanea di studi in onore di Vittore Branca*, 5 vols. (Florence, Olschki, 1983), I, 459–78
  'Alle origini della filologia e della grammatica italiana: il Fortunio', in G. Bolognesi and V. Pisani (eds.), *Linguistica e filologia: Atti del VII Convegno internazionale di linguisti* (Brescia, Paideia, 1987), pp. 189–204

and Trovato, Paolo, 'Sul commento al Petrarca di Sebastiano Fausto e sull' "Introduttione alla lingua volgare" di Domenico Tullio Fausto', *Rivista di letteratura italiana*, 7 (1989), 249–88

'Les Commentaires de Pétrarque', in *Les Commentaires*, pp. 147–55

Beloch, K. J. *Bevölkerungsgeschichte Italiens*, Vols. II and III (Berlin, De Gruyter, 1965, 1961)

Bembo, Pietro, *Opere in volgare*, edited by Mario Marti (Florence, Sansoni, 1961)
  Prose e rime, edited by Carlo Dionisotti, 2nd edition (Turin, UTET, 1966)

Biagi, Guido, *Aneddoti letterari* (Milan, Treves, 1887)

Biagiarelli, Berta Maracchi, 'Il privilegio di stampatore ducale nella Firenze medicea', *Archivio storico italiano*, 123 (1965), 304–70
  'Editori di incunabuli fiorentini', in *Contributi ... Donati*, pp. 211–20

Blunt, Anthony, *Artistic Theory in Italy 1450–1600* (Oxford, Clarendon Press, 1940)

Boccaccio, Giovanni, *Commedia delle ninfe fiorentine (Ameto)*, edited by A. E. Quaglio (Florence, Sansoni, 1963)

Bongi, Salvatore, *Annali di Gabriel Giolito de' Ferrari da Trino di Monferrato stampatore in Venezia*, 2 vols. (Rome, Ministero della pubblica istruzione, 1890–5)

Borghini, Vincenzio, *Scritti inediti o rari sulla lingua*, edited by J. R. Woodhouse (Bologna, Commissione per i testi di lingua, 1971)

Branca, Vittore (ed.), *Rinascimento europeo e rinascimento veneziano* (Florence, Sansoni, 1967)
  'Introduzione', in G. Boccaccio, *Decameron: edizione critica secondo l'autografo Hamiltoniano* (Florence, Accademia della Crusca, 1976)

Brotto, Giovanni and Zonta, Gasparo, *La facoltà teologica dell'Università di Padova, Parte I (secoli XIV e XV)* (Padua, Tipografia del Seminario, 1922)

Brown, Horatio F., *The Venetian Printing Press 1469–1800* (Amsterdam, van Heusden, 1969: reprint of London, 1891 edition)

Brown, Peter M., *Lionardo Salviati: A Critical Biography* (London, Oxford University Press, 1974)

Brunello, Giorgio and Lodo, Antonio (eds.), *Luigi Groto e il suo tempo: Atti del Convegno di studi, Adria, 27–29 aprile 1984*, 2 vols. (Rovigo, Minelliana, 1987)

Bruni, Roberto and Zancani, Diego, *Antonio Cornazzano: la tradizione testuale* (Florence, Olschki, 1992)

Bruscagli, Riccardo (ed.), *I libri di 'Orlando innamorato'* (Modena, Panini, 1987)

Bussi, Giovanni Andrea, *Prefazioni alle edizioni di Sweynheim e Pannartz prototipografi romani*, edited by M. Miglio (Milan, Il Polifilo, 1978)

Carrai, Stefano and Madricardo, Silvia, 'Il "Decameron" censurato: preliminari alla "rassettatura" del 1573', *Rivista di letteratura italiana*, 7 (1989), 225–47

Carrara, Enrico, *Studi petrarcheschi ed altri scritti* (Turin, Bottega d'Erasmo, 1959)

Carter, Tim, 'Another promoter of the 1582 "rassettatura" of the *Decameron*', *Modern Language Review*, 81 (1986), 893–9

Cavalcanti, Guido, *Le rime*, edited by Guido Favati (Milan and Naples, Ricciardi, 1957)

Ceruti, Antonio (ed.), *Lettere inedite di dotti italiani del secolo XVI* (Milan, Tipografia e libreria arcivescovile, 1867)

Cicogna, Emmanuele, *Delle inscrizioni veneziane*, 6 vols. (Venice, Orlandelli, 1824–53)

Clubb, Louise George and Clubb, William G., 'Building a lyric canon: Gabriel Giolito and the rival anthologists. Part I', *Italica*, 68 (1991), 332–44

Les Commentaires et la naissance de la critique littéraire. Textes réunis et présentés par G. Mathieu-Castellani et M. Plaisance (Paris, Aux Amateurs de Livres, 1990)

SELECT BIBLIOGRAPHY

Contributi alla storia del libro italiano: miscellanea in onore di Lamberto Donati (Florence, Olschki, 1969)
De la Mare, Albinia C., 'New research on humanistic scribes in Florence', in A. Garzelli (ed.), *Miniatura fiorentina del Rinascimento 1440–1525: un primo censimento*, 2 vols. (Florence, La Nuova Italia, 1985), I, 393–600
De Robertis, Domenico, 'Il canzoniere Escorialense e la tradizione "veneziana" delle rime dello stil novo', *GSLI*, supplement no. 27 (1954)
(ed.), *Sonetti e canzoni di diversi antichi autori toscani*, 2 vols. (Florence, Le Lettere, 1977)
Editi e rari: studi sulla tradizione letteraria tra Tre e Cinquecento (Milan, Feltrinelli, 1978)
Della Torre, Arnaldo, 'Per la storia della "toscanità" del Petrarca', in A. Della Torre and P. L. Rambaldi (eds.), *Miscellanea di studi critici in onore di Guido Mazzoni*, 2 vols. (Florence, Tipografia Galileiana, 1907), I, 185–223
Di Filippo Bareggi, Claudia, *Il mestiere di scrivere: lavoro intellettuale e mercato librario a Venezia nel Cinquecento* (Rome, Bulzoni, 1988)
Dionisotti, Carlo, 'Girolamo Claricio', *SB*, 2 (1964), 291–341
'Dante nel Quattrocento', in *Atti del Congresso* ... *(1965)*, pp. 333–78
Gli umanisti e il volgare fra Quattro e Cinquecento (Florence, Le Monnier, 1968)
'Fortuna del Petrarca nel Quattrocento', *IMU*, 17 (1974), 61–113
Machiavellerie (Turin, Einaudi, 1980)
Doni, Anton Francesco, *I Marmi*, edited by E. Chiòrboli, 2 vols. (Bari, Laterza, 1928)
Fabrizio-Costa, Silvia, and La Brasca, Frank, 'De l'âge des auteurs à celui des polygraphes: les commentaires de la Divine Comédie de C. Landino (1481) et A. Vellutello (1544)', in *Les Commentaires*, pp. 175–93
Fahy, Conor, *Saggi di bibliografia testuale* (Padua, Antenore, 1988)
L'"Orlando furioso" del 1532: profilo di una edizione (Milan, Vita e Pensiero, 1989)
Federici, Domenico Maria, *Memorie trevigiane sulla tipografia del secolo XV* (Venice, Andreola, 1805)
Folena, Gianfranco, *La crisi linguistica del Quattrocento e l'"Arcadia" di I. Sannazaro* (Florence, Olschki, 1952)
'Filologia testuale e storia linguistica', in *Studi e problemi*, pp. 17–34
'Überlieferungsgeschichte der altitalienischen Literatur', in *Geschichte der Textüberlieferung der antiken und mittelalterlichen Literatur*, 2 vols. (Zurich, Atlantis, 1961–4), II, 319–537
'La tradizione delle opere di Dante', in *Atti del Congresso* ... *(1965)*, pp. 1–76
Fowler, Mary, *Catalogue of the Petrarch Collection Bequeathed by Willard Fiske* (London, Oxford University Press, 1916)
Frasso, Giuseppe, 'Sebastiano Fausto, editore e volgarizzatore di storici medioevali e umanistici', *Aevum*, 64 (1990), 363–74
Frey, K. (ed.), *Der literarische Nachlass Giorgio Vasaris*, Vol. I (Munich, Müller, 1923)
Fulin, R., 'Documenti per servire alla storia della tipografia veneziana' and 'Nuovi documenti ...', *Archivio veneto*, 23 (1882), 84–212, 390–405
Gargan, Luciano, *Lo studio teologico e la biblioteca dei Domenicani a Padova nel Tre e Quattrocento* (Padua, Antenore, 1971)
Garzoni, Tomaso, *La piazza universale di tutte le professioni del mondo* (Venice, Somasco, 1585)
Gentile, Salvatore, *Repatriare Masuccio al suo lassato nido: contributo filologico e linguistico*. Atti del Convegno Nazionale di Studi su Masuccio Salernitano, Salerno, 9–10 maggio 1976, Vol. II (Galatina, Congedo, 1979)

Gerber, Adolph, *Niccolò Machiavelli. Die Handschriften, Ausgaben und Übersetzungen seiner Werke im 16. und 17. Jahrhundert*, 3 vols. and facsimiles (Gotha, Perthes, 1912-13)

Ghinassi, Ghino, 'Correzioni editoriali di un grammatico cinquecentesco', *SFI*, 19 (1961), 33-93

'L'ultimo revisore del "Cortigiano"', *SFI*, 21 (1963), 217-64

Gigli, Ottavio (ed.), *Studi sulla Divina commedia di Galileo Galilei, Vincenzo Borghini ed altri* (Florence, Le Monnier, 1855)

Giovio, Paolo, *Lettere*, edited by G. G. Ferrero, 2 vols. (Rome, Libreria dello Stato, 1956-8)

Grafton, Anthony, *Joseph Scaliger: A Study in the History of Classical Scholarship. I: Textual Criticism and Exegesis* (Oxford, Clarendon Press, 1983)

Grendler, Paul F., *Critics of the Italian World (1530-1560)* (Madison, Milwaukee, and London, University of Wisconsin Press, 1969)

  The Roman Inquisition and the Venetian Press 1540-1605 (Princeton University Press, 1977)

  Schooling in Renaissance Italy: Literacy and Learning 1300-1600 (Baltimore and London, Johns Hopkins University Press, 1989)

Guidi, José, 'Reformulations de l'idéologie aristocratique au xvie siècle: les différentes rédactions et la fortune du "Courtisan"', in *Réécritures I: commentaires, parodies, variations dans la littérature italienne de la Renaissance* (Paris, Université de la Sorbonne Nouvelle, 1983), pp. 121-84

Harris, Neil, 'L'avventura editoriale dell' "Orlando innamorato"', in Bruscagli, *I libri*, pp. 35-100

  Bibliografia dell' 'Orlando Innamorato', Vol. 1 (Modena, Panini, 1988)

Hellinga, Lotte, 'Manuscripts in the hands of printers', in J. B. Trapp (ed.), *Manuscripts in the Fifty Years after the Invention of Printing* (London, Warburg Institute, 1983), pp. 3-11

Hirsch, Rudolf, *Printing, Selling and Reading, 1450-1550* (Wiesbaden, Harrassowitz, 1967)

Javitch, Daniel, *Proclaiming a Classic: The Canonization of 'Orlando furioso'* (Princeton University Press, 1991)

Kenney, E. J., *The Classical Text: Aspects of Editing in the Age of the Printed Book* (Berkeley and London, University of California Press, 1974)

Lee, R. W., '*Ut pictura poesis*: the humanistic theory of painting', *Art Bulletin*, 22 (1940), 197-269

Lepschy, Anna Laura, '"Quel libro del Mandavila ... che me aveva tuto travaliato". A presentation of the Italian incunables', in Lepschy and others, *Book Production*, pp. 210-19

  and others (eds.), *Book Production and Letters in the Western European Renaissance: Essays in Honour of Conor Fahy* (London, Modern Humanities Research Association, 1986)

Lowry, Martin, *The World of Aldus Manutius* (Oxford, Blackwell, 1979)

  Nicholas Jenson and the Rise of Venetian Publishing (Oxford, Blackwell, 1991)

Luiso, Francesco Paolo, 'Le edizioni della "Cronica" di Giovanni Villani', *Bullettino dell'Istituto storico italiano e Archivio muratoriano*, 49 (1933), 279-315

Machiavelli, Niccolò, *Discorso intorno alla nostra lingua*, edited by P. Trovato (Padua, Antenore, 1982)

Mambelli, Giuliano, *Gli annali delle edizioni dantesche* (Bologna, Zanichelli, 1931)

Mardersteig, Giovanni, 'La singolare cronaca della nascita di un incunabolo: il commento di Gentile da Foligno stampato da Pietro Maufer nel 1477', *IMU*, 8 (1965), 249–67

Martinelli, Lucia Cesarini, 'Contributo all'epistolario di Pier Vettori (lettere a don Vincenzio Borghini, 1546–1565)', *Rinascimento*, 2nd series, 19 (1979), 189–227

Mestica, Giovanni, 'Il "Canzoniere" del Petrarca nel codice originale a riscontro col ms. del Bembo e con l'edizione aldina del 1501', *GSLI*, 21 (1893), 300–34

Mordenti, Raul, 'Le due censure: la collazione dei testi del *Decameron* "rassettati" da Vincenzio Borghini e Lionardo Salviati', in *Le Pouvoir et la plume*, pp. 253–73

Moro, Giacomo, 'A proposito di antologie epistolari cinquecentesche', *Studi e problemi di critica testuale*, 38 (1989), 71–107

Mortimer, Ruth, *Harvard College Library, Department of Printing and Graphic Arts: Catalogue of Books and Manuscripts. Part II: Italian 16th Century Books*, 2 vols. (Cambridge, Mass., Belknap Press of Harvard University Press, 1974)

Nesi, Emilia, *Il diario della stamperia di Ripoli* (Florence, Seeber, 1903)

Niccolai, Francesco, *Pier Vettori (1499–1585)* (Florence, Seeber, 1912)

Novo libro di lettere scritte da i più rari auttori e professori della lingua volgare italiana (ristampa anastatica delle edd. Gherardo, 1544 e 1545), edited by Giacomo Moro (Bologna, Forni, 1987)

Nuovo, Angela, *Alessandro Paganino (1509–1538)* (Padua, Antenore, 1990)

Ossola, Carlo, '"Il libro del Cortegiano": esemplarità e difformità', in *La corte e il 'Cortegiano'*, edited by C. Ossola and A. Prosperi, 2 vols. (Rome, Bulzoni, 1980), I, 15–82

Panofsky, Erwin, *Idea: A Concept in Art Theory*, translated by J. Peake (Columbia, University of South Carolina Press, 1968)

Parent, Annie, *Les Métiers du livre à Paris au XVIe siècle (1535–1560)* (Geneva, Droz, 1974)

Pesenti, Tiziana, 'Le edizioni veneziane dell'umanista tedesco Friedrich Nausea', in M. C. Billanovich and others (eds.), *Viridarium floridum: studi di storia veneta offerta dagli allievi a Paolo Sambin* (Padua, Antenore, 1984), pp. 295–316

Petrocchi, Giorgio, *Introduzione*, in Dante Alighieri, *La commedia secondo l'antica vulgata*, 4 vols. (Milan, Mondadori, 1966–7), Vol. 1

Pettas, William A., *The Giunti of Florence: Merchant Publishers of the Sixteenth Century* (San Francisco, Rosenthal, 1980)

Pillinini, Stefano, 'Traguardi linguistici nel Petrarca bembino del 1501', *SFI*, 39 (1981), 57–76

Plaisance, Michel, 'Une première affirmation de la politique culturelle de Côme Ier: la transformation de l'Académie des "Humidi" en Académie Florentine (1540–1542)', in A. Rochon and others, *Les Ecrivains et le pouvoir en Italie à l'époque de la Renaissance* (Paris, Université de la Sorbonne Nouvelle, 1973), pp. 361–438

'Culture et politique à Florence de 1542 à 1551: Lasca et les *Humidi* aux prises avec l'Académie Florentine', in A. Rochon and others, *Les Ecrivains et le pouvoir en Italie à l'époque de la Renaissance*, 2nd series (Paris, Université de la Sorbonne Nouvelle, 1974), pp. 149–242

Poggiali, Cristoforo, *Memorie per la storia letteraria di Piacenza*, 2 vols. (Piacenza, Orcesi, 1789)

Porta, Giuseppe, 'Censimento dei manoscritti delle Cronache di Giovanni, Matteo e Filippo Villani', *SFI*, 34 (1976), 61–129, 37 (1979), 93–117, 44 (1986), 65–7

'L'ultima parte della "Nuova cronica" di Giovanni Villani', *SFI*, 41 (1983), 17–36

SELECT BIBLIOGRAPHY

Le Pouvoir et la plume: incitation, contrôle et répression dans l'Italie du XVIe siècle (Paris, Université de la Sorbonne Nouvelle, 1982)

Pozzi, Mario, 'Il pensiero linguistico di Vincenzio Borghini', *GSLI*, 148 (1971), 216–94, and 149 (1972), 207–68

'Vincenzio Borghini e la lingua del "Decameron"', *SB*, 7 (1973), 271–304

Procaccioli, Paolo, *Filologia ed esegesi dantesca nel Quattrocento: l'"Inferno" nel 'Comento sopra la Comedia' di Cristoforo Landino* (Florence, Olschki, 1989)

Prose fiorentine, collected by Carlo Dati, Vol. IV, Part IV (Florence, Tartini e Franchi, 1745)

Quaglio, Antonio Enzo, 'Per il testo della *Fiammetta*', *SFI*, 15 (1957), 5–205

Scienza e mito nel Boccaccio (Padua, Liviana, 1967)

Quarta, Nino, 'I commentatori quattrocentisti del Petrarca', *Atti della Reale Accademia di archeologia, lettere e belle arti di Napoli*, 23 (1905), 269–324

Quondam, Amedeo, 'La letteratura in tipografia', in *Letteratura italiana*, directed by A. Asor Rosa, Vol. II (Turin, Einaudi, 1983), pp. 555–686

Renouard, Antoine Auguste, *Annales de l'imprimerie des Alde*, 3rd edition (Paris, Renouard, 1834)

Reynolds, L. D. and Wilson, N. G., *Scribes and Scholars: A Guide to the Transmission of Greek and Latin Literature*, 3rd edition (Oxford, Clarendon Press, 1991)

Rhodes, Dennis E., *Gli annali tipografici fiorentini del XV secolo* (Florence, Olschki, 1988)

Richardson, Brian (ed.), *Trattati sull'ortografia del volgare 1524–1526* (University of Exeter, 1984)

'An editor of vernacular texts in sixteenth-century Venice: Lucio Paolo Rosello', in Lepschy, *Book Production*, pp. 246–78

'Editing the *Decameron* in the Cinquecento', *IS*, 45 (1990), 13–31

'The first edition of Jacopone's *Laude* (Florence, 1490) and the development of vernacular philology', *IS*, 47 (1992), 26–40

'The two versions of the "Appendix Aldina" of 1514', *The Library*, 6th series, 13 (1991), 115–25

Ridolfi, Roberto, *Studi guicciardiniani* (Florence, Olschki, 1978)

Rigoni, Erice, 'Stampatori del sec. xv° a Padova', *Atti e memorie della R. Accademia di scienze, lettere ed arti in Padova*, n. s., 50 (1933–4), 277–333

Rizzo, Silvia, *Il lessico filologico degli umanisti* (Rome, Edizioni di Storia e letteratura, 1973)

Roaf, Christina, 'The presentation of the *Decameron* in the first half of the sixteenth century with special reference to the work of Francesco Sansovino', in P. Hainsworth and others (eds.), *The Languages of Literature in Renaissance Italy* (Oxford, Clarendon Press, 1988), pp. 109–21

Rogledi Manni, Teresa, *La tipografia a Milano nel XV secolo* (Florence, Olschki, 1980)

Rotondò, Antonio, 'Nuovi documenti per la storia dell'"Indice dei libri proibiti" (1572–1638)', *Rinascimento*, 2nd series, 3 (1963), 145–211

Salviati, Lionardo, *Degl'avvertimenti della lingua sopra 'l Decameron*, 2 vols. (Venice, D. and G. B. Guerra, 1584 and Florence, Giunta, 1586)

Samuels, R., 'Benedetto Varchi, the *Accademia degli Infiammati* and the origins of the Italian academic movement', *Renaissance Quarterly*, 29 (1976), 599–633

Sandal, Ennio, *Editori e tipografi a Milano nel Cinquecento*, 3 vols. (Baden-Baden, Koerner, 1977–81)

(ed.), *I primordi della stampa a Brescia 1472–1511: Atti del Convegno internazionale ... 1984* (Padua, Antenore, 1986)

Santagata, Marco and Quondam, Amedeo (eds.), *Il libro di poesia dal copista al tipografo* (Modena, Panini, 1989)

Sartori, Antonio, 'Documenti padovani sull'arte della stampa nel sec. xv', in *Libri e stampatori a Padova: miscellanea di studi storici in onore di mons. G. Bellini, tipografo editore libraio* (Padua, Tipografia Antoniana, 1959), pp. 111–231

Scholderer, Victor, *Fifty Essays in Fifteenth- and Sixteenth-Century Bibliography*, edited by D. E. Rhodes (Amsterdam, Hertzberger, 1966)

Solerti, Angelo, *Le vite di Dante, Petrarca e Boccaccio* (Milan, Vallardi, 1904)

Sorrentino, Andrea, *La letteratura italiana e il Sant'Uffizio* (Naples, Perrella, 1935)

Studi e problemi di critica testuale (Bologna, Commissione per i testi di lingua, 1961)

Sturel, René, 'Recherches sur une collection in-32 publiée en Italie au début du XVIe siècle', *Revue des livres anciens*, 1 (1913), 50–73

Tapella, Claudia and Pozzi, Mario, 'L'edizione del "Decameron" del 1573: lettere e documenti sulla rassettatura', *GSLI*, 165 (1988), 54–84, 196–227, 366–98, 511–44

Timpanaro, Sebastiano, *La genesi del metodo del Lachmann* (Florence, Le Monnier, 1963)

Trattati d'arte del Cinquecento, edited by P. Barocchi, 3 vols. (Bari, Laterza, 1960–2)

Trinkaus, Charles, 'A humanist's image of humanism: the inaugural orations of Bartolommeo della Fonte', *Studies in the Renaissance*, 7 (1960), 90–147

Trovato, Paolo, 'Primi appunti sulla norma linguistica e la stampa tra Quattrocento e Cinquecento', *Lingua nostra*, 48 (1987), 1–7

'Per un censimento dei manoscritti di tipografia in volgare', in Santagata and Quondam, *Il libro di poesia*, pp. 43–81

Con ogni diligenza corretto: la stampa e le revisioni editoriali dei testi letterari italiani (1470–1570) (Bologna, Il Mulino, 1991)

'Serie di caratteri, formato e sistemi di interpunzione nella stampa dei testi in volgare (1501–1550)', in E. Cresti and others (eds.), *Storia e teoria dell'interpunzione* (Rome, Bulzoni, 1992), pp. 89–110

Vasari, Giorgio, *Le vite de' più eccellenti pittori, scultori e architettori nelle redazioni del 1550 e 1568*, edited by R. Bettarini and P. Barocchi, 8 vols. (Florence, Studio per edizioni scelte, 1966–87)

Vattasso, Marco, 'Introduzione', in *L'originale del Canzoniere di Francesco Petrarca* (Milan, Hoepli, 1905)

Weiss, Roberto, *Un inedito petrarchesco: la redazione sconosciuta di un capitolo del Trionfo della Fama* (Rome, Edizioni di Storia e Letteratura, 1950)

Wilkins, Ernest H., 'The fifteenth-century editions of the Italian poems of Petrarch', *Modern Philology*, 40 (1943), 225–39

INDEX OF ITALIAN EDITIONS, 1470–1600

s. n. t. (Hain 12750)
c. 1474 Petrarch, *Canz.* and *Trionfi*, 198

ANCONA
Guerralda, Bernardino
1516 Fortunio, *Regole*, 58, 66–7
1518 Peranzone, *Opusculum*, 198
[1523] Peranzone, *Vaticinium*, 198

BOLOGNA
s. t.
c. 1525 *Ciento novelle antike (Novellino)*, 230
Benedetti, Girolamo de'
1525 *Ciento novelle antike (Novellino)*, 62, 88, 144, 159, 168
Benedetti, Platone de'
1494 Poliziano, *Stanze* etc., 70
Griffo, Francesco
1516 Petrarch, *Canz.* and *Trionfi*, 11, 57–8
Henricus de Colonia
1483 Arienti, *Porretane*, 65
Malpigli, Annibale
1475–6 Petrarch, *Canz.* and *Trionfi*, 32, 34–6, 198

BRESCIA
Bonini, Bonino
1487 Dante, *Commedia*, 38, 199
1487 *Esopo*, 4

FANO
Soncino, Girolamo
1503 Petrarch, *Canz.* and *Trionfi*, 53–4, 57–8

FERRARA
Rosso, Francesco
1532 Ariosto, *Orlando furioso*, 93, 118, 120, 122, 189

FLORENCE
s. t.
c. 1520 *Falconetto*, 80–1
Bonaccorsi, Francesco
1485/6 Petrarch, *Trionfo della Fama*, 44
1486 Usuardus, *Martyrologium*, 43

1490 Dante, *Convivio*, 45
1490 Jacopone, *Laude*, 45–7, 87, 92, 178, 186, 195
Bonacorso, Zuanne
1508 *Falconetto*, 80–1
Cennini, Bernardo and **Domenico**
1471–2 Servius, *Commentarius*, 42
Doni, Anton Francesco
1547 *Orationi diverse*, 136
1547 *Prose antiche*, 136
Francesco di Dino
1482/3 Pulci, Luigi, *Morgante*, 105–6, 220
Giunta, Benedetto
1537 Machiavelli, *Historie*, 225
1544 Ariosto, *Orlando furioso*, 98, 118, 133–4, 183, 226
Giunta, Bernardo
1532 Machiavelli, *Il principe* etc., 132
1532 Machiavelli, *Historie*, 225
1540 Machiavelli, *Il principe* etc., 132
1547 Sophocles, *Tragoediae*, 189
1548 Firenzuola, *Prose*, 132
1548 Martelli, *Opere*, 133
1548 Berni etc., *Opere burlesche*, 135, 136
1549 Firenzuola, *Rime*, 132
Giunta, Filippo
1504 Petrarch, *Canz.* and *Trionfi*, 54, 82, 83, 214
1506 Dante, *Commedia*, 58, 82–3, 173, 219
1510 Petrarch, *Canz.* and *Trionfi*, 54–5
1514 Cei, *Sonetti*, 210
1515 Petrarch, *Canz.* and *Trionfi*, 54, 57–8, 70
1515 Sannazaro, *Arcadia*, 60, 84
1516 Boccaccio, *Decameron*, 4, 12, 83, 86, 111, 112, 162, 208
1516 Boccaccio, *Corbaccio* and *Epistola a Pino de' Rossi*, 83, 212
1517 Boccaccio, *Fiammetta*, 72, 77, 84, 212
Giunta, Filippo, heirs of
1518 Boccaccio, *Ninfale fiesolano*, 214
1519 Benivieni, *Opere*, 71
1520 Cornazzano, *De re militari*, 84–5
1521 Boccaccio, *Ameto*, 72, 84, 110
1522 Petrarch, *Canz.* and *Trionfi*, 85, 101, 146, 173, 214–15
1524 Boccaccio, *Fiammetta*, 72

1526 Bruni, *Prima guerra*, 107
1527 Boccaccio, *Decameron*, 86, 89, 99, 111, 112, 113, 162-3, 169, 186-7, 195, 218
1527 *Sonetti e canzoni*, 62, 86-9, 132, 168, 173, 176, 185, 186
Giunta, Filippo and **Iacopo**
1587 Villani, G., *Cronica*, 175-6
Giunta, Filippo, *the younger*
1589 Maffei, *Istorie delle Indie orientali*, 233
1594 Boccaccio, *Fiammetta*, 184
1594 Boccaccio, *Corbaccio*, 183
1595 Galeotti, *Della varia dottrina*, 233
1596 Boccaccio, *Delle donne illustri*, 178, 233
1598 Boccaccio, *De gl'huomini illustri*, 178, 233
1598/9 Bracciolini, *Istoria*, 178, 233
Giunti, *house of*
1552 Burchiello, *Sonetti*, 136, 226
1556 Virgil, *Opere*, 147
1560 Aristotle, *Poetica*, 131
1561 Grazzini, *Spiritata*, 227
1565 Piovano Arlotto, *Facezie*, 180
1568 Buonaccorsi, *Diario*, 177
1568 Malispini, *Historia*, 177-8, 181, 233
1572 *Libro di novelle (Novellino)*, 159-61, 166, 189
1572 Pulci, Luca, *Ciriffo Calvaneo*, 179, 234
1573 Boccaccio, *Decameron*, 143, 161-5, 169, 171, 176
1574 Deputati, *Annotationi*, 163, 165
1577 Villani, M. and F., *Cronica*, 176
1578 *Istorie pistolesi*, 165-6
1581 Villani, M., *Cronica*, 176-7
1582 Boccaccio, *Decameron*, 17, 168-73
1586 Salviati, *Secondo volume degli Avvertimenti*, 168-71
Libri, Bartolomeo de'
c. 1485 Boccaccio, *Ninfale fiesolano*, 44-5
Manzani, Domenico
1595 Dante, *Commedia*, 83, 173-4, 187
Miscomini, Antonio
c. 1485 Cavalca, *Disciplina*, 179
[before 1490] Pulci, Luca, *Ciriffo Calvaneo*, 179, 234
1492 Landino, *Formulario*, 24
1492 Panciera, *Trattati*, 45
Morgiani, Lorenzo and **Johann Petri**
1492 Mandeville, *Trattato*, 202
1492 Panciera, *Trattati*, 45
Niccolò di Lorenzo
1481 Dante, *Commedia*, 38, 41, 43-4, 52-3, 59, 199, 223
Peri, Lorenzo
1546 Burchiello, *Sonetti*, 136, 226
Petri, Johann
1472 Boccaccio, *Filocolo*, 197
Sant'Iacopo di Ripoli
1477 St Antoninus, *Specchio di coscienza*, 44

[before 2 Jan. 1483?] *Fiore di virtù*, 44
[1483 (Sept.)?] *Fiore di virtù*, 44
[1483] Boccaccio, *Decameron*, 44, 83
Sermartelli, Bartolomeo
1569 Cavalca, *Disciplina*, 179, 184
1574 Pulci, Luigi, *Morgante*, 179-80
1576 Dante, *Vita nuova*, 89, 173
1580? Passavanti, *Specchio di vera penitenza*, 172
1582 Dell'Uva, *Le vergini prudenti*, 26
1584 Nardi, *Storie*, 178-9
1585 Passavanti, *Specchio di vera penitenza*, 172
Torrentino, Lorenzo
1548 *Facetie e motti*, 135
1549 Varchi, *Due lezzioni*, 133
1549-50 Fórnari, *Spositione*, 134
1550 Vasari, *Vite*, 133, 155-6
1553 Justinian, *Digesta*, 130-1, 155, 168
1554 Villani, G., *Seconda parte della Cronica*, 137-9
1554 Villani, M., *Prima parte della Cronica*, 137-8, 176
1558 Cicero, *Epistolae familiares*, 131
1559 *Tutti i trionfi*, 136-7
1561 Guicciardini, *Storia d'Italia* (1-16, 2°), 151, 156, 175
1561 Guicciardini, *Storia d'Italia* (1-16, 8°), 151, 156, 175
Vescovado, presso al
1560 Alberti (attrib.), *Historia de Hippolito*, 137, 226
Zucchetta, Bernardo
c. 1510 Ariosto, *Cassaria*, 79
c. 1510 Ariosto, *Suppositi*, 79
c. 1515 Arlotto, *Facetie*, 79-80
1525 Cristoforo Fiorentino, *Sonetti*, 212

FOLIGNO
Angelini, Evangelista and **Johann Numeister**
1472 Dante, *Commedia*, 219

MANTUA
Butzbach, Georg and **Paul**
1472 Dante, *Commedia*, 8, 37, 38
Micheli, Pietro Adamo de'
1472 Boccaccio, *Decameron*, 32, 60

MILAN
Agostino da Vimercate
1518 Dante, *Canzoni*, 208
Calvo, Andrea
1542 Boiardo, *Orlando innamorato*, 149
Castiglione, Giovanni
1521 Boccaccio, *Amorosa visione*, 67-9
Domenico da Vespolate
1476 Boccaccio, *Filocolo*, 31, 197

INDEX OF EDITIONS 1470–1600

Filippo da Lavagna
1478 Boccaccio, *Filocolo*, 31, 197
Lodovico and **Alberto Pedemontani**
1478 Dante, *Commedia*, 37, 43, 199
Meda, Valerio and **Girolamo**
1558 Ser Giovanni, *Pecorone*, 138
Minuziano, Alessandro
1503 Corio, *Historia*, 152
1520 Boccaccio, *Ameto*, 67–9, 110–11
Pietro da Corneno
1480 Mandeville, *Trattato*, 41
Rocco da Valle
c. 1520 Fregoso, *Dialogo de Fortuna*, 71
Scinzenzeler, Ulrich
1494 Petrarch, *Canz.* and *Trionfi*, 3, 10, 11, 27, 33, 54, 198
1496 Mandeville, *Trattato*, 41
1497 (Oct.) Mandeville, *Trattato*, 41
1497 (Dec.) Mandeville, *Trattato*, 41
1498 Petrarch, *De vita solitaria*, 9
Valdarfer, Cristoforo
1483 Masuccio, *Novellino*, 40
Zarotto, Antonio
1473 Petrarch, *Canz.* and *Trionfi*, 198
1476 Boccaccio, *Decameron*, 44
1489 Manilius, *Astronomicon*, 25
1490 Simonetta, *Commentarii*, 221
1494 Petrarch, *Canz.* and *Trionfi*, 198

NAPLES
s. n. t. [Naples?]
c. 1470 Boccaccio, *Decameron*, 163, 169, 231
s. t.
1477 Dante, *Commedia*, 219
[1507] Sannazaro, *Arcadia*, 59
Arnoldus de Bruxella
1477 Petrarch, *Canz.* and *Trionfi*, 197–8
Frezza, Antonio
1518 Del Pozzo, *Duello*, 74
1521 della Valle, *Vallo*, 72–3
Mayr, Sigismondo
1504 Sannazaro, *Arcadia*, 59, 64
[1508] Pontano, *De prudentia*, 191

ORTONA
Soncino, Girolamo
1518 Cornazzano, *De re militari*, 84

PADUA
Percacino, Grazioso
1557 Robortello, *De arte sive ratione*, 110
Valdezocco, Bartolomeo and **Martinus de Septem Arboribus**
1472 Petrarch, *Canz.* and *Trionfi*, 32, 33–4, 198

PESARO
Soncino, Girolamo
1507 Cornazzano, *De re militari*, 85

POIANO (*Verona*)
Antiquario (Feliciano), Felice and **Innocente Ziletti**
1476 Petrarch, *Degli homini famosi*, 9

ROME
Guillery, Etienne
1515 Tacitus, *Annales*, 10–11, 25, 62
Han, Ulrich
1470? Plutarch, *Vitae*, 31
Han, Ulrich and **Simone Cardella**
1471 Tortelli, *De orthographia*, 22
La Legname, Giovanni Filippo
1473 Petrarch, *Canz.* and *Trionfi*, 198
Lauer, Georg
1471 Petrarch, *Canz.* and *Trionfi*, 198
Puecher, Vitus
1475? Petrarch, *Trionfo della Fama*, 44
Sweynheim, Konrad and **Arnold Pannartz**
1469 Aulus Gellius, *Noctes Atticae*, 15, 192
1470 Pliny, *Naturalis Historia*, 22
1471 Bible (Lat.), 197

SIENA
Simeone di Niccolò
1517 Petrarch, *El secreto*, 70

TOSCOLANO
Paganini, Paganino and **Alessandro**
1520? Dante, *Commedia*, 58

TREVISO
Manzolo, Michele
1479 Boccaccio, *Ameto*, 31
1480 Caesar, *Commentarii*, 24

TURIN
Cravotto, Martino and **Francesco Robi**
1536 Ariosto, *Orlando furioso*, 96

VENICE
s. t.
1472 Bruni, *Epistolae*, 9
[1472–81] Boccaccio, *Fiammetta*, 197, 214
1477? Boccaccio, *Ninfale fiesolano*, 44, 210
c. 1495 *Virtutes psalmorum*, 41, 201
1501? *Hystoria del mondo falace*, 41
1511 *Danese Ugieri*, 74
s. t. (**Bernardino Benagli?**)
[1493] *Monte de la oratione*, 40–1
s. t. (**Manfredo Bonelli** or **Matteo Capcasa?**)

INDEX OF EDITIONS 1470–1600

c. 1495 *Monte de la oratione*, 40–1
Albertino da Vercelli (da Lissona)
1501 Pliny, *Storia naturale*, 221
1503 Petrarch, *Canz.* and *Trionfi*, 55
Angelieri, Giorgio
1574 Guicciardini, *Storia d'Italia* (1–20), 151
1585 Petrarch, *Canz.* and *Trionfi*, 146
1586 Petrarch, *Canz.* and *Trionfi*, 146, 228
Antonio da Gussago
1497 Boccaccio, *Filocolo*, 31
Antonio da Strada
1481 Boccaccio, *Decameron*, 32
Arrivabene, Andrea
1545 Boccaccio, *I casi de gli huomini illustri*, 178
Avanzi, Lodovico
1565 *Rime di diversi*, 150
Ballarino, Tommaso, di Ternengo
1534 Pliny, *Storia naturale*, 106, 221
Barezzi, Barezzo
1592 Petrarch, *Canz.* and *Trionfi*, 146
Basa, Bernardo
1584 Castiglione, *Cortegiano*, 142–3
Benagli, Bernardino
1516? Boccaccio, *Corbaccio*, 66
Benagli, Bernardino and Matteo Capcasa (Codecà)
1491 Dante, *Commedia*, 29, 38, 59, 199, 200
Bevilacqua, Nicolò
1559 Villani, G., *Cronica*, 125, 138–9, 175–6
1562 Petrarch, *Canz.* and *Trionfi*, 145
1563 Guicciardini, *Storia d'Italia* (1–16), 151
1563 Petrarch, *Canz.* and *Trionfi*, 145
1565 Guicciardini, *Storia d'Italia* (1–16), 151
1565 Petrarch, *Canz.* and *Trionfi*, 145
1568 Guicciardini, *Storia d'Italia* (1–16), 151
Bindoni, Francesco and Maffeo Pasini
1524 Olimpo da Sassoferrato, *Camilla*, 74
1527 Boiardo, *Orlando innamorato*, 74
1530 Boccaccio, *Filocolo*, 211
1532 Petrarch, *Canz.* and *Trionfi*, 25, 93–4, 97
1535 Ariosto, *Orlando furioso*, 95–6, 97
1537 Pulci, Luigi, *Morgante*, 106, 220
1541 Boccaccio, *Decameron*, 12, 99–100, 112, 113
1542 Ariosto, *Orlando furioso* (8°), 97, 218
1542 Ariosto, *Orlando furioso* (4°), 97, 218
Blavi, Tommaso de'
1481 Cecco d'Ascoli, *Acerba*, 200
Bonelli, Giovanni Maria
1552 Collenuccio, *Compendio*, 12, 122–3, 126, 184–5
1554 Corio, *Historia*, 152
1562 Guicciardini, *Storia d'Italia* (1–16), 151
Bonelli (Bon), Manfredo (Manfrino)
1494 Pulci, Luigi, *Morgante*, 105–6
[1495–8] *Salve Regina*, 41, 201
1496 Mandeville, *Trattato*, 41
1498 Boccaccio, *Decameron*, 32

1507 Pulci, Luigi, *Morgante*, 106, 220
1515 Cavalca, *Specchio di Croce*, 60, 154
Bonelli, Michele
1574 Boiardo, *Orlando innamorato*, 149
1576 Boiardo, *Orlando innamorato*, 149–50
Brucioli, Alessandro and brothers
1548 Petrarch, *Canz.* and *Trionfi*, 114–15
Burgofranco, Iacob (Pocatela) del
1529 Dante, *Commedia*, 59
Butrici, Massimo
1491 Boccaccio, *Fiammetta*, 31
Capcasa (Codecà), Giovanni
1493 Petrarch, *Canz.* and *Trionfi*, 33
Capcasa (Codecà), Matteo
1493 Dante, *Commedia*, 38–9
1494 St Catherine, *Dialogo*, 29
Cavalli, Giorgio
1565 Corio, *Historia*, 152–3
Cesano, Bartolomeo
1551 Boccaccio, *Filocolo*, 222
Ciotti, Giovanni Battista
1583 Ariosto, *Rime* and *Satire*, 149
Codecà, *see* Capcasa
Comin da Trino
1544 *Novo libro di lettere*, 108
1545 Boccaccio, *Libro delle donne illustri*, 178
1545 *Nuovo libro di lettere*, 108
1545–6 Pulci, Luigi, *Morgante*, 105–6, 220
1562 Petrarch, *Canz.* and *Trionfi*, 146
1567 Ariosto, *Orlando furioso*, 146
1571 Ariosto, *Orlando furioso*, 146
Conti, Federico de'
1472 Dante, *Commedia*, 37, 205
Costantini, Baldassare
1542 Sansovino, *Lettere*, 110
de Gregori, Giovanni and Gregorio
1492 Boccaccio, *Decameron*, 32
de Gregori, Gregorio
1508 Petrarch, *Canz.* and *Trionfi*, 54, 55, 195
1516 Boccaccio, *Decameron*, 7, 60–1, 83, 86, 100, 111, 185, 190, 218
1521 Del Pozzo, *Duello*, 74
1522 Masuccio, *Novellino*, 75
1523 Del Pozzo, *Duello*, 74
1524 della Valle, *Vallo*, 73
1525 Boccaccio, *Decameron*, 61
1525 Del Pozzo, *Duello*, 74
1525? Boccaccio, *Corbaccio*, 75
1525 Arienti, *Porretane*, 74
1526 Boccaccio, *Ameto*, 211
1526 Liburnio, *Le tre fontane*, 191
Deuchino (Dehuchino), Pietro, *heirs of*
1581 Goselini, *Rime*, 18
Farri, Domenico
1569 Dante, *Commedia*, 145
1572 Dante, *Commedia*, 145
1575 Dante, *Commedia*, 145

245

INDEX OF EDITIONS 1470–1600

1578 Dante, *Commedia*, 145
Filippo di Pietro
1477 *Atila*, 200
1478 Dante, *Commedia*, 37–8
1478 Cecco d'Ascoli, *L'Acerba*, 39
1481 Boccaccio, *Fiammetta*, 31, 214
1481 Boccaccio, *Filocolo*, 31
1482 Petrarch, *Canz.* and *Trionfi*, 32, 198
Fino, Pietro, *see* Pietro da Fino
Franceschi, Domenico De
1567 Ariosto, *Rime*, 149
Franceschi, Francesco De
1584 Ariosto, *Orlando furioso*, 148–9
1588 Goselini, *Rime*, 18
Francesco Veneziano
1517 *Danese Ugieri*, 74
Gabriele di Pietro
1477 Bible (It.), 197
Gabriele and Filippo di Pietro
1472 Boccaccio, *Filocolo*, 31, 197
1473 Petrarch, *Canz.* and *Trionfi*, 36
Gherardo, Paolo
1550 Petrarch, *Canz.* and *Trionfi*, 115
Giglio, Domenico
1587 Castiglione, *Cortegiano*, 227–8
Giolito de Ferrari, Gabriele
1541 Castiglione, *Cortegiano*, 124
1542 Ariosto, *Orlando furioso*, 96–7, 110
1542 Boccaccio, *Decameron*, 100
1542 Boccaccio, *Fiammetta*, 218
1543 Ariosto, *Orlando furioso*, 97, 226
1543 Cavalca, *Specchio di Croce*, 154
1543 Pliny, *Storia naturale*, 106, 221
1544 Ariosto, *Orlando furioso*, 97–8, 134, 218, 226
1544 Petrarch, *Canz.* and *Trionfi*, 101
1545 Boccaccio, *Ameto*, 110, 222
1545 Boccaccio, *Corbaccio*, 183, 218, 222
1545 Bruni, *Prima guerra*, 107
1545 *Rime diverse*, 108
1546 Ariosto, *Cassaria* (verse), 121, 224
1546 Ariosto, *Orlando furioso*, 118
1546 Boccaccio, *Decameron*, 25, 110–14, 144, 222
1547 Petrarch, *Canz.* and *Trionfi*, 114
1548 Ariosto, *Orlando furioso*, 118
1549 Ariosto, *Orlando furioso*, 223
1550 Ariosto, *Satire*, 121
1550 Cavalca, *Specchio di Croce*, 154
1550 Dolce, *Osservationi*, 115
1551 Ariosto, *Lena*, 121, 224
1551 Ariosto, *Negromante*, 121
1551 Ariosto, *Suppositi* (verse), 121
1552 Ariosto, *Orlando furioso*, 118, 120
1552 Boccaccio, *Decameron*, 112, 223
1552 Castiglione, *Cortegiano*, 124
1552 Dolce, *Osservationi*, 114

1553 Petrarch, *Canz.* and *Trionfi* (12°), 115
1553 Petrarch, *Canz.* and *Trionfi* (4°), 114
1554 Equicola, *Libro de natura de Amore*, 153
1554 Petrarch, *Canz.* and *Trionfi*, 114
1555 Dante, *Commedia*, 117, 118
1555 Tasso, B., *Amori*, 14
1556 Castiglione, *Cortegiano*, 125, 142
1557 Ariosto, *Rime* and *Satire*, 122
1558 Ariosto, *Rime* and *Satire*, 122
1558 Boccaccio, *Ameto*, 222
1559 Ariosto, *Orlando furioso*, 121
1559 Castiglione, *Cortegiano*, 125
1560 Ariosto, *Rime* and *Satire*, 149
1560 Tasso, B., *Amadigi*, 14
1560 Tasso, B., *Rime*, 192
1561 Equicola, *Libro de natura de Amore*, 153
1562 Dolce, *Osservationi*, 223
1562 Domenichi, *Dialoghi*, 191
1563 Boccaccio, *Corbaccio*, 148
1564 Guicciardini, *Storia d'Italia* (17–20), 151
1565 Cavalca, *Specchio di Croce*, 154
1566 Sannazaro, *Arcadia*, 228
1567 Ariosto, *Rime* and *Satire*, 149
1567 Rota, *Sonetti et canzoni*, 193
1567/8 Guicciardini, *Storia d'Italia* (1–20), 151
1568 Ariosto, *Rime* and *Satire*, 149
1568 Cavalca, *Specchio di Croce*, 154
1571 Bembo, *Asolani*, 9, 153–4
Giolito de Ferrari, Giovanni
1538 Boccaccio, *Decameron*, 98–9
1539 Franco, *Dialogi*, 96
Girardengo, Nicolò
1480 Guido de Cauliaco, *Chirurgia*, 40
Giunta, Bernardo, *the younger* and brothers
1582 Grazzini, *Comedie*, 142, 227
Giunta, Filippo and Iacopo
1582 Boccaccio, *Decameron*, 17, 168–73
Giunta, Lucantonio
1534–7 Cicero, *Opera*, 103, 129
Gobbi, Orazio de'
1580 Ariosto, *Orlando furioso*, 148
Griffio, Alessandro
1581–2 Petrarch, *Canz.* and *Trionfi*, 146
Griffio, Giovanni
1546 *Orationi diverse*, 136, 221
1549 Boccaccio, *Decameron*, 222
1554 Petrarch, *Canz.* and *Trionfi*, 4, 116
1573 Petrarch, *Canz.* and *Trionfi*, 145–6
Guerra, Domenico and Giovanni Battista
1560 Cappello, *Rime*, 2, 191
1561 *Rime in morte di Irene di Spilimbergo*, 18
1562 Zane, *Rime*, 18
1562 Villani, M., *Cronica*, 156, 176
1568 Ariosto, *Orlando furioso* (4°), 112, 148
1568 Ariosto, *Orlando furioso* (12°), 112, 148
1570 Ariosto, *Orlando furioso*, 148
1584 Salviati, *Avvertimenti*, 168–71

Guglielmo da Fontaneto
1518 Dante, *Canzoni*, 61-2, 208
Jenson, Nicholas
1470 Cicero, *Epistolae ad Brutum* etc., 194
1471 Caesar, *Commentarii*, 22
1472 *Scriptores rei rusticae*, 22
1474 Gratian, *Decretum*, 10
1475 Gregory IX, *Decretales*, 10
1475 Iacopo da Varazze, *Legenda aurea*, 2
1478 *Breviarium Romanum*, 26
Johann of Cologne and **Johann Manthen**
[1477] Asconius, *Commentarii*, 15
Le Rouge (Rubeus), Jacques
1474 Herodotus, 5
Locatelli, Boneto
1491 Durand, *Rationale*, 5
Manuzio, Aldo
1495 Lascaris, *Erotemata*, 22
1496 Bembo, *De Aetna*, 204
1497 Aristotle, *Opera*, IV, 21
1498 Aristotle, *Opera*, V, 21
1499 Perotti, *Cornucopiae*, 191
1500 St Catherine, *Epistole*, 23, 29
1501 Petrarch, *Canz.* and *Trionfi*, 9, 48-52, 53-6, 77, 85, 93, 94, 102, 111, 114, 115, 116, 173, 186, 195, 202-4, 210, 216
1501 Manuzio, *Rudimenta*, 204
1502 Dante, *Commedia*, 9, 48, 52-3, 59, 82, 102, 186, 195, 204-6
1502 Sophocles, *Tragoediae*, 21
1502 Interiano, *Vita et sito*, 2, 59
1503 Euripides, *Tragoediae*, 21
1505 Bembo, *Asolani*, 66
1508 Pliny the Younger, *Epistolae*, 21
1509 Plutarch, *Opuscula*, 21
1509 Horace, *Opera*, 51
1513 Perotti, *Cornucopiae*, 17
1514 Hesychius, *Dictionarium*, 22
1514 Petrarch, *Canz.* and *Trionfi*, 5, 56-8, 61, 77, 83, 85, 94, 98, 101, 116, 145
1514 Sannazaro, *Arcadia*, 59-60
Manuzio, Aldo and **Andrea (Torresani) da Asola**, *house of*
1515 Dante, *Commedia*, 58, 117, 128, 157
1521 Petrarch, *Canz.* and *Trionfi*, 58, 101
1522 Budé, *De asse*, 204
1522 Boccaccio, *Decameron*, 25, 61, 86, 161
1528 Castiglione, *Cortegiano*, 62-3, 124, 191, 208-9
Manuzio, Aldo and **Andrea (Torresani) da Asola**, *heirs of*
1533 Castiglione, *Cortegiano*, 124
1533 Cicero, *Epistolae familiares*, 103
1533 Petrarch, *Canz.* and *Trionfi*, 26, 94-5, 101, 114, 217
Manuzio, Aldo, *sons of*
1539 Ramberti, *Libri tre delle cose de' Turchi*, 221

1540 Cicero, *Epistolae familiares*, 103, 113, 130
1540 Cicero, *Epistolae ad Atticum* etc., 130
1541 Castiglione, *Cortegiano*, 124
1541 Cicero, *De officiis* etc., 221
1542 *Lettere volgari*, 108
1543 *Viaggi fatti da Vinetia*, 221
1545 *Lettere volgari*, 108
1545 Ariosto, *Orlando furioso*, 98, 118
1546 Petrarch, *Canz.* and *Trionfi*, 101, 114
1547 Castiglione, *Cortegiano*, 124
1551 Conti, *De venatione*, 109
Manuzio, Aldo, *the younger*
1581 Tasso, T., *Rime* (8°), 150
1581-2 Tasso, T., *Rime* (12°), 150
Manuzio, Paolo
1555 Livy, 116
Marcolini, Francesco
1538 Aretino, *Lettere*, 107
1538 Bembo, *Prose della volgar lingua*, 115, 203
1539 Petrarch, *Canz.* and *Trionfi*, 102, 115
1544 Dante, *Commedia*, 9, 102-3
1553 Burchiello, *Rime*, 125
Miscomini, Antonio
1477 Bible (It.), 28-9
Muschio, Andrea
1575 *Lettere facete*, 228
Nalli, Stefano and **Bernardino di**
1494 Seneca, *Pistole*, 9, 65
Navò, Curzio Troiano
1537 Berni and Mauro, *Terze rime*, 135
1542? *Letere de diversi*, 108
Nicolini, Domenico
1563 *Rhetoricorum ... tabulae*, 191
Nicolini da Sabbio, Giovanni Antonio and brothers
1525 Petrarch, *Canz.* and *Trionfi*, 77-8, 93, 102, 145
1526 Boccaccio, *Decameron*, 61
1532 *Sonetti e canzoni*, 89
1533 Petrarch, *Canz.* and *Trionfi*, 94, 146
1536 Equicola, *Libro de natura de Amore*, 219
1541 Petrarch, *Canz.* and *Trionfi*, 101, 219
Nicolini da Sabbio, Pietro
1534 Virgil, *Opera*, 9
1539 Boiardo, *Orlando innamorato*, 219
Nicolini da Sabbio, Pietro and **Giovanni Antonio**
1549 Petrarch, *Canz.* and *Trionfi*, 101
Paganini, Alessandro
1515 Petrarch, *Canz.* and *Trionfi*, 24, 57
1515 Sannazaro, *Arcadia*, 57
[1515] Dante, *Commedia*, 58
[1515] Dante, *Commedia* (another edition), 58
1516 Boccaccio, *Corbaccio*, 12, 66, 183
Paganini, Girolamo
1492 Bible (Lat.), 29, 197

Pasquali, Pellegrino
1488 Boccaccio, *Filocolo*, 31
Pasquali, Pellegrino and **Domenico Bertocchi**
1486 Petrarch, *Canz.* and *Trionfi*, 32
Penzio, *see also* Pincio
Penzio, Girolamo
1528 Boccaccio, *Pistola a Pino de' Rossi*, 212
1528 Boccaccio, *Teseida*, 76, 208, 212
Penzio, Iacopo
1527 Boccaccio, *Filocolo*, 76
Piasi, Piero de
1484 Petrarch, *Canz.* and *Trionfi*, 10, 24, 32, 35
1490 Petrarch, *Canz.* and *Trionfi*, 33, 36, 38, 198
1491 Dante, *Commedia*, 29, 38, 56, 61, 199
1491-2 Petrarch, *Canz.* and *Trionfi*, 29, 33, 36, 198
Pietrasanta, Plinio
1553 Ruscelli, *Tre discorsi*, 109, 119-20, 223
1554 Ariosto, *Satire*, 121-2, 185
1554 *Commedie elette*, 123-4
1554 Petrarch, *Canz.* and *Trionfi*, 116-17, 118
Pietro da Fino
1568 Dante, *Commedia*, 145
Pincio, *see also* Penzio
Pincio, Aurelio
1549 Seneca, *Epistole*, 107
Pincio, Donino
1502 Plutarch, *Vitae*, 31
1503 Boccaccio, *Filocolo*, 31
Pocatela, *see* Burgofranco
Portonari, Francesco
1560 Cornazzano, *Vita di Pietro Avogadro*, 125
Quarengi, Piero
1492-3 Ruffo, *Arte de cognoscere*, 29-30
1493 Guido de Cauliaco, *Chirurgia*, 40
1494 Petrarch, *Canz.* and *Trionfi*, 33
1497 Dante, *Commedia*, 38-9, 200
Rampazetto, Francesco
1549 Ariosto, *Orlando furioso* (4°), 146
1549 Ariosto, *Orlando furioso* (8°), 146
1559 Sannazaro, *Arcadia*, 117-18
1561-2 Grazzini, *Spiritata*, 142, 227
1562 Ariosto, *Orlando furioso*, 146-7
1564 Ariosto, *Orlando furioso*, 146-7
1564 Ariosto, *Rime*, 149
1566 *Cento novelle scelte*, 144
1566 Grazzini, *Spiritata*, 227
1566 Pallavicino, *Delle Lettere*, 189
1566 Ruscelli, *Imprese illustri*, 192
Ravani, Pietro, *heirs of*
1546 Petrarch, *Canz.* and *Trionfi*, 114
Rizzo, Bernardino

1487 Cecco d'Ascoli, *Acerba*, 200
1488 Petrarch, *Canz.* and *Trionfi*, 33, 36
Rosso (Rubeus), Giovanni, da Vercelli
1495 Caracciolo, *Specchio della fede*, 28
1512 Sannazaro, *Arcadia*, 59-60
Rubeus, *see* Le Rouge, Rosso
Ruffinelli, Venturino
1543 Simonetta, *Sforziade*, 107
1545 Cavalca, *Specchio di Croce*, 154
Rusconi, Giorgio
1503 Boccaccio, *Ameto*, 64-5, 111
1516 Arlotto, *Facetie*, 70-1, 79
1518 Boccaccio, *Ninfale fiesolano*, 210
1518 Cei, *Opera nova*, 210
1520 Caviceo, *Peregrino*, 71
Rusconi, Giovanni Francesco and **Giovanni Antonio**
1524 Caviceo, *Peregrino*, 72
Sansovino, Francesco
1561 Ariosto, *Rime*, 149
1561 Bruni, *Historia*, 151
1561 *Cento novelle scelte*, 144
1562 *Cento novelle scelte*, 144
1562 Guicciardini, *Storia d'Italia* (1-16), 151
1562 Nicetas Choniates, *Historia*, 151
1563 *Cento novelle scelte*, 144
Scoto, Brandino and **Ottaviano**
1539 Savonarola, *Prediche per tutto l'anno*, 106
1539 Savonarola, *Prediche sopra 'Quam bonus'*, 106
Scoto, Girolamo
1545 Boiardo, *Orlando innamorato*, 104, 117, 122-3, 184
1545 Pulci, Luigi, *Morgante*, 104-5, 220
Scoto, Gualtiero
1552 Aeschylus, *Tragoediae*, 110
Scoto, Ottaviano
1484 Dante, *Commedia*, 199
Sessa, Giovanni Battista
1500 Falconetto, 80
1501 Boiardo, *Sonetti*, 209
1502 Pulci, Luigi, *Morgante*, 6, 200
Sessa, Giovanni Battista and **Giovanni Bernardo**
1596 Dante, *Commedia*, 145
Sessa, Giovanni Battista and **Melchiorre,** *the younger*
1559 Ruscelli, *Del modo di comporre*, 12, 13
1560 Giovio, *Lettere volgari*, 142
1564 Dante, *Commedia*, 144
1578 Dante, *Commedia*, 145
Sessa, Melchiorre
1531 Arienti, *Porretane*, 74
Sessa, Melchiorre and **Pietro Ravani**
1516 Pliny, *Storia naturale*, 221

Sessa, Melchiorre, *heirs of*
1558 *Fiori delle rime*, 191
1571 *Cento novelle scelte*, 144, 159
Siliprandi, Domenico
1477 Petrarch, *Canz.* and *Trionfi*, 32, 35, 36, 198
Soardi, Lazaro de'
1511 Petrarch, *Canz.* and *Trionfi*, 55
Speranza, al segno della
1547 Savonarola, *Della semplicità*, 221
Stagnino, Bernardino
1512 Dante, *Commedia*, 59
1513 Petrarch, *Canz.* and *Trionfi*, 1, 8, 10, 15, 24, 55–6
1520 Dante, *Commedia*, 59
Tacuino, Giovanni
1525 Bembo, *Prose della volgar lingua*, 61, 86
Theodore of Reynsburch and **Rainaldus de Novimagio**
1478 Petrarch, *Canz.* and *Trionfi*, 32, 35, 198
Torti, Alvise
1536 Ariosto, *Orlando furioso* (4°), 96
1536 Ariosto, *Orlando furioso* (8°), 96
Torti, Battista
1484 Masuccio, *Novellino*, 40
1487 *Articella*, 195
Tramezzino, Michele
1539 Collenuccio, *Compendio*, 122
Valdarfer, Cristoforo
1471 Boccaccio, *Decameron*, 31–2
Valgrisi, Vincenzo
1546 Speroni, *Canace*, 223
1549 Petrarch, *Canz.* and *Trionfi*, 114
1552 Boccaccio, *Decameron*, 10, 12, 112–14, 118, 143, 223
1556 Ariosto, *Orlando furioso*, 118–21, 147
1558 Ariosto, *Orlando furioso*, 147, 223
1565 Ariosto, *Orlando furioso*, 146
Valla, Bruno and **Tommaso de' Blavi**
1477 Boccaccio, *Ninfale fiesolano*, 210
Valvassori, Giovanni Andrea, *known as* 'Guadagnino'
1549 Ariosto, *Orlando furioso*, 118
1553 Ariosto, *Orlando furioso*, 118, 147
1554 Ariosto, *Orlando furioso*, 118
1566 Ariosto, *Orlando furioso*, 141, 147–8
Varisco, Giovanni
1563 Ariosto, *Orlando furioso*, 147
1565 Gamucci, *Dell'antichità*, 153
1566 Ariosto, *Orlando furioso*, 147
1568 Ariosto, *Orlando furioso*, 147
1569 Gamucci, *Le antichità*, 153
Viani, Bernardino
1533 Savonarola, *Della semplicità*, 220–1
1536 Crescenzi, *Libro della agricultura*, 106

Vidali, Giacomo
1574 Petrarch, *Canz.* and *Trionfi*, 146
Vitali, Bernardino
1524 Boccaccio, *Fiammetta*, 25, 77, 212
1528 Petrarch, *Canz.* and *Trionfi*, 93
1532 Petrarch, *Canz.* and *Trionfi*, 93
1535 Boccaccio, *Decameron*, 98
Windelin of Speyer
1470 Livy, 192
1470 Petrarch, *Canz.* and *Trionfi*, 198
1477 Dante, *Commedia*, ix, 35, 37, 43, 82, 117, 182, 219
Zanetti, Bartolomeo
1537 Villani, G., *Cronica*, 99, 125, 137–8
1538 Petrarch, *Canz.* and *Trionfi*, 9, 93
Zanni, Agostino
1515 Petrarch, *Canz.* and *Trionfi*, 55
1518 Boccaccio, *Decameron*, 208
Zanni, Bartolomeo
1497 Petrarch, *Canz.* and *Trionfi*, 33
1500 Petrarch, *Canz.* and *Trionfi*, 1, 11, 33, 36, 48, 58
1503 Masuccio, *Novellino*, 209
1504 Arienti, *Porretane*, 65, 112, 209
1507 Dante, *Commedia*, 59
1508 Petrarch, *Canz.* and *Trionfi*, 1
1510 Arienti, *Porretane*, 209
1510 Boccaccio, *Decameron*, 208
1510 Masuccio, *Novellino*, 209
Zenaro, Damiano
1581 Ruscelli, *Commentarii*, 119
Ziletti, Giordano
1556 Giovio, *Ragionamento*, 224
Zoppini, Fabio and **Agostino,** and **Onofrio Farri**
1588 Boccaccio, *Decameron*, 143–4
Zoppino, Nicolò d'Aristotile, *known as*
1524 Sannazaro, *Arcadia*, 73
1525 Boccaccio, *Fiammetta*, 72
1525 Cornazzano, *Proverbi*, 73
1529 della Valle, *Vallo*, 72–3
1529 Lucian, *Dialoghi*, 6
1530 Boccaccio, *Filocolo*, 73, 211
1530 Sannazaro, *Arcadia*, 73, 211
1530 Tebaldeo, *Opere d'amore*, 73, 211
1531 Cornazzano, *Vita di Cristo*, 73
1536 Ariosto, *Orlando furioso*, 96
Zoppino, Nicolò d'Aristotile, *known as*, and **Vincenzo di Paolo**
1518 Arlotto, *Facetie*, 71
1518 Cornazzano, *Vita di Cristo*, 70, 73
1520 Petrarch, *El secreto*, 70
1521 Fregoso, *Dialogo de Fortuna*, 71
1521 Petrarch, *Canz.* and *Trionfi*, 58, 70
1521 Petrarch, *Trionfi*, 58
1521 Poliziano, *Stanze*, 70

1521 Sannazaro, *Arcadia*, 73
1522 Benivieni, *Opere*, 6, 71, 211
1524 Boccaccio, *Ameto*, 72
VERONA (*see also Poiano*)
Alvise, Giovanni and **Alberto**
1479 *Aesopus* (Lat. and It.), 4

VICENZA
Achates, Leonhard
1474 Petrarch, *Canz.* and *Trionfi*, 32, 198
Johannes Renensis
1478 Boccaccio, *Decameron*, 32

INDEX OF MANUSCRIPTS AND ANNOTATED COPIES

EL ESCORIAL
Biblioteca de San Lorenzo
Lat. e.III.23, 193

FLORENCE
Accademia della Crusca
53 ('Raccolta Bartoliniana'), 89, 216
Biblioteca Marucelliana
R.O.304, 158, 230
Biblioteca Medicea Laurenziana
Codex Pisanus, 130
32.9, 155
40.42, 173
42.1 (Mn), 83, 86, 158, 162–3, 167–70, 181, 214, 231
42.11, 166, 232
42.12, 166, 232
42.38, 136
49.9, 129–30, 131, 162
Ashburnham 409, 63
Ashburnham 443, 19–20
Ashburnham 510, 178, 233
Mediceo Palatino 119, 215
Redi 9, 19
Biblioteca Nazionale Centrale
2.1.108, 178, 233
2.1.289, 233
2.3.350, 165, 231
2.4.28, 178, 233
2.10.66, 158
6.112, 166, 232
22.A.4.1, 161, 231
22.A.4.2, 163, 215
22.A.5.18, 163
25.28, 165, 231
Landau Finaly stampe 262, 230
Palatino 1081, 138
Palatino E.B.10.3, 176, 233
Panciatichiano 32, 159–61, 230–1
Biblioteca Riccardiana
707, 198
2731, 137

LONDON
British Library
177.g.16, 175, 233

VATICAN CITY
Biblioteca Apostolica Vaticana
Lat. 3195 (V), 32, 49–50, 77, 115, 116, 203–4
Lat. 3197, 49–53, 203
Lat. 3199 (Vat), 52–3, 157, 174, 204–5
Lat. 3214, 62

VENICE
Biblioteca Nazionale Marciana
It. IX.191 (= 6754), 62, 208
It. Z.34 (= 4772), 175

GENERAL INDEX

Acarisio, Alberto, 111
Accademia degli Infiammati, 127, 129, 224
Accademia degli Umidi, 127-8, 135
Accademia dei Fantastici, 146
Accademia della Crusca, 173-4, 187
Accademia fiorentina, 127-9, 131, 161, 173, 224
accents, diacritic
  rejection of, 58, 167
  use of, by editors and printers, x, 2, 26, 51, 54, 59, 69, 82, 87, 95, 101, 116, 120, 123, 131, 139, 156, 167, 178, 204, 210, 214, 215, 217, 223, 226, 232; *see also* diaeresis
Acciaiuoli, Donato, 151
Accorso, Buono, *see* Bonaccorso
Adamo da Montaldo, 22
Adriani, Giovambattista, 161
Aeschylus, 110
Aesop, 4, 21, 134
Ageno, Franca, 200, 202, 232
Agnelli, Giuseppe, 217, 224, 227, 228
Alberti, Leon Battista, 137, 185, 234
Albignani (Trecius), Pietro, 10, 26, 28, 190
Albignani, Giovanni, 190
Albini, Enrico, 193
Albricci, Matteo, 16
Alessandro de' Medici, Duke of Florence, 127
Alexander VI, Pope, 153
Alexandria, Museum in, x
Alfieri, Francesco, 54, 57, 85, 184
Alighieri, Dante, *see* Dante Alighieri
Alighieri, Iacopo, 37
allegorie, *see* moral glosses
Allenspach, Joseph, 190, 192, 197, 199
Alunno, Francesco, 102, 111, 115, 154
Ambrosoli, Mauro, 220
Ammirato, Scipione, 119
Angeleri, Carlo, 229
Angeli da Barga, Pietro, 167
Angelieri, Giorgio, 146, 151, 228
Anguillara, Giovanni Andrea, 147, 228
annotations, use of, by editors, x, xii, 3, 4, 183; on Ariosto, 95, 119-21, 146-9, 223; on Bembo, 154; on Boccaccio, 68, 99, 100, 110-11, 112-13, 144, 163, 165, 171, 172; on Castiglione, 143; on Dante, 117, 145; on Guicciardini, 151; on Machiavelli (*Mandragola*), 124; on Petrarch, 95, 101, 114, 115, 145-6; on Pulci, 105-6; *see also* margins, regional Italian forms
anthologies, 107-8, 134-7, 144, 150, 183, 221
Antinori, Bastiano, 161
Antoninus, St, *see* Pierozzi
Antonio da Ferrara, 20, 193
Antonio da Strada, 32
Antonio da Tempo (pseudo-), 35-6, 55, 199
Antonius Andreae, 196
apocope, 50-1, 54, 71, 82, 120, 167, 204, 207, 209, 212, 214, 226
Apollonio Campano, *see* Clario
apostrophe, 51, 55, 58, 69, 82, 87, 102-3, 116, 120, 156, 160, 167, 170, 205-6, 207, 210, 214, 221, 223
Apuleius, 138
Aramei, 127, 131, 133, 135, 136
Arbib, Lelio, 179, 233
Arbizzoni, Guido, 211
Aretino, Pietro, 2, 92-3, 100, 102, 107, 216, 219
Argilagnes, Franciscus, 195
argomenti, *see* summaries
Arienti, Sabadino degli, 65, 74, 112
Ariosto, Galasso, 120
Ariosto, Lodovico, 90, 93, 100, 124
  biography of, 119, 146, 147, 148
  Cinque canti, editions of, 98, 118, 146-7
  comedies, editions of, 79, 121, 224
  Orlando furioso, editions of, 4, 95-8, 102, 112, 117, 118-21, 132, 133-4, 141, 145, 146-9, 183, 189, 217-18, 223-4, 226
  Rime and *Satire*, editions of, 121-2, 149, 184-5, 224
Ariosto, Virginio, 98, 118, 121, 223
Aristarchus, 2
Aristotle, 16, 21, 131
Arlenio, Arnoldo, 133, 134
Arlia, C., 230
Arlotto Mainardi, Piovano, 70-1, 135, 180
Ascarelli, Fernanda, 216

252

Asconius Pedianus, 15
Asolano, *see* Torresani
Astemio, Lorenzo, 54, 205
Astemio, Marco, 26, 61, 195, 208
asterisks, use of, 110, 130, 131, 164, 168, 171
Atanagi, Dionigi, 2, 8, 11, 17, 141, 148, 150, 156–7, 159, 176, 191, 192, 193, 228
Augurello, Giovanni Aurelio, 48, 202
Aulus Gellius, 15, 192
Ausonius, 23
Avicenna, 16

Badalin, T., 218
Baldacci, Luigi, 212, 216, 219, 228
Baldelli, Francesco, 8
Balduino, Armando, 202, 210, 214
Ballarino, Tommaso, 106
Ballistreri, G., 196
Balsamo, Luigi, 190, 192, 207
Barbarigo, Lodovico, 55
Barbaro, Daniele, 141
Barbaro, Ermolao, 23, 67, 116, 194
Barbi, Michele, 61–2, 173, 202, 208, 216, 219, 224, 231, 232, 233
Barezzi, Barezzo, 146
Barnes, John C., 233
Bartoli, Cosimo, 133
Bartolini, Lorenzo, 89
Bartolomeo da Alzano, fra, 29
Basa, Bernardo, 142
Basilico, 198
Basso, Jeannine, 221
Baudrier, Henri and Julien, 228
Beccadelli, Lodovico, 163, 231
Beer, Marina, 211
Belloni, Gino, 190, 199, 204, 207, 209, 212, 216, 217, 218, 219
Beloch, K. J., 200, 201
Bembo, Bernardo, 52–3
Bembo, Carlo, 9
Bembo, Pietro, 12, 60, 62–3, 77, 99, 101, 102, 127, 129, 131, 143, 159, 208, 217, 221, 225
  as author, 53, 66, 90, 92
  as editor, x, 7, 8, 9, 48–63, 64, 67, 82, 85, 86, 87, 94, 103, 111, 115, 144, 157, 167, 174, 186, 187, 195, 202–7, 210, 214, 216, 219
  as grammarian, xii, 53, 61, 63, 86, 88, 91–2, 94, 102, 115, 118, 121, 132, 135, 137, 145, 157, 166, 170, 182, 208, 211, 212, 214, 216, 217, 228
  editions of works, 9, 66, 153–4, 204
Benagli, Bernardino, 38, 66, 90
Benci, Antonio, 178, 233
Benedictines, 7, 155
Benivieni, Antonio, 161

Benivieni, Girolamo, 6, 71, 82–3, 211, 213, 214, 220
Benivieni, Lorenzo, 213
Bentley, J. H., 194
Benvenuto da Imola, ix, 37
Benzoni, fra Francesco, 196
Benzoni, G., 232
Berardi, Cristoforo, ix, xii, 37, 43, 82, 117, 182, 198
Beretin Convento (Frari), press, 7, 196
Berni, Francesco, 104, 135, 136, 149–50, 226
Beroaldo, Filippo, the younger, 10, 25, 62
Bertoli, Gustavo, 195
Bertolo, Fabio, 208
Bessi, Rossella, 199
Bettarini, Rosanna, 225
Betussi, Giuseppe, 178
Bevilacqua, Nicolò, 145, 151, 228
Biagi, Guido, 230, 231
Biagiarelli, Berta Maracchi, 201, 202, 212, 224, 225, 229
Biblioteca Laurenziana, 173
Billanovich, Myriam, 190
Bindoni, Francesco, 74, 93, 95, 97, 99, 104, 112, 113, 134, 211, 218
Bindoni, press, 90
Black, M. H., 197
Black, Robert, 216
Blado, Antonio, 124
Blunt, Anthony, 234
Boccaccio, Giovanni, xi, 20, 52, 59, 61, 64, 66, 90, 92, 96, 124, 186
  as biographer and editor of Dante, x, 20, 37, 117, 199, 223
  Ameto (Commedia delle ninfe fiorentine), editions of, 31, 64–5, 67–9, 72, 84, 110–11, 211, 222
  Amorosa visione, editions of, 67–9
  biography of, 31, 32, 111
  Corbaccio, editions of, 12, 66, 75, 83, 148, 161–2, 167–8, 183, 218, 222, 232
  Decameron, editions of, 4, 7, 10, 12, 17, 25, 31–2, 42, 44, 60–1, 83, 86, 89, 98–100, 102, 103, 110–14, 117, 141, 143–4, 158, 161–5, 168–73, 176, 185, 186–7, 190, 195, 207–8, 218, 222–3, 231; Fortunio and text of, 67; manuscript tradition of, 21
  Fiammetta, editions of, 25, 31, 72, 77, 84, 184, 197, 212, 214
  Filocolo, editions of, 31, 42, 73, 76, 197, 211, 222
  imitation of, 66, 69, 71, 72, 92–3, 113, 166, 167, 183
  letters, editions of, 64–5, 75, 83, 136, 212, 226
  Ninfale fiesolano, editions of, 44–5, 210, 214
  Teseida, editions of, 76, 208, 211, 212

GENERAL INDEX

Boiardo, Matteo Maria, 74, 97, 104, 149–50, 209, 219–20
Bologna, Corrado, 218
Bologni, Girolamo, 24, 31, 197
Bonaccorsi, Francesco, 43, 45
Bonaccorsi, Lodovico, 129
Bonaccorso, 17, 189, 195
Bonacorso, Zuanne, 80–1
Bonagiunta da Lucca, 87
Bond, W. H., 191
Bonelli, Giovanni Maria, 122, 151, 152
Bonelli (Bon), Manfredo (Manfrino), 105
Bonelli, Michele, 149–50
Bonelli, Sebastiano, 151
Bongi, Salvatore, 154, 190, 193, 224, 227, 229
Bonini, Bonino, 4
Bono Gallo, 16
Bononome, Giuseppe, 148
book-privileges, 5, 9, 15, 61, 74–5, 96, 101, 195, 208, 211, 221
Bordegazzi, Giacomo, 9, 16
Borghini, Vincenzio, x, 5, 7, 111, 130, 133, 155–66, 168–9, 171, 173, 175, 176, 177, 181, 187, 229–32, 233
Bosso, Matteo, 23
Botticelli, Sandro, 44
Bracciolini, Iacopo di Poggio, 44, 178
Bracciolini, Poggio, 178
Branca, Vittore, 21, 194, 210, 214, 231, 232
Brandolini, Ettore, 29, 197
Brandolini, Giancoate, 29, 197
Bresciano, G., 190
Brevio, Francesco, 16
British Library, 81, 109, 140
Brocardo, Antonio, 100
Bronzino, Agnolo, 129
Brotto, Giovanni, 196
Brown, Horatio F., 211, 227
Brown, Peter M., 171, 231, 232, 234
Brucioli, Antonio, 8, 98–101, 106–7, 111, 115, 125, 127, 189, 218
Brugnoli, Benedetto, 5, 28
Bruni, Leonardo, 9, 107, 151
Bruni, Roberto, 210, 218, 224
Bruno, fra Gabriele, 29–30, 33, 38, 196
Bryce, Judith, 225
Budé, Guillaume, 204
Bujanda, J. M. de, 227
Buonaccorsi, Biagio, 71, 177
Burchiello, Domenico di Giovanni, *known as il*, 125, 136
Burgofranco, Iacob (Pocatela) del, 59
Bussi, Giovanni Andrea, 6, 15, 22, 190, 191, 192, 194, 195

Cachey, Theodore J., Jr, 203
Caesar, C. Julius, 22, 24

Caimi, Francesco, 9
Calfurnio, Giovanni, 28
Camaldolese order, 7, 179
Cambi Opportuni, Alfonso, 146, 228
Camerini, Luigi S., 233
Camerini, Paolo, 213
Camilli, Camillo, 148
Camillo Delminio, Giulio, 101, 115, 218
Campolo, fra Giovanni, 193
Canigiani, Bernardo, 162
Cannata Salamone, Nadia, 221
Capcasa (Codecà), Matteo, 29, 38
capitalization, 26, 51, 116, 120, 131, 139, 156, 167, 178
Capoduro, Luisa, 22, 194
Cappello, Bernardo, 2, 191
Caracciolo, fra Roberto, 28
Carari, Giovanni Pietro, 9, 16, 28
Carbone, Lodovico, 28
Cardini, Roberto, 202
Carducci, Niccolò, 173
Carlini, Antonio, 222
Carmelites, 7, 149
Caro, Annibale, 222
Carrai, Stefano, 230, 231
Carrara, Enrico, 199
Carter, Tim, 193, 231, 232
Casciano, Paola, 194
Casella, Angela, 212, 224
Cassiodoro Ticinese, 6, 71, 75, 184, 211
Castellani, G., 205
Castelvetro, Lodovico, 176, 223
Castiglione, Baldesar, 76, 105, 111, 112, 186
  Cortegiano, editions of, 62–3, 124–5, 142–3, 183, 191, 208–9, 227–8
Castiglione, Camillo, 142
Castiglione, Giovan Battista da, 92, 216
Castiglione, Giovanni, 69
Catherine of Siena, St, 23, 28
Cavalca, fra Domenico, 60, 154, 179, 184
Cavalcanti, Bartolomeo, 135
Cavalcanti, Guido, 20, 56, 61, 87, 194, 215
Cavalli, Giorgio, 152
Caviceo, Iacopo, 71, 72
Cecco d'Ascoli, 39
Cei, Francesco, 210
Cellini, Benvenuto, 225
Cennini, Bernardo, 43
Cennini, Domenico, 43
Cennini, Pietro, 43, 201
censorship, 80, 135, 136, 138, 140–4, 146, 149–50, 156, 159, 164–5, 166, 168, 171–3, 175, 180, 181, 213, 227, 230, 232, 233; *see also* Indexes of Prohibited Books, Inquisition
Centone, Girolamo, 7, 33, 198
Ceruti, Antonio, 190, 192, 193, 219, 227, 230

Charles V, Emperor, 127
Chiòrboli, Ezio, 135
Cian, Vittorio, 208, 227
Ciccarelli, Antonio, 142-3
Cicero, 15, 103, 107, 110, 113, 116, 129-30,
    131, 162, 177, 194, 221
Cicogna, Emmanuele A., 207, 212, 217, 222
Ciento novelle antike, see *Novellino*
Cinico, Giovan Marco, 190
Cino da Pistoia, 56, 61, 87
Cioni, A., 228
Cipolla, Carlo, 192
Claricio, Girolamo, 67-9, 76, 84, 87, 92, 95,
    110-11, 210
Clario, Giovanni Antonio (Apollonio
    Campano), 114, 136, 221, 223
Clement VII, Pope, 127
Clement VIII, Pope, 142
clergy, and editing, 7, 8, 82
Clubb, Louise George and William G., 222
Coccia, Paola, 223
Coccio, Francesco, 219
Codecà, see Capcasa
Coglievina, Lionella, 229
collana ('series'), 150-1, 229
Collenuccio, Pandolfo, 12, 122-3, 126, 152,
    184
Colombino Veronese, 8, 37, 38
Colonne, Guido delle, 87
Colucia, Francesco, 10
Columella, 17
Comin da Trino, 90, 105-6, 146, 220
commentaries, 3, 6
  on Burchiello, 125
  see also Dante, Petrarch
'common' language or usage, 66, 68, 76-7,
    96, 112, 113, 123, 147, 217
Compagni, Dino, 20
compositors, 10, 16, 17, 26, 27, 156, 192, 195,
    196
Concini, Bartolomeo, 175
Congregation of the Index, 142, 165
conjecture, use of, xi, 20, 21-3, 33, 38, 46,
    50, 52, 84, 87, 89, 110, 113, 129-31,
    157, 158, 160, 169, 175, 176, 194, 199
consonants, see Latinisms, phonology
Contarini, Iacopo, 175
Conti, A., 234
Conti, Giusto de', 168
Conti, Natale de', 109
copyists, see scribes
Coraldo, Livio, 146
Corbinelli, Iacopo, 159, 161-2, 166-8, 169,
    181, 232
Corbon, Jean, 168
Corio, Bernardino, 152
Cornazzano, Antonio, 70, 73, 84-5, 125

corrections, stop-press, 25, 195, 205
'correctors' (It. *correttori*, Lat. *correctores*), ix, 4,
    9, 10, 12, 13, 16, 17, 22, 26, 33, 189, 191
Corti, Maria, 193
Cosimo I de' Medici, Duke of Florence
    (1537-69), Grand Duke of Tuscany
    (1569-74), 127, 128, 130, 131, 133, 134,
    138, 151, 155, 161, 162, 175, 177
Costabili, fra Paolo, 165
Costantini, Baldassare, 222
Costo, Tomaso, 146
Cotton, J. Hill, 210
Cox, Virginia, 224
Craverio, Antonio, 8
Crescenzi, Pietro, 106, 158
Cresci, Pietro, 146
Cristoforo Fiorentino, 212
Croce, Benedetto, 193

D'Alessandro, Alessandro, 218
Danese Ugieri, 74
Daniello, Bernardino, 101, 103, 145, 219
Dante Alighieri, xi, 20, 90, 96, 127, 177
  biography of, ix, 37, 103
  Commedia: commentaries on, see Benvenuto
    da Imola, Daniello, della Lana,
    Landino; editions of, ix, 8, 9, 29, 35,
    36-9, 41, 43-4, 48, 52-3, 58-9, 61, 82-3,
    102-3, 117, 118, 128, 144-5, 157, 173-4,
    182, 186, 187, 195, 199-200, 204-6, 207,
    219, 223; Fortunio and text of, 67;
    manuscript tradition and text of, 20, 128,
    157, 173, 224-5
  Convivio, editions of, 45
  De vulgari eloquentia, editions of, 168, 232
  letters, editions of, 136, 226
  lyric poetry, editions of, 38, 56, 61-2, 87, 215
  Vita nuova, editions of, 89, 173
Dante da Maiano, 87
Davis, Charles T., 233
Debenedetti, Santorre, 88, 215
De Caro, G., 197
de Gregori, Gregorio, 8, 55, 60, 61, 73-5, 75,
    90, 195, 211
de la Mare, Albinia, 194, 201
Delfin(o) (Dolfin), Nicolò, 7, 60-1, 83, 86,
    99-100, 112, 114, 185, 186, 190, 207,
    208, 210, 218
della Fonte (Fonzio), Bartolomeo, 43, 197,
    201, 202
della Lana, Iacopo, 37, 38
della Rovere, Francesco Maria, 213
Della Torre, Arnaldo, 197, 199
della Valle, Battista, 72-3
Dell'Uva, Benedetto, 26
Del Pozzo, Paride, 73-4
del Rosso, Paolo, 168

GENERAL INDEX

De Maldé, Vania, 229
Demetrius, 185, 186
'Deputati' (for *Decameron* of 1573), 161–5, 171, 172, 231
De Robertis, Domenico, 193, 199, 207, 214, 215, 216
Diacceto, Francesco, Bishop of Fiesole, 172
diaeresis
　diacritic sign, 207
　in verse, 65, 150, 214, 220
dialoephe, 46, 69, 76, 88, 106
dictionaries, *see* glossaries
Di Filippo Bareggi, Claudia, 190, 191, 193, 228
Di Francia, Letterio, 230
Dinali, *see* Nalli
Dionisotti, Carlo, 45, 189, 190, 194, 197, 198, 199, 202, 203, 205, 207, 208, 209, 210, 211, 213, 214, 215, 216, 218, 219, 223, 228
Dolce, Lodovico, 140, 185–6
　as editor, 2, 3, 8, 12, 13, 95–7, 99–101, 104, 109, 112–22, 124–6, 134, 142, 144, 146, 147–8, 149, 153, 183, 187, 191, 217, 218, 222, 223, 226, 228
　as grammarian, 114, 115, 223
　as translator, 147, 221
Dolcino, Stefano, 25, 195
Dolfin, *see* Delfino
Domenichi, Lodovico, 3, 7, 8, 101, 104–8, 109, 116, 117, 122, 132, 135, 138, 139, 142, 146–50, 176, 183, 184, 187, 191, 218, 220, 222, 226
Dominicans, 7, 21, 28–9, 125
Donati, Lamberto, 192, 233
Donatus, 91–2
Doni, Anton Francesco, 8, 11, 107, 109, 121, 125, 131–2, 136, 189, 191, 222, 224
Dortelata, Neri, 105
Dragoncino, Giovanni Battista, 74
Ducas, Demetrius, 21
Duns Scotus, 196
Durand, Guillaume, 5, 16, 29

Echard, J., 224
editors
　and layout of text, 11
　and printers, 3–12
　as agents, 14
　as publishers, 9
　background of, 7–8, 28–30
　demand for, 1–7
　employment and payment of, 15–18, 63
　Florentine, characteristics of, 42–3, 45
　identification of, 1
　methods of, factors influencing, 19–27;
　　aesthetic principles, 185–6;
　　comprehensibility of text, xi, 40, 44, 69, 72, 77, 80, 84, 88; conformity with rules, 64, 67, 71, 76, 99, 113, 118–23, 125, 129, 137, 157, 160, 163, 166, 168, 177, 180, 184, 209; haste, xi, 24–5, 35, 111, 144, 195; lack of control, 26; 'openness' of text, 25–6, 61, 175; treatment and choice of manuscripts, 21, 23–4, 26–7
　motives of, 15
　opinions of, among men of letters, 13–15
　portraits of, 1
　printers as, 5
　see also clergy, 'correctors', lexis, manuscripts, morphology, paraphrase, phonology, portraiture, scribes, statues, syntax
elision, 50–1, 54, 82, 167, 210, 216; *see also* apostrophe
Elizabeth I of England, 144
engravings, *see* illustrations
epithets, 91, 93, 97, 102, 111, 115, 116, 148
Equicola, Mario, 153, 219
Erasmo da Valvasone, 148
errata, lists of, 11, 12, 24, 25–6, 45, 50, 166, 175, 195, 214
Esposito, Enzo, 227
Eugenico, Nicolò, 118, 119, 147, 223
expurgation, *see* censorship

Fabrizio-Costa, Silvia, 202, 219, 228
Fahy, Conor, 189, 191, 195, 196, 227
Falaschi, Enid, 223
Falconetto, 80–1
Farri, Domenico, 145
Fascitello, Onorato, 219
Fatini, Giuseppe, 217, 225
Fausto, Domenico Tullio, da Longiano, 97, 118, 119, 134, 147, 217–18
Fausto, Sebastiano, da Longiano, 25, 93–4, 107, 151, 221
Fava, M., 190
Favati, Guido, 215
Favretti, E., 193
Federici, Domenico Maria, 195, 197
Federigo da Montefeltro, 213
Fedi, Roberto, 222
Ferrari, Mirella, 207
Ferroni, G., 224
Filelfo, Francesco, 32, 34–5, 44, 55
Filippo da Lavagna, 17, 195
Filippo di Pietro, 39
Fino, Pietro, *see* Pietro da Fino
Fiore di virtù, 44, 193
Fiorelli, Piero, 220
Firenzuola, Agnolo, 132, 138, 144
Flaminio, Marco Antonio, 108
Florence
　output of presses, 42, 127

population, 42, 201
Studio of, 42
Foà, S., 207
Foffano, T., 197
Folena, Gianfranco, 193, 194, 204, 207, 213, 225, 226, 230
foliation, 36, 199
folio, *see* formats
Fontanini, Giusto, 205
Fontes-Baratto, Anna, 226
formats
    duodecimo, 114, 145, 148
    folio, 36, 55, 59, 121, 130, 144, 151, 152, 175, 183, 208
    octavo, 48, 58, 59, 61, 82, 96, 97, 124, 148, 151, 175, 218
    quarto, 55, 96, 97, 100, 101, 105, 114, 121, 148, 151, 154, 218
    sixteens, 100, 154
    twenty-fours, 57, 58, 66
Fórnari, Simon, 119, 134, 147, 148
Fortunio, Gian Francesco, xii, 58, 66–71, 71, 80, 84, 90, 92, 207, 211, 212, 213, 214
Foscolo, Ugo, 170–1
Fowler, Mary, 92, 189, 195, 198, 205, 207, 215
Francesco da Montefeltro, fra, 196
Francesco de' Medici, Grand Duke of Florence, 155, 168
Francesco di Dino, 105, 220
Francesio, Giovan de, 107
Francini, Antonio, 86, 127, 129
Franciscans, 7, 21, 28–30, 38, 39
Franco, Niccolò, 96
Frasso, Giuseppe, 190, 192, 197, 199, 202, 208, 219, 221, 229
Frati, C., 208
Fregoso, Antonio Fileremo, 71
Frey, K., 191, 225
Frova, Carla, 194
Fulin, R., 190, 192, 195

Gabriele, Trifon, 101, 145
Gaetano di Thiene, 16
Gaetano, Tizzone, 7, 8, 25, 60, 73, 75–8, 86, 87, 184, 208, 212, 218, 222
Gamucci, Bernardo, 153
Ganda, Arnaldo, 190
Garanta, Nicolò, 61, 74, 208
Gargan, Luciano, 196
Garofalo, Giacomo, 148
Garzoni, Tomaso, 10, 191
Gaskell, Philip, 196
Gelli, Giambattista, 127
Gentile, Salvatore, 200
Gerber, Adolph, 221, 224, 225
Gesualdo, Giovanni Andrea, 94–5, 114, 146, 216–17

Gherardo, Paolo, 108
Ghinassi, Ghino, 190, 191, 208, 209, 211, 212
Giacomo da Lentini, 87
Giambullari, Pierfrancesco, 105, 127, 133
Giannotti, Donato, 127
Gigli, Ottavio, 224, 230
Giglio, Domenico, 228
Giglio, Giacomo, 14, 192
Giocondo, fra Giovanni, 15, 17, 192
Giolito, Gabriele, 3, 8, 10, 14, 25, 90, 96–8, 100, 101, 106, 107, 108, 109–18, 121–2, 124, 128, 134, 140, 145, 146, 147, 149, 150–1, 153–4, 192, 218, 222, 223, 224, 226, 228
Giolito, Giovanni, the elder, 8, 9, 90, 96, 98, 190
Giolito, Giovanni, the younger, 140
Giolito, Giovanni Paolo, 140
Giovanni, ser, 138, 144
Giovanni di Dio, 29
Giovanni Antonio da Padova, fra, 196
Giovanni Francesco da Venezia, fra, 196
Giovio, Paolo, 13, 133, 142, 191, 224, 227, 228
Giunta family (Giunti), presses of, x, 5, 58, 79, 81–9, 107, 110, 125, 127, 131, 136, 147, 155, 165, 175, 178, 179, 180, 184, 220, 225
Giunta, Benedetto, 133
Giunta, Bernardo, 17, 72, 82–4, 86, 127, 128, 132, 133, 135, 142, 155, 189, 214
Giunta, Filippo, the elder, 4, 7, 54, 57, 79, 81–3, 213, 214
Giunta, Filippo, the younger, 17, 128, 138, 156, 159, 161, 168, 175, 176, 177, 179, 183
Giunta, Iacopo, 17, 128, 138, 159, 168, 175, 176, 177, 179
Giunta, Lucantonio, 29, 59, 82, 127, 129, 213
Giunta, Tommaso, 128
Giuntini, Francesco, 146, 179
glossaries, xii, 3, 45, 68, 97, 98, 100, 107, 111, 116, 118, 119, 144, 147, 148, 160, 183; *see also* regional Italian forms
Gobbi, Orazio de', 148
Gonzaga, Federigo, 7
Gonzaga, Pirro, 7
Goselini, Giuliano, 18
Grafton, Anthony, 194, 208, 225
Graziano da Brescia, fra, 196
Grazzini, Anton Francesco, 128, 132, 135, 136–7, 139, 142, 156, 223, 226, 227, 230
Grazzini, Giovanni, 227
Grendler, Paul F., 140, 190, 216, 217, 221, 222, 223, 224, 226, 227, 229
Griffio, Giovanni, 4, 10, 12, 90, 116, 146, 221, 222
Griffo, Francesco, 11, 57–8, 204

Grimani, Giovanni, 185
Grohovaz, V., 218
Grolier, Jean, 124
Groto, Luigi, 143–4, 146, 227
Guadagnoni, Piero Donato, 9
Gualteruzzi, Carlo, 62, 88, 144, 159–61, 168, 208
Guarnieri, Francesco, 191
Guazzo, Marco, 7, 73, 96, 211
Guerra, Domenico, 148, 176
Guerra, Giovanni Battista, 148, 176
Guicciardini, Agnolo di Girolamo, 151, 175, 233
Guicciardini, Francesco, 151, 156, 168, 174–5, 230
Guicciardini, Lodovico, 151
Guicciardini, Paolo, 229, 233
Guidi, José, 224, 227
Guido de Cauliaco, 40
Guidotti, Ascanio, 8
Guinizelli, Guido, 87
Guittone d'Arezzo, 19, 87–8, 165

h, use of, 50, 68, 113, 116, 117, 120, 133, 156, 169, 171, 202, 205, 216, 223, 225
Harris, Neil, 208, 210, 211, 219, 228, 229
Hellinga, Lotte, 189, 194, 196
Henri, dauphin of France, 97, 218
Herbort, Johannes, 16
Herodotus, 5
Hesychius, 22
Hinman, Charlton, 196
Hirsch, Rudolf, 190, 200, 201
Holy Office, 140–1, 227
Homer, x, 97, 119, 134
Horace, 48, 51
Hystoria del mondo falace, 41

Iacopo da Varazze, 2
Ilicino, Bernardo, 34–6, 55, 56
illustrations, 3
  engraved, 44, 148
  woodcut, 32, 36, 58, 59, 82, 103, 104, 105, 119, 134, 208; maps, 77–8
imprimaturs, 109, 140
indexes, 2–3, 15, 29, 32, 36, 38, 39, 45, 95, 97, 99, 104, 105, 115, 117, 124, 125, 133, 134, 142, 147, 151, 183, 199, 218, 221, 226
Indexes of Prohibited Books, 141, 142, 143, 149, 155; *see also* Congregation of the Index
Inquisition, 29, 140–1, 142, 143, 149, 150, 155, 156, 159, 164, 165, 172, 178, 180, 181, 187
Interiano, Giorgio, 2, 59
Ioannes Florentinus, 41, 201
Istoria di Eneas, 20, 193

Istorie pistolesi, 165
italic, *see* typefaces

Jacopone da Todi, 45–7, 178
Javitch, Daniel, 149, 192, 217, 218, 223, 228
Jenson, Nicholas, xi, 5, 7, 10, 22, 26, 28, 29, 194, 196
John of Cologne, 10
John of Speyer, 7
Justinian, 130, 168, 177

k, use of, 62, 160, 162, 205
Kenney, E. J., x, 190, 192, 194, 195, 222
koinè, 30, 65, 217; *see also* 'common' language

La Brasca, Frank, 202, 219, 228
Laelius, C. Lucius, 37–8
Landi, A., 218
Landino, Cristoforo
  as author and translator, 24, 106–7, 151, 221
  as editor and commentator, 38, 39, 43–4, 52–3, 58, 59, 82, 92, 102, 117, 144, 145, 199, 204, 205, 219
Lascaris, Constantine, 22
Latini, Brunetto, 167
Latinisms
  lexical: eliminated, 40, 65, 72, 123, 132, 211, 221; introduced, 20, 65, 81, 219
  orthographic and phonetic, 63, 65, 68, 113; eliminated, 33, 39, 50, 65, 67, 71, 72, 74, 75, 77, 97, 123, 153, 169, 200, 201, 205, 207, 210, 211, 212, 213, 214, 221, 224, 225; introduced, 38, 39, 71, 82, 197, 201, 202, 210, 213; restored, 118; unchanged, 40, 50, 70, 73, 74, 208, 209, 210, 213; *see also h*
Laurario, Bartolomeo, 70, 210
Laurario, Castorio, 7, 12, 60, 66, 70, 76, 92, 183
Laurent, M. H., 195, 196
Lavezuola, Alberto, 148–9
lectio difficilior, 46, 158
lectio facilior, 32, 157, 158, 159, 174
Lee, R. W., 234
Lenhart, J. M., 200
Lentzen, Manfred, 202
Lenzoni, Carlo, 127, 133
Leo X, Pope, 60, 213
Lepschy, Anna Laura, 201, 202
Le Rouge (Rubeus), Jacques, 5, 16, 35
Lesca, G., 231
Le Voirrier, Pierre, 168
lexis
  changes: to archaic or obscure forms, 41, 44, 64, 65, 69, 77, 98, 112, 113–14, 156, 161–3, 169, 182; to everyday or popular forms, 52, 69, 80, 112, 132, 135, 200,

211; to northern forms, 40, 44, 104, 202; to southern forms, 40, 211; *tornare* replaced, 76
   strengthening of, 40, 65, 75, 80
   see also Latinisms
Libri, Bartolomeo de', 44, 214
Libri, Sigismondo de', 34
Libro de la destructione de Troya, 193
Liburnio, Nicolò, 13, 91–3, 98, 100, 216
Livy, x, 16, 65, 116, 192
Locatelli, Boneto, 5
Lodovico da Canossa, 63
Löslein, Peter, 9, 189
Longinus, 110
Longo, Nicola, 227
Lorio, Lorenzo, 61, 208
Lotti, Lorenzetto, 185
Lowry, Martin, x–xi, 21, 190, 191, 192, 194, 195, 196, 199, 202, 214
Luca da Suvereto, fra, 196
Lucian, 190
lui, lei, loro as subject pronouns, *see* syntax
Luiso, Francesco Paolo, 226, 227

Machiavelli, Bernardo, 16, 192
Machiavelli, Niccolò, 53, 76, 107, 123–4, 132, 137, 141, 224, 225, 233
Madricardo, Silvia, 230, 231
Mainardi, Arlotto, *see* Arlotto
Maler, Bernhard, 9, 189
Malermi (Malerbi), Nicolò, 2, 28, 189, 197
Malispini, Ricordano, 177–8
Malpigli, Annibale, 34
Mambelli, Giuliano, 228
Manacorda, Guido, 224
Mandeville, John, 41, 202
Manilio, Sebastiano, 8, 9, 65–6, 74, 75, 76, 107, 112
Manilius, 25, 195
Mannelli, Francesco, 83, 158, 162–3, 167, 168–70
Manrique, fra Tommaso, 161–2, 164–5, 171, 231
Mantese, Giovanni, 190
manuscripts, transmission of texts in, 19–21
Manuzio, Aldo, the elder, x, xi, 7, 9, 13, 17, 21, 22, 23, 29, 48–9, 59, 60–1, 66, 83, 85, 94, 103, 204, 213
   as editor, 2, 5, 56–9, 77, 95, 101, 116
Manuzio, Aldo, the younger, 142, 150, 228, 229
Manuzio, Antonio, 108, 221
Manuzio, Paolo, 90, 95, 98, 103, 108, 109, 110, 113, 114, 118, 124, 128, 129–30, 142, 145, 221, 230
Manzani, Domenico, 173
Manzolo, Michele, 197

Marcolini, Francesco, 90, 102, 107
Mardersteig, Giovanni, 190, 192
margins, use of, 142, 183
   for annotations, 107, 113, 117, 139, 143, 145, 154, 157, 171, 172, 180, 204
   for commentaries, 36
   for corrections and notes by reader, 25, 204
   for cross-references, 37
   for *notabilia*, 38–9, 59, 125, 151, 221, 222, 229
   for variants, 25, 65, 111, 112, 113, 117, 174, 227
Marino, fra, 29, 196
Mariotto da Pistoia, fra, 196
Marsili-Libelli, C. Ricottini, 226
Marsilio, Pre', 3, 8, 10, 11, 15, 24, 55–6, 190
Martelli, Lodovico, 133, 207
Martelli, Ugolino, 129
Martinelli, Lucia Cesarini, 225, 229
Martini, Luca, 128, 137, 225
Masi, Giorgio, 191, 224
Massetti, Niccolò, 6, 200
Master of the Sacred Palace, 141, 142, 150, 161, 165, 168
Masuccio Salernitano (Tommaso Guardati), 39–40, 75, 144, 200, 209
Maufer, Pierre, 9, 16
Mauro, Giovanni, 135
Mayr, Sigismondo, 59
Mazzanti, fra Pietro, 38, 59, 199
Mazzarello, Lazaro, 29
Medici, family, 127, 135, 180
Medici, Alessandro de', *see* Alessandro de' Medici
Medici, Caterina de', 97, 110
Medici, Cosimo de', *see* Cosimo I de' Medici
Medici, Francesco de', *see* Francesco de' Medici
Medici, Lorenzo de' ('il Magnifico'), 108, 136, 177, 213
Medici, Lorenzo de', Duke of Urbino, 213
Medici, Lucrezia di Lorenzo de', 79
Melchiori, Francesco, 18
Menato, Marco, 216
Menegazzo, Emilio, 192
Meneghin, Vittorino, 189
Mengaldo, Pier Vincenzo, 209, 229
Merato, Bernardino, 98, 118, 218
Mercati, Giovanni, 191
Merula, Giorgio, 22, 28, 192
Mestica, Giovanni, 202, 203, 212
metaphony, *see* phonology
metre, 30, 167, 197
   revisions concerning, 20, 33, 44–6, 50–2, 71, 74, 76, 81, 85, 88, 104, 106, 135, 137, 179, 198, 200, 202, 205, 213, 214, 226, 234; *see also* apocope, diaeresis, dialoephe, elision, synaloephe

Mezzabarba, Antonio Isidoro, 62, 208
Miglio, Massimo, 194, 195
Migliorini, Bruno, 212, 216
Minerbi, Lucilio, 98, 218
Minturno, Antonio, 13
Minuziano, Alessandro, 189
Miscomini, Antonio, 28
Mocenigo, family, 18, 190
Mocenigo, Iacopo, 191
Molino, Girolamo, 219
Molza, Francesco Maria, 108
Mombrizio, Bonino, 31, 197
Monfasini, John, 191
Monforte, Vincenzo, 228
Monte de la oratione, 40–1
Montemurlo, battle of, 127
moral glosses and *allegorie*, 66, 97, 112, 117, 118, 144, 146, 148, 149, 172
Morato, *see* Moreto
Mordenti, Raul, 231, 232
Morel, Fédéric, 168, 232
Moreni, Domenico, 224
Moreto (Morato), Pellegrino, 92–3
Moretti, Matteo, 17
Moretto, Antonio, 9
Morgiani, Lorenzo, 45
Moro, Giacomo, 221, 223, 225, 227
morphology
 articulated prepositions, 26, 115, 120, 121, 122, 153, 208, 210, 211
 definite articles, 63, 70, 71, 75, 77, 104, 117, 118, 119–21, 123, 124, 200, 201, 205, 213, 214, 220, 221, 225
 noun and adjective plurals, 70, 71, 85, 121, 166, 207, 210, 213, 214, 216, 220, 221, 226
 popular Florentine forms: eliminated, 80, 105, 107, 124, 136, 137, 211, 213, 214, 220, 221, 225, 226; introduced, 135, 201, 213, 214; restored, 106; unchanged, 70–1, 210
 possessives, 211, 225
 pronouns, 70, 104, 116, 118, 123, 208, 213, 214, 216, 220, 226, 229
 verb forms: present indicative, 63, 73, 104, 120, 137, 153, 220; imperfect indicative, 166, 208, 221, 232; past historic, 104, 119, 120, 123, 166, 175, 200, 208, 211, 214, 216, 220, 221; future, 104, 213, 220, 221; conditional, 200, 208, 211; subjunctive, 76, 123, 135, 175, 205, 207, 212, 213, 216, 220, 221, 225, 226, 229
Mortimer, Ruth, 205, 218
Motta, E., 192
Musuro, Marco, 22
Mutini, C., 193, 205, 228

Nalli (Dinali), Stefano and Bernardino di, 9
Nannini, fra Remigio, 125–6, 138–9, 150–1, 154, 157–8, 175, 227, 229, 230, 231, 233
Nardi, Bruno, 197
Nardi, Iacopo, 108, 127, 178, 221
Nausea, Fredericus, 195
Navagero, Andrea, 5, 129
Navò, Curzio, 107, 108, 135, 218
Nencioni, Giovanni, 225
Nesi, Emilia, 201
Niccolai, Francesco, 225, 229
Niccolò da Tobia, fra, 43
Niccolò di Lorenzo, 43
Nicetas Choniates, 151
Nicolini, press, 90, 94, 101
Nicolini, Domenico, 191
Nicolini, Giovanni Antonio, 89, 104
Nidobeato (Nibia), Martino Paolo, 37, 43, 199
Nievo, Alessandro, 10, 11, 28, 190
Nigro, Joannes Dominicus, 190
Noakes, Susan, 195
Nolhac, Pierre de, 192
notes, *see* annotations
Novellino, 62, 88, 111, 144, 159–61, 163, 166, 189, 228, 230–1
Nuovo, Angela, 191, 195, 207, 209

octavo, *see* formats
Ognibene da Lonigo (Leoniceno), 28
Olimpo degli Alessandri, Caio Baldassarre, da Sassoferrato, 74
Ong, Walter, 25
Orologgi, Giuseppe, 147
orthography, *see* spelling
Ossola, Carlo, 224, 227
Ovid, 97, 119, 147

Pacini, Bernardo, 70, 79–81, 212
Pacini, Piero, 79
Padua, University of, 7, 28, 42, 65, 66, 75
Paganini, Alessandro, 24, 57, 58, 66, 207
Paganini, Girolamo, 197
Paganini, Paganino, 58, 195
Paganucci, Bartolomeo, 108
Pallavicino, Giuseppe, 2, 189
Panciera, fra Ugo, 45
Panella, Antonio, 229
Panigada, Costantino, 233
Pannartz, Arnold, 15
Panofsky, Erwin, 234
Panvinio, Onofrio, 230
Parabosco, Girolamo, 144
paraphrase, 75, 77, 80, 153, 158, 179, 180, 212, 234
Parent, Annie, 193
Parmegiano, Lanfranco, 116
Paruta, Giovanni Giacomo, 147

Pasini, Maffeo, 74, 93, 95, 97, 99, 104, 112, 113, 134, 211, 218
Pasquilini, Eugenio, 211
Passano, Giovanni Battista, 228
Passavanti, Iacopo, 172
Paul III, Pope, 140
Paul IV, Pope, 141, 142, 155
Penzio, *see also* Pincio
Penzio, Girolamo, 75
Penzio, Iacopo, 75
Peranzone, Nicolò, 1, 8, 11, 33-4, 36, 48, 55, 198
Peri, Lorenzo, 136, 226
Perini, Leandro, 189
Perotti, Niccolò, 13, 17, 22, 23, 191
Perry, M., 234
Pertile, Lino, 207, 228
Pesenti, Tiziana, 195, 208
Pettas, William A., 192, 214
Petrarch (Petrarca), Francesco, xi, 20, 52, 59, 69, 86, 90, 96, 113, 124, 125, 127, 156, 165, 186, 212, 217
  as editor of Livy, x
  biography of, 34, 36, 55, 77, 115, 116, 199
  Canzoniere and *Trionfi*: appendices of *rime disperse* etc., 54-5, 56-8, 61, 78, 85, 93, 94, 101, 115, 116; commentaries, *see* Antonio da Tempo, Daniello, Sebastiano Fausto, Filelfo, Gesualdo, Ilicino, Squarzafico, Vellutello; division and ordering, 34, 49, 51-2, 53-6, 58, 78, 85, 93, 94, 101, 114, 115, 116, 207, 214; editions, 1, 3, 4, 8, 9, 10, 11, 15, 24-7, 29, 32-6, 39, 44, 48-59, 61, 70, 77-8, 82, 83, 85, 93-5, 97, 98, 100-2, 103, 111, 114-17, 118, 144, 145-6, 173, 184, 186, 195, 197-9, 202-4, 206-7, 210, 214, 216-17, 218-19, 228; Fortunio and text of, 67
  De viris illustribus, editions of, 9
  De vita solitaria, editions of, 9
  imitation of, 92-3, 101, 119, 135, 167, 183
  Secretum, editions of, 70
Petri, Johann, 45
Petrocchi, Giorgio, 174, 194, 200, 204, 225
Philip II of Spain, 13
phonology
  northern forms: eliminated, 65, 81, 104, 201, 211, 213, 214, 219; introduced, 40, 65, 73, 200, 201; unchanged, 40, 41, 70, 72-4, 200, 201, 209, 210, 211, 213
  popular Florentine forms: eliminated, 80, 137, 210, 212, 221; introduced, 135, 214; unchanged, 210
  southern forms: eliminated, 200, 207; unchanged, 40, 73, 74, 84, 207
  vowels: diphthongs, 63, 65, 73, 74, 76, 84, 97, 120, 124, 202, 205, 209, 210, 212, 213, 214, 216, 226; final, 63, 74, 205; metaphony, 46, 76; pretonic *ar / er*, 71, 84, 213, 214; pretonic *i / e*, 33, 50, 113, 200, 213; stressed, 74; syncope, 207, 208
  consonants: single / double, 25, 66, 72-4, 97, 116, 124, 170, 175, 200, 201, 202, 207, 208, 210, 213, 214, 216
  see also apocope, elision, Latinisms
Piasi, Piero de, 33, 35, 56
Picone, Michelangelo, 215
Pierozzi, Antonio (St Antoninus), 43, 44
Pietrasanta, Plinio, 9, 109, 116, 122, 123, 227
Pietro Angelo di Montolmo, fra, 29
Pietro da Fino, 145
Pietro da San Cancian, fra, 196
Pigna, Giovan Battista, 119
Pillinini, Stefano, 192, 202, 203, 204
Pincio, *see also* Penzio
Pincio, Aurelio, 98, 107
Piovan, Francesco, 225
Pirotti, Umberto, 225
Pisa, Studio of, 42, 127
Piscini, A., 218
Pius II, Pope, 221
Plaisance, Michel, 156, 224, 225, 226, 229, 230, 232
Plautus, 79
Pliny the Elder, 22, 66, 106
Pliny the Younger, 21, 189
Plutarch, 31
Pocatela, *see* Burgofranco
Poggiali, Cristoforo, 191, 218
Polica, Sante, 220
poligrafi, 90-1
Poliziano, Angelo
  as author, 69, 70, 76, 135
  as scholar, x, 22-3, 46-7, 53, 116, 128, 129, 181, 194
Polletti, Giovanni Nicolò, 16
Pontano, Giovanni, 191
Poppi, A., 196
Porcacchi, Tomaso, 8, 9, 147-8, 150-4, 184, 226, 229
Porro, Antonietta, 225
Porro, Girolamo, 148
Porta, Giuseppe, 175, 226-7, 233
Portonari, Francesco, 125
portraiture, analogies with, 162, 185-6, 187
Pozone, Angela, 199
Pozzi (Puteus), Bartolomeo, 10
Pozzi, Mario, 157, 230, 231, 232
Price, Hereward T., 196
print culture, characteristics of, xi
printers
  avarice of, 3, 11-12, 24
  concern for accuracy, 6

concern for being up to date, 4
see also editors
privileges, *see* book-privileges
Procaccioli, Paolo, 202, 204, 219
proof-correction, 10–11, 24, 26, 129, 196; *see also* 'correctors'
proverbs, explanations and lists of, 98, 99, 100, 105, 106, 111, 115, 134
Pulci, Giovanni, 105–6, 220
Pulci, Luca, 179, 220
Pulci, Luigi, 6, 104–6, 179–80, 200, 202, 220
punctuation, added or revised by editors, x, 1, 2, 14, 26, 32, 51, 56, 59, 62, 63, 71, 74, 87, 95, 112, 116, 120, 123, 131, 138, 139, 150, 156, 160, 167, 177–8, 180, 195, 204, 214, 221, 223
see also apostrophe
Pusterla, Giovanni Francesco, 167
Puteo, Paris de, *see* Del Pozzo
Puteus, *see* Pozzi

Quaglio, Antonio Enzo, 65, 197, 209, 210, 211, 212, 214, 222
Quaquarelli, Leonardo, 232
Quarengi, Piero, 29, 38
Quarta, Nino, 199, 202
quarto, *see* formats
Quétif, J., 224
Quintilian, 67, 185
Quondam, Amedeo, 189, 190, 200, 221, 222, 227

Rabitti, Giovanna, 193
'Raccolta aragonese', x, 46–7, 87
'Raccolta Bartoliniana', 89, 216
Raimondi, Ezio, 199, 210, 219
Rajna, Pio, 193, 232
Ramberti, Benedetto, 108, 221
Rampazetto, Francesco, 117, 142, 146
Ramusio, Giovambattista, 225
Raphael (Raffaello Sanzio), 186
Rapirio, Giovita, 222
Ratdolt, Erhard, 9, 29, 189
Ravani, Pietro, 74, 114
Ravegnani, Giuseppe, 217, 224, 227, 228
Razzi, Silvano, 179, 184, 229
regional or dialectal Italian forms
northern, 33, 38, 59, 65, 81
southern, 40, 59–60
Umbrian, 45–6
used in annotations and glossaries, 94, 97, 98, 99, 106, 107, 111, 114, 116, 144, 145, 154, 164, 183, 217, 218, 220, 221, 222, 228, 229
see also lexis, morphology, phonology
regulation of printing and editing by state, 5, 12, 13, 74–5; *see also* book-privileges

Renouard, Antoine Auguste, 228
Reynolds, L. D., 194, 196
Rhodes, Dennis E., 201, 208, 212, 226, 234
rhyme, 46, 81, 88, 114, 115, 119, 137, 209, 211, 219
rhyming dictionaries (*rimari*), 92, 93, 97, 100, 115, 116, 134, 145, 147
Ricasoli, Braccio, 161
Ricci, Giuliano de', 176, 233
Riccio, Angelo, 18
Richardson, Brian, 195, 202, 207, 208, 209, 211, 223
Ridolfi, Lucantonio, 145, 146, 228
Ridolfi, Roberto, 213, 229, 233
Rigoni, Erice, 192
rimari, *see* rhyming dictionaries
Rizzo, Silvia, 189, 208
Roaf, Christina, 207, 222, 223
Robortello, Francesco, 109–10, 126
Rocchi, Ivonne, 209
Roddewig, Marcella, 205
Rogledi Manni, Teresa, 189, 192, 201
roman, *see* typefaces
Romei, G., 217
Ronchi, G., 224
Rosello, Lucio Paolo, 7, 8, 75, 78, 86, 184
Rossi, Bastiano de', 173–4, 181, 187
Rossi, Nicolò de', 193
Rossi Pinelli, O., 234
Rosso, Giovanni, da Vercelli, 28, 59–60
Rota, Berardino, 193
Rota, Giovan Battista, 148
Rotondò, Antonio, 227, 229, 231
Rouillé, Guillaume, 145, 146
Rüdiger, Wilhelm, 225
Ruffo, Giordano, 29
Rufino, fra, 196
Ruscelli, Girolamo, 140, 191, 223, 227
  as author, 12, 13, 119, 223
  as editor, 2, 4, 8, 9, 10, 12, 13–15, 109, 112–14, 116–24, 126, 139, 143, 146, 147–8, 152, 156, 162, 169, 172, 181, 184–5, 187, 192, 223, 224, 230, 231
Rusconi, Giorgio, 7, 64, 70–2, 90, 104, 210
Rutilius Taurus, 10

Sabbadini, Remigio, 194, 209
Sabellico, Marcantonio, 28
Sacchetti, Franco, 166
Salve Regina, 41, 201
Salviati, Iacopo, 79
Salviati, Lionardo, 7, 17, 168–75, 178, 181, 187, 230, 231, 232, 233
Salviati, Pietro, 79, 213
Sambin, Paolo, 203
Samuels, R., 224, 225
Sandal, Ennio, 201, 210

Sanguinacci, Iacopo, 65, 209
Sannazaro, Iacopo, 19, 57, 59–60, 73, 84, 90, 92, 117–18, 228
Sansone, G. E., 197
Sansovino, Francesco, 2, 7, 8, 110–12, 114, 117–18, 144–5, 146, 148, 149, 150–2, 159–60, 189, 222, 223, 224, 228, 231
Sansovino, Iacopo, 8, 110
Santasofia, family, 49
Sant'Iacopo di Ripoli, press, 7, 43, 44, 201
Santoro, C., 192
Sanudo, Marino, 190
Sapori, Giuliana, 233
Sarti, Alessandro, 70, 210
Sartori, Antonio, 190, 192
Saviotto, Anne, 224
Savonarola, Girolamo, 106, 220
Scala, Lorenzo, 132, 135, 225
Scapecchi, Piero, 195
Scardeone, B., 198
Scholderer, Victor, 28, 189, 196, 200, 201
Schutte, Anne Jacobson, 193, 201, 221
Scinzenzeler, Ulrich, 3, 33
Sclarici dal Gambero, Tommaso, 11, 57–8, 191
Scoto, press, 90
Scoto, Brandino, 106
Scoto, Girolamo, 104–5
Scoto, Ottaviano, 106
scribes
   as editors, 8, 43, 190
   methods of, 156, 158–61, 163, 166
   see also manuscripts
scrittura, 14, 62, 64, 112, 123, 161, 169, 170, 232
Securo, fra Francesco, 196
Segni, Bardo, 86–9, 186
Segre, Cesare, 224
semigothic, *see* typefaces
Seneca, 9, 66, 107
Senni, fra Francesco, 29
Serdonati, Francesco, 178
series, *see* collana
Sermartelli, Bartolomeo, 172, 173, 178–80
Servius, 42, 95
Sessa, press, 90, 144, 145, 191
Sessa, Giovanni (Zuan) Battista, the elder, 80
Sessa, Giovanni Battista, the younger, 12
Sessa, Melchiorre, the elder, 74
Sessa, Melchiorre, the younger, 4, 7, 12
Severo, Girolamo, 74
Sforza, Lodovico, 229
Sicilian school of poetry, 19
Sigonio, Carlo, 116
Siliprandi, Domenico, 35
Siliprandi, Gaspare, 35
Simone (Fidati) da Cascia, 76

Simonetta, Giovanni, 107, 151
simplicity of language, 125, 161, 169, 178, 179
Simpson, Percy, 196
Sixtus IV, Pope, 15
Smith, Margaret M., 199
Soardi, Lazaro de', 55
Soldati, Paolo, 232
Solerti, Angelo, 199, 212
Soncino, Girolamo, 53–4, 57
Sophocles, 155, 189
Sorbelli, Albano, 191, 192
Sorrentino, Andrea, 227, 228, 231, 232, 234
Spadari, Giorgio de, 26
spelling, relation to pronunciation, 61, 87, 116, 169–71, 214, 232; *see also* h, k, Latinisms
Speroni, Sperone, 100, 108, 175, 223, 233
Spini, Giorgio, 218
Squarzafico, Gerolamo, 8, 9, 10, 15, 16, 24, 31–2, 35, 43, 55, 75, 76, 77, 92, 192, 197, 214, 221
Stadter, P. A., 201
Stagnino, Bernardino, 3, 10, 15, 55, 59, 90
Stampa, Gaspara, 110–11
Statius, 134, 147, 148, 228
statues, analogies with cleaning and restoration of, 87, 132, 139, 176, 185, 234
Stefani, Ottavio degli, 24, 57
Stefano da Capua, fra, 28
Stoppelli, Pasquale, 209
Storie pistolesi, see Istorie pistolesi
Stramazzo da Perugia, 78
Sturel, René, 207
Stussi, Alfredo, 207
summaries of cantos, books etc. (*argomenti*), ix, 37, 104, 105, 106, 115, 117, 119, 125, 142, 144–9, 151, 153, 179, 180, 224
Summonte, Pietro, 59, 191
Sweynheim, Konrad, 15
synaloephe, 50, 88, 135, 204, 216, 220
syntax
   changes to, 40, 65, 71, 75, 77, 80, 81, 104, 137, 200, 205, 207, 212, 213, 219; *see also* paraphrase
   lui, lei, loro as subject pronouns, 67, 68, 94, 123, 157, 180, 211, 213, 220, 221, 225

Tacitus, 10, 25, 62
Taddei, Giovanni, 225
Tambara, G., 224, 228
Tanzio, Francesco, 3, 10, 11, 27, 33, 54, 198
Tapella, Claudia, 231, 232
Tasso, Bernardo
   as author, 14
   and editors, 2, 13–15, 109, 191–2

Tasso, Torquato
  as author, 150, 229
  and editors, 13–15
Tebaldeo, Antonio, 73
Tedeschi, John A., 229
Teo, Bartolomeo, 39, 200
Teofilo da Cremona, fra, 196
Terence, 35, 53, 79
Terzago, Guido, 37
Thomas Aquinas, St, 196
Thorndike, Lynn, 198
Timpanaro, Sebastiano, 27, 194, 196
Tiraboschi, Cristoforo, 208
Tolomei, Claudio, 108
Tomasoli, Nicolò, 34
Torelli, Francesco, 130–1
Torelli, Lelio, 130, 155, 168, 177, 181, 225
Tornaquinci, family, 233
Tornaquinci, Benedetto, 175
Torrentino, Lorenzo, 128, 130–6, 138, 155, 175
Torresani, Andrea, 61, 90
Torresani, Francesco, da Asola (Asolano), 25, 61, 124
Tortelli, Giovanni, 22
Torti, Alvise, 96, 97, 104
Torti, Battista, 40, 195
Tournes, Jean de, 145, 167
Tramezzino, Michele, 90, 122, 126
Trecius, *see* Albignani
Trent, Council of, 141; *see also* Indexes of Prohibited Books
'Tre Savii sopra eresia', 140
Triclinius, x
Trinkaus, Charles, 201, 202
Trissino, Gian Giorgio, 76, 87, 215
Tromba, Francesco, 211
Trombetta, fra Antonio, 196
Trovato, Paolo, 32, 58, 109, 189, 190, 192, 195–203 *passim*, 205, 207–28 *passim*, 231, 232
truncation, *see* apocope
Tschiesche, Jacqueline, 228
Tucci, Mariano, 17
Turchi, fra Francesco, 149, 154, 228
typefaces
  gothic or semigothic, 55, 71, 96, 97
  italic, 48, 55, 57, 59, 82, 97, 100, 154, 164
  roman, 59, 71, 96, 97, 100, 154, 164, 214

Uberti, Fazio degli, 111
Ulivi, Pietro, 133–4, 147, 149, 183, 226
Ullman, B. L., 201
Usher, Jonathan, 232
Usuardus, 43

Vaccari, A., 197

Valdarfer, Cristoforo, 32
Valenti, T., 190
Valerio (Valier), Giovan Francesco, 63, 191, 208
Valgrisi, Vincenzo, 10, 90, 118, 143, 146, 147, 223
Valla, Lorenzo, 5
Valori, Baccio, 175
Valori, Niccolò, 177
Valvassori, Clemente, 118, 147, 223
Valvassori, Giovanni Andrea, 90, 118, 141, 147–8
Vandelli, Giuseppe, 225
Varasi, Elena, 212
Varchi, Benedetto, 127–30, 131, 133, 134, 173, 219, 225
Varisco, Giovanni, 147, 153
Varisco, Paolo, 9, 40
Vasari, Giorgio, 13, 133, 156, 185, 186
Vattasso, Marco, 202, 203
Vedova, G., 210
Vellutello, Alessandro, 7, 8, 9, 77–8, 92, 93, 94, 101, 102–3, 115, 116, 117, 144, 145, 217, 219, 223
Veneziani, Paolo, 210
Venice
  output of presses, 28, 39, 42, 109, 140, 200
  population, 39, 42, 200
  Scuola di San Marco, 28
Verdezotti, Giovanni Mario, 147
Vernarecci, A., 190
Vernia, Nicoletto, 16, 28, 192
Vespucci, Giorgio Antonio, 43, 201
Vettori, Piero, x, 86, 103, 116, 128–31, 139, 155, 156, 161, 162, 177, 181, 225, 230, 231
Viani, Bernardino, 106
Villani, Filippo, 176
Villani, Giovanni, 99, 111, 125, 137–9, 157–8, 175–6, 177, 228
Villani, Matteo, 137–8, 156–7, 176–7
Virgil, 9, 35, 48, 52, 95, 97, 119, 147
Virgili, A., 226
Virtutes psalmorum, 41, 201
Visconti, Gasparo, 34
Vitali, Bernardino, 75, 90, 98
vocabulary, *see* lexis
vowels, *see* Latinisms, phonology

Watt, Ian, 182
Weaver, Elissa, 219
Weiss, Roberto, 202, 203, 204, 219
Wilkins, E. H., 32, 197–8, 199
Williamson, Edward, 192
Wilson, N. G., 194, 196
Windelin of Speyer, ix, 31, 192, 198
woodcuts, *see* illustrations
Woodhouse, H. F., 228

Woodhouse, John R., 156, 234
word division, 25, 26, 51, 100, 167, 170, 177, 178, 205, 212, 216; *see also* apostrophe, elision
word lists, *see* glossaries

Zanato, Tiziano, 226
Zancani, Diego, 210, 224
Zane, Iacopo, 18
Zanetti, Bartolomeo, 98, 99
Zanni, Agostino, 55, 90, 208
Zanni, Bartolomeo, 1, 33, 59, 65, 90, 208, 209
Zappella, Giuseppina, 189
Zarotto, Antonio, 198
Zenaro, Damiano, 227
Zeuxis, 60, 185–6
Ziletti, Giordano, 224
Ziletti, Innocente, 9
Zilius, 65, 209
Zonta, Gasparo, 196
Zoppino, Nicolò d'Aristotile, *known as*, 6, 58, 70–3, 79, 90, 96, 104, 210
Zovenzoni, Raffaele, 28
Zucchetta, Bernardo, 79

Printed in the United Kingdom
by Lightning Source UK Ltd.
102159UKS00002B/208-210